Newne

Physi

Pocke

for engineers

J O Bird

Bsc (Hons), AFIMA, TEng (CEI), MIElecIE

P J Chivers

BSc (Hons), PhD

Newnes Technical Books

Newnes Technical Books
is an imprint of the Butterworth Group
which has principal offices in
London, Boston, Durban, Singapore, Sydney, Toronto, Wellington

First published 1983

British Library Cataloguing in Publication Data

Bird, J. O.
Newnes physical science pocket book.
1. Science
I. Title II. Chivers, P. J.
500.2 Q158.5

ISBN 0-408-01343-5

Typeset by Mid-County Press
Printed in England by The Thetford Press Ltd,
Thetford, Norfolk

Preface

This Physical Science pocket book is intended to provide students, technicians, scientists and engineers with a readily available reference to the essential physical sciences formulae, definitions and general information needed during their studies and/or work situation. The book is divided, for convenience of reference, into four sections embracing general science, physics, electrical science and chemistry.

The text assumes little previous knowledge and is suitable for a wide range of courses. It will be particularly useful for students studying for Technician certificates and diplomas, and for CSE and 'O' and 'A' levels.

The authors would like to express their appreciation for the friendly co-operation and helpful advice given to them by the publishers and by the editor, Tony May. Thanks are also due to Simon Pascoe for his agreeing to use of some material from *Physics 2 Checkbook*, and to Mrs Elaine Woolley and Mrs Sandra Chivers for the excellent typing of the manuscript.

Finally the authors would like to add a word of thanks to their wives, Elizabeth and Sandra for their patience, help and encouragement during the preparation of this book.

J O Bird and P J Chivers
Highbury College of Technology
Portsmouth

Contents

1 SI units

1 The systems of units used in engineering and science is the **Système Internationale d'Unites** (International system of units), usually abbreviated to SI units, and is based on the metric system. This was introduced in 1960 and is now adopted by the majority of countries as the official system of measurement.

2 The basic units in the SI system are given in *Table 1.1.*

Table 1.1

Quantity	*Unit*
length	metre, m
mass	kilogram, kg
time	second, s
electric current	ampere, A
thermodynamic temperature	kelvin, K
luminous intensity	candela, cd
amount of substance	mole, mol

3 SI units may be made larger or smaller by using **prefixes** which denote multiplication or division by a particular amount. The eight most common multiples, with their meaning, are listed in *Table 1.2.*

Table 1.2

Prefix	*Name*	*Meaning*
T	tera	multiply by 1 000 000 000 000 (i.e. $\times 10^{12}$)
G	giga	multiply by 1 000 000 000 (i.e. $\times 10^{9}$)
M	mega	multiply by 1 000 000 (i.e. $\times 10^{6}$)
k	kilo	multiply by 1 000 (i.e. $\times 10^{3}$)
m	milli	divide by 1 000 (i.e. $\times 10^{-3}$)
μ	micro	divide by 1 000 000 (i.e. $\times 10^{-6}$)
n	nano	divide by 1 000 000 000 (i.e. $\times 10^{-9}$)
p	pico	divide by 1 000 000 000 000 (i.e. $\times 10^{-12}$)

4 (i) **Length** is the distance between two points. The standard unit of length is the **metre**, although the **centimetre, cm, millimetre, mm** and **kilometre, km,** are often used.

$1\ \text{cm} = 10\ \text{mm};\ 1\ \text{m} = 100\ \text{cm} = 1000\ \text{mm};$
$1\ \text{km} = 1000\ \text{m}.$

(ii) **Area** is a measure of the size or extent of a plane surface and is measured by multiplying a length by a length. If the lengths are in metres then the unit of area is the **square metre, m^2.**

$$1\ \text{m}^2 = 1\ \text{m} \times 1\ \text{m} = 100\ \text{cm} \times 100\ \text{cm} = 10\,000\ \text{cm}^2 \text{ or } 10^4\ \text{cm}^2$$
$$= 1000\ \text{mm} \times 1000\ \text{mm} = 1\,000\,000\ \text{mm}^2 \text{ or } 10^6\ \text{mm}^2$$

Conversely, $1\ \text{cm}^2 = 10^{-4}\ \text{m}^2$ and $1\ \text{mm}^2 = 10^{-6}\ \text{m}^2$.

(iii) **Volume** is a measure of the space occupied by a solid and is measured by multiplying a length by a length by a length. If the lengths are in metres then the unit of volume is in **cubic metres, m^3**.

$$1\ \text{m}^3 = 1\ \text{m} \times 1\ \text{m} \times 1\ \text{m} = 100\ \text{cm} \times 100\ \text{cm} \times 100\ \text{cm} = 10^6\ \text{cm}^3$$
$$= 1000\ \text{mm} \times 1000\ \text{mm} \times 1000\ \text{mm} = 10^9\ \text{mm}^3$$

Conversely, $1\ \text{cm}^3 = 10^{-6}\ \text{m}^3$ and $1\ \text{mm}^3 = 10^{-9}\ \text{m}^3$

Another unit used to measure volume, particularly with liquids, is the litre (l) where 1 litre = 1000 cm^3.

(iv) **Mass** is the amount of matter in a body and is measured in **kilograms, kg**.

1 kg = 1000 g (or conversely, $1\ \text{g} = 10^{-3}\ \text{kg}$) and
1 tonne (t) = 1000 kg.

5 **Derived SI units** use combinations of basic units and there are many of them. Two examples are:

velocity	metres per second, (m/s)
acceleration	metres per second square, (m/s^2).

(a) The unit of **charge** is the coulomb, (C), where one coulomb is one ampere second. (1 coulomb = 6.24×10^{18} electrons). The coulomb is defined as the quantity of electricity which flows past a given point in an electric circuit when a current of one ampere is maintained for one second. Thus

charge in coulombs, $\boldsymbol{Q = It}$

where I is the current in amperes and t is the time in seconds.

(b) The unit of **force** is the newton, (N), where one newton is one kilogram metre per second squared. The newton is defined as the force which, when applied to a mass of one kilogram, gives it an acceleration of one metre per second squared. Thus

force in newtons, $\boldsymbol{F=ma}$,

where m is the mass in kilograms and a is the acceleration in metres per second squared. Gravitational force, or weight, is mg, where $g=9.81$ m/s^2.

(c) The unit of **work or energy** is the joule, (J), where one joule is one newton metre. The joule is defined as the work done or energy transferred when a force of one newton is exerted through a distance of one metre in the direction of the force. Thus

work done on a body in joules, $\boldsymbol{W=Fs}$,

where F is the force in newtons and s is the distance in metres moved by the body in the direction of the force. Energy is the capacity for doing work.

(d) (i) The unit of **power** is the watt, (W), where one watt is one joule per second. Power is defined as the rate of doing work or transferring energy. Thus:

power in watts, $P=\dfrac{W}{t}$,

where W is the work done or energy transferred in joules and t is the time in seconds.

Hence, energy in joules, $W=Pt$.

(e) The unit of **electric potential** is the volt (V) where one volt is one joule per coulomb. One volt is defined as the difference in potential between two points in a conductor which, when carrying a current of one ampere dissipates a power of one watt.

$$\left(\text{i.e. volts}=\frac{\text{watts}}{\text{amperes}}=\frac{\text{joules/second}}{\text{amperes}}=\frac{\text{joules}}{\text{ampere seconds}}=\frac{\text{joules}}{\text{coulomb}}\right)$$

A change in electric potential between two points in an electric circuit is called a potential difference. The electromotive force (e.m.f.) provided by a source of energy such as a battery or a generator is measured in volts.

2 Density

1 (i) **Density** is the mass per unit volume of a substance. The symbol used for density is ρ (Greek letter rho) and its units are kg/m^3.

$$\textbf{Density} = \frac{\textbf{mass}}{\textbf{volume}}, \text{ i.e., } \boxed{\rho = \frac{m}{v}} \quad \text{or}$$

$$\boxed{m = \rho V} \quad \text{or} \quad \boxed{V = \frac{m}{\rho}}$$

where m is the mass in kg, V is the volume in m^3 and ρ is the density in kg/m^3.

(ii) Some typical values of densities include:

aluminium 2 700 kg/m^3, copper 8 900 kg/m^3,
lead 11 400 kg/m^3, cast iron 7 000 kg/m^3,
steel 7 800 kg/m^3, water 1 000 kg/m^3,
cork 250 kg/m^3, petrol 700 kg/m^3.

2 (i) The **relative density** of a substance is the ratio of the density of the substance to the density of water,

$$\text{i.e. relative density} = \frac{\text{density of substance}}{\text{density of water}}$$

Relative density has no units, since it is the ratio of two similar quantities.

(ii) Typical values of relative densities can be determined from para. 1, (since water has a density of 1000 kg/m^3, and include: aluminium 2.7, copper 8.9, lead 11.4, cast iron 7.0, steel 7.8, cork 0.25, petrol 0.7.

(iii) The relative density of a liquid (formerly called the 'specific gravity') may be measured using a **hydrometer**.

3 Scalar and vector quantities

Quantities used in engineering and science can be divided into two groups:

(a) **Scalar quantities** have a size or magnitude only and need no other information to specify them. Thus, 10 cm, 50 sec, 7 litres and 3 kg are all examples of scalar quantities.

(b) **Vector quantities** have both a size or magnitude and a direction, called the line of action of the quantity. Thus, a velocity of 50 km/h due east, on acceleration of 9.8 m/s^2 vertically downwards and a force of 15 N at an angle of 30° are all examples of vector quantities.

4 Standard quantity symbols and units

Quantity	Quantity symbol	Unit	Unit symbol
Acceleration:			
gravitational	g	metres per second squared	m/s^2
linear	a	metres per second squared	m/s^2
Angular acceleration	α	radians per second squared	rad/s^2
Angular velocity	ω	radians per second	rad/s
Area	A	square metres	m^2
Area, second moment of	I	$(metre)^4$	m^4
Capacitance	C	farad	F
Capacity	V	litres	l
Coefficient of friction	μ	No unit	
Coefficient of linear expansion	α	per degree Celsius	/°C
Conductance	G	seimens	S
Cubical expan sion, coeffi-cient of	γ	per degree Celsius	/°C
Current	I	ampere	A
Density	ρ	kilogram per cubic metre	kg/m^3
Density, relative	d	no unit	
Dryness fraction	x	no unit	
Efficiency	η	no unit	
Elasticity, modulus of	E	Pasçal ($1\ Pa = 1\ N/m^2$)	Pa
Electric field strength	E	volts per metre	V/m

Quantity	*Quantity symbol*	*Unit*	*Unit symbol*
Electric flux density	D	coulomb per square metre	C/m^2
Energy	W	joules	J
Energy, internal	U, E	joules	J
Energy, specific internal	u, e	kilojoules per kilogram	kJ/kg
Enthalpy	H	joules	J
Enthalpy, specific	h	kilojoules per kilogram	kJ/kg
Entropy	S	kilojoules per kelvin	kJ/K
Expansion: coefficient of cubical	γ	per degree Celsius	/°C
coefficient of linear	α	per degree Celsius	/°C
coefficient of superficial	β	per degree Celsius	/°C
Field strength: electric	E	volts per metre	V/m
magnetic	H	ampere per metre	A/m
Flux: electric	D	coulomb per square metre	C/m^2
magnetic	B	tesla	T
Flux:			
electric	ψ	coulomb	C
magnetic	Φ	weber	Wb
Force	F	newtons	N
Frequency	f	hertz	Hz
Heat capacity, specific	c	kilojoules per kilogram kelvin	kJ/(kg K)
Impedance	Z	ohm	Ω
Inductance: self	L	henry	H
mutual	M	henry	H
Internal energy	U, E	joules	J
specific	u, e	kilojoules per kilogram	kJ/kg
Inertia, moment of	I, J	kilogram metre squared	$kg\ m^2$
Length	l	metre	m

Quantity	Quantity symbol	Unit	Unit symbol
Luminous intensity	I	candela	cd
Magnetic field strength	H	ampere per metre	A/m
Magnetic flux	Φ	weber	Wb
density	B	tesla	T
Magnetomotive force	F	ampere	A
Mass	m	kilogram	kg
Mass, rate of flow	V	cubic metres per second	m^3/s
Modulus of elasticity	E	Pascal	Pa
rigidity	G	Pascal	Pa
Moment of force	M	newton metre	N m
Moment of inertia	I, J	kilogram metre squared	kg m^2
Mutual inductance	M	henry	H
Number of turns in a winding	N	no unit	
Periodic time	T	second	s
Permeability:			
absolute	μ	henry per metre	H/m
absolute of free space	μ_o	henry per metre	H/m
relative	μ_r	no unit	
Permittivity:	ε		
absolute		farad per metre	F/m
of free space	ε_o	farad per metre	F/m
relative	ε_r	no unit	
Polar moment of area	J	$(\text{metre})^4$	m^4
Power: apparent	S	volt ampere	VA
active	P	watt	W
reactive	Q	volt ampere reactive	VAr
Pressure	p	Pascal ($1 \text{ Pa} = 1 \text{ N/m}^2$)	Pa
Quantity of heat	Q	joule	J

Quantity	*Quantity symbol*	*Unit*	*Unit symbol*
Quantity of electricity	Q	coulomb	C
Reactance	X	ohm	Ω
Reluctance	S	per henry or ampere per weber	/H or A/Wb
Resistance	R	ohm	Ω
Resistivity	ρ	ohm metre	Ω m
Second moment of area	I	$(\text{metre})^4$	m^4
Shear strain	γ	no unit	
stress	τ	Pascal	Pa
Specific gas constant	R	kilojoules per kilogram kelvin	kJ/(kg K)
Specific heat capacity	c	kilojoules per kilogram kelvin	kJ/(kg K)
Specific volume	v	cubic metres per kilogram	m^3/kg
Strain, direct	ε	no unit	
Stress, direct	σ	Pascal	Pa
Shear modulus of rigidity	G	Pascal	Pa
Temperature coefficient of resistance	α	per degree Celsius	/°C
Temperature, thermodynamic	T	kelvin	K
Time	t	second	s
Torque	T	newton metre	N m
Velocity	v	metre per second	m/s
angular	ω	radian per second	rad/s
Voltage	V	volt	V
Volume	V	cubic metre	m^3
Volume, rate of flow	V	cubic metre per second	m^3/s
Wavelength	λ	metre	m
Work	W	joule	J
Young's modulus of elasticity	E	Pascal	Pa

5 Speed and velocity

1 Speed is the rate of covering distance and is given by:

$$\textbf{speed} = \frac{\textbf{distance travelled}}{\textbf{time taken}}$$

The usual units for speed are metres per second (m/s or m s^{-1}), or kilometres per hour (km/h or km h^{-1}). Thus if a person walks 5 kilometres in 1 hour, the speed of the person is $\frac{5}{1}$, that is, 5 kilometres per hour.

The symbol for the SI unit of speed and velocity is written as 'm s^{-1}', called the 'index notation'. However, engineers usually use the symbol m/s, called the 'oblique notation', and it is this notation which is largely used in this chapter and other chapters on mechanics. One of the exceptions is when labelling the axes of graphs, when two obliques occur, and in this case the index notation is used. Thus for speed or velocity, the axis markings are speed/m s^{-1} or velocity/m s^{-1}.

2 One way of giving data on the motion of an object is graphically. A graph of distance travelled (the scale on the vertical axis of the graph), against time (the scale on the horizontal axis of the graph), is called a **distance-time graph**. Thus if a plane travels 500 km in its first hour of flight and 750 km in its second hour of flight, then after 2 h, the total distance travelled is (500 + 750) kilometres, that is, 1250 km. The distance-time graph for this flight is shown in *Figure 5.1*.

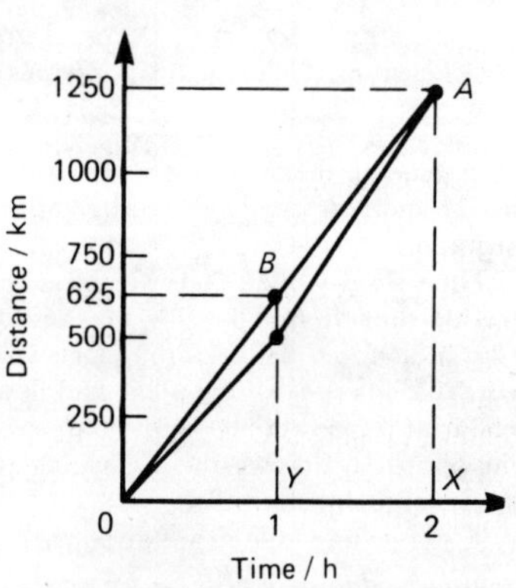

Figure 5.1

3 The **average speed** is given by

$$\frac{\text{total distance travelled}}{\text{total time taken}}$$

Thus, the average speed of the plane in para. 2 is:

$$\frac{(500+750)\text{ km}}{(1+1)\text{ h}}, \text{ i.e. } \frac{1250}{2} \text{ or } 625 \text{ km/h.}$$

If points O and A are joined in *Figure 5.1*, the slope of line OA is defined as

$$\frac{\text{change in distance (vertical)}}{\text{change in time (horizontal)}}$$

for any two points on line OA.

For point A, the change in distance is AX, that is 1250 km, and the change in time is OX, that is, 2 h. Hence the average speed is

$$\frac{1250}{2}, \text{ i.e. } 625 \text{ km/h}$$

Alternatively, for point B on line OA, the change in distance is BY, that is, 625 km and the change in time is OY, that is, 1 h, hence the average speed is

$$\frac{625}{1}, \text{ that is, } 625 \text{ km/h}$$

In general, the average speed of an object travelling between points M and N is given by the slope of line MN on the distance-time graph.

4 The **velocity** of an object is the speed of the object **in a specified direction**. Thus, if a plane is flying due south at 500 km/h, its speed is 500 km/h, but its velocity is 500 km/h **due south**. It follows that if the plane had flown in a circular path for one hour at a speed of 500 km/h hour, so that one hour after taking off it is again over the airport, its average velocity in the first hour of flight is zero.

5 The **average velocity** is given by:

$$\frac{\text{distance travelled in a specific direction}}{\text{time taken}}$$

If a plane flies from place O to place A, a distance of 300 km in

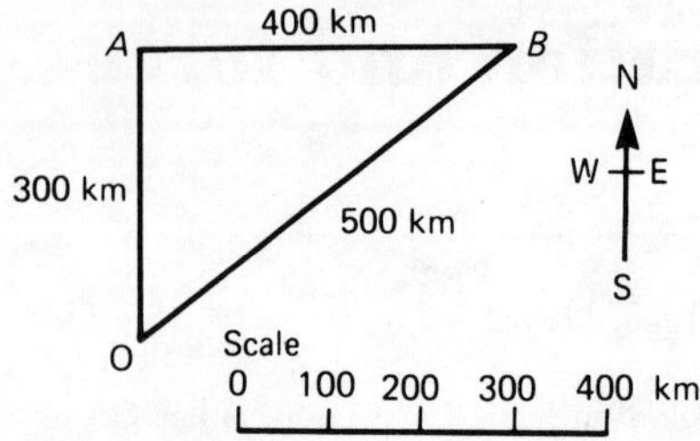

Figure 5.2

1 h, A being due north of O, then OA in *Figure 5.2* represents the first hour of flight. It then flies from A to B, a distance of 100 km during the second hour of flight, B being due east of A, thus AB in *Figure 5.2* represents its second hour of light. Its average velocity for the two hour flight is

$$\frac{\text{distance OB}}{\text{2 hours}}, \text{ that is, } \frac{500 \text{ km}}{2 \text{ h}}$$

or 250 km/h in direction OB.

6 A graph of velocity (scale on the vertical axis), against time, (scale on the horizontal axis), is called a **velocity-time graph**. The graph shown in *Figure 5.3* represents a plane flying for 3 h at a constant speed of 600 km/h in a specified direction. The shaded area represents velocity (vertically), multiplied by time (horizontally), and has units of kilometres/hours × hours, i.e. kilometres, and represents the distance travelled in a specific direction.

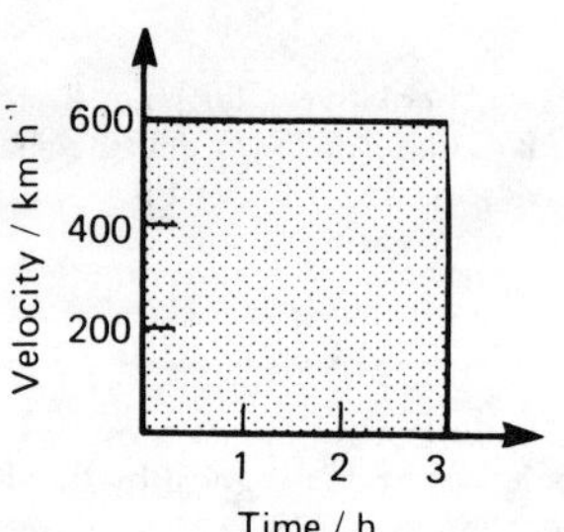

Figure 5.3

Another method of determining the distance travelled is from:

distance travelled = average velocity × time

Thus if a plane travels due south at 600 km/h for 20 minutes, the distance covered is

$$\frac{600 \text{ km}}{1 \text{ h}} \times \frac{20}{60} \text{ h, that is, 200 km.}$$

6 Acceleration and force

1 **Acceleration** is the rate of change of speed or velocity with time. The average acceleration, a, is given by:

$$a = \frac{\textbf{change in velocity}}{\textbf{time taken}}$$

The usual units are metres per second squared (m/s^2 or m s^{-2}). If u is the initial velocity of an object in m/s, v is the final velocity in m/s and t is the time in seconds elapsing between the velocities of u and v, then

$$\textbf{average acceleration, } a = \frac{v-u}{t} \textbf{ m/s}^2.$$

2 A graph of speed (scale on the vertical axis), against time (scale on the horizontal axis) is called a **speed-time graph**. For the speed-time graph shown in *Figure 6.1*, the slope of line OA is given by AX/OX. AX is the change in velocity from an initial

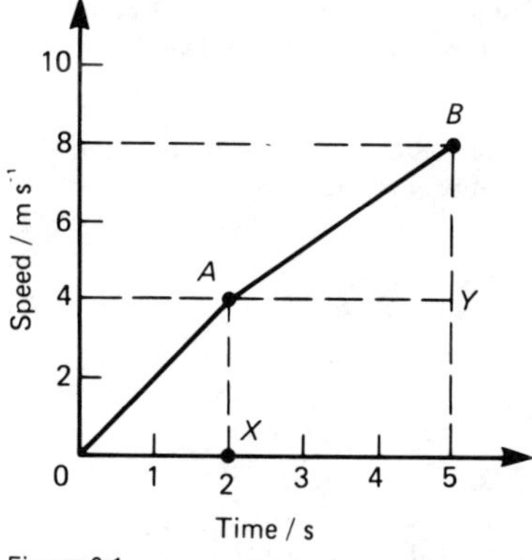

Figure 6.1

velocity u of zero to a final velocity, v, of 4 metres per second. OX is the time taken for this change in velocity, thus

$$\frac{AX}{OX} = \frac{\text{change in velocity}}{\text{time taken}}$$

$$= \text{the acceleration in the first two seconds.}$$

From the graph:

$$\frac{AX}{OX} = \frac{4 \text{ m/s}}{2 \text{ s}} = 2 \text{ m/s}^2,$$

i.e. the acceleration is 2 m/s^2.

Similarly, the slope of line AB in *Figure 6.1* is given by BY/AY, that is, the acceleration between 2 and 5 s is

$$\frac{8-4}{5-2} = \frac{4}{3} = 1\tfrac{1}{3} \text{ m/s}^2$$

In general, the slope of a line on a speed-time graph gives the acceleration.

3 If a dense object such as a stone is dropped from a height, called **free fall**, it has a constant acceleration of approximately 9.8 metres per second squared. In a vacuum, all objects have this same constant acceleration vertically downwards, that is, a feather has the same acceleration as a stone. However, if free fall takes place in air, dense objects have the approximately constant acceleration of 9.8 metres per second squared over short distances, but objects which have a low density, such as feathers, have little or no acceleration.

4 For bodies moving with a constant acceleration, the average acceleration is the constant value of the acceleration, and since from para. 1,

$$a = \frac{v-u}{t}, \text{ then } a \times t = v - u \text{ or } \boldsymbol{v = u + at},$$

where
u is the initial velocity in m/s,
v is the final velocity in m/s,
a is the constant acceleration in m/s^2, and
t is the time in s.
When symbol 'a' has a negative value, it is called **deceleration** or

retardation. The equation, $v=u+at$ is called an **equation of motion**.

5 When an object is pushed or pulled, a force is applied to the object. This force is measured in **newtons**, (**N**). The effects of pushing or pulling an object are:

(i) to cause a change in the motion of the object, and
(ii) to cause a change in the shape of the object.

If a change occurs in the motion of the object, that is, its speed changes from u to v, then the object accelerates. Thus, it follows that acceleration results from a force being applied to an object. If a force is applied to an object and it does not move, then the object changes shape, that is, deformation of the object takes place. Usually the change in shape is so small that it cannot be detected by just watching the object. However, when very sensitive measuring instruments are used, very small changes in dimensions can be detected.

6 A force of attraction exists between all objects. The factors governing the size of this force are the masses of the objects and the distances between their centres,

$$\left(F \propto \frac{m_1 m_2}{d^2}\right).$$

Thus, if a person is taken as one object and the earth as a second object, a force of attraction exists between the person and the earth. This force is called the **gravitational force** and is the force which gives a person a certain weight when standing on the earth's surface. It is also this force which gives freely falling objects a constant acceleration in the absence of other forces.

7 To make a stationary object move or to change the direction in which the object is moving requires a force to be applied externally to the object. This concept is known as **Newton's first law of motion** and may be stated as:

> '*an object remains in a state of rest, or continues in a state of uniform motion in a straight line, unless it is acted on by an externally applied force*'.

8 Since a force is necessary to produce a change of motion, an object must have some resistance to a change in its motion. The force necessary to give a stationary pram a given acceleration is far less than the force necessary to give a stationary car the same acceleration. The resistance to a change in motion is called the **inertia** of an object and the amount of inertia depends on the mass of the object. Since a car has a much larger mass than a pram, the inertia of a car is much larger than that of a pram.

9 **Newton's second law of motion** may be stated as:

'the acceleration of an object acted upon by an external force is proportional to the force and is in the same direction as the force.'

Thus, force α acceleration

or force = a constant × acceleration, this constant of proportionality being the mass of the object, i.e.

force = mass × acceleration.

The unit of force is the newton (N) and is defined in terms of mass and acceleration. One newton is the force required to give a mass of 1 kilogram an acceleration of 1 metre per second squared. Thus:

$F = ma$, where

F is the force in newtons (N), m is the mass in kilograms (kg) and a is the acceleration in metres per second squared (m/s^2), i.e.

$$\mathbf{1\ N = 1\ \frac{kg\ m}{s^2}}$$

10 **Newton's third law of motion** may be stated as:

'for every force, there is an equal and opposite reacting force'.

Thus, an object on, say, a table, exerts a downward force on the table and the table exerts an equal upward force on the object, known as a **reaction force** or just a **reaction**. When an object is accelerating, the force due to the inertia of the body ($F = ma$) is a reaction force acting in the opposite direction to the motion of the object.

11 When an object moves in a circular path at constant speed, its direction of motion is continually changing and hence its velocity (which depends on **both magnitude and direction**) is also continually changing. Since acceleration is the

$$\frac{\text{change in velocity}}{\text{change in time}},$$

the object has an acceleration. Let the object be moving with a constant angular velocity of ω and a tangential velocity of magnitude v and let the change of velocity for a small change of angle of θ ($=\omega t$) be V.

Then, $\boldsymbol{v_2} - \boldsymbol{v_1} = \boldsymbol{V}$.

The vector diagram is shown in *Figure 6.2(b)* and since the magnitudes of $\boldsymbol{v_1}$ and $\boldsymbol{v_2}$ are the same, i.e. $\boldsymbol{v}$, the vector diagram is also an isosceles triangle.

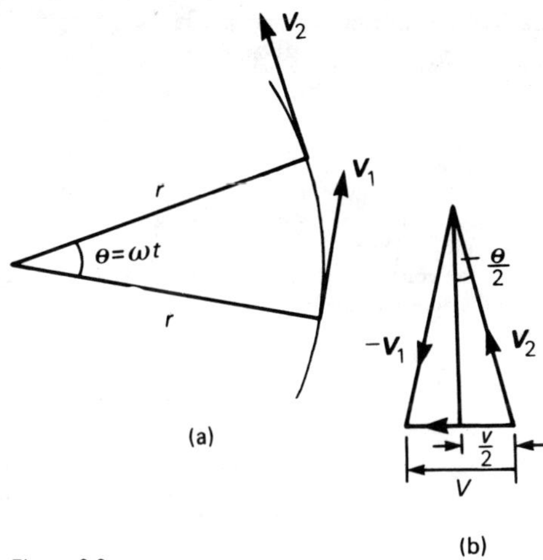

Figure 6.2

Bisecting the angle between $\boldsymbol{v_2}$ and $\boldsymbol{v_1}$ gives:

$$\sin\frac{\theta}{2}=\frac{V/2}{v_2}=\frac{V}{2v}$$

i.e. $V=2v\sin\frac{\theta}{2}$ (1)

Since $\theta=\omega t$, $t=\frac{\theta}{\omega}$ (2)

Dividing (1) by (2) gives:

$$\frac{V}{t}=\frac{2v\sin\frac{\theta}{2}}{\frac{\theta}{\omega}}=\frac{v\omega\sin\frac{\theta}{2}}{\frac{\theta}{2}}$$

For small angles, $\dfrac{\sin\frac{\theta}{2}}{\frac{\theta}{2}}$ is very nearly equal to unity.

Hence, $\frac{V}{t} = \frac{\text{change of velocity}}{\text{change of time}} = \text{acceleration}, a = v\omega$

But, $\omega = \frac{v}{r}$, thus $v\omega = v \cdot \frac{v}{r} = \frac{v^2}{r}$

That is, the acceleration a is v^2/r and is towards the centre of the circle of motion (along V). It is called the **centripetal acceleration**. If the mass of the rotating object is m, then by Newton's second law the **centripetal force** is mv^2/r, and its direction is towards the centre of the circle of motion.

7 Linear momentum and impulse

1 (i) The **momentum** of a body is defined as the product of its mass and its velocity,

i.e. **momentum** $= mu$
where
m = mass (in kg) and
u = velocity (in m/s).

The unit of momentum is kg m/s.

(ii) Since velocity is a vector quantity, **momentum is a vector quantity**, i.e. it has both magnitude and direction.

2 (i) **Newton's first law of motion** states:

> *'a body continues in a state of rest or in a state of uniform motion in a straight line unless acted on by some external force.'*

Hence the momentum of a body remains the same provided no external forces act on it.

(ii) **The principle of conservation of momentum** for a closed system (i.e. one on which no external forces act) may be stated as:

> *'the total linear momentum of a system in any given direction is a constant'.*

(iii) The total momentum of a system before collision in a given direction is equal to the total momentum of the system after collision in the same direction. In *Figure 7.1*, masses m_1 and m_2 are travelling in the same direction with velocity $u_1 > u_2$. A collision

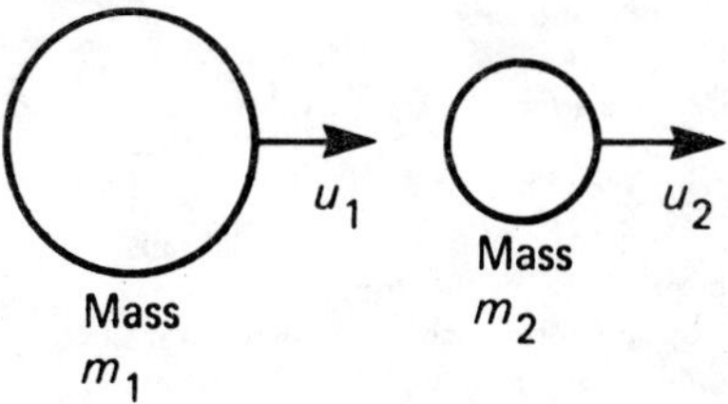

Figure 7.1

will occur, and applying the principle of conservation of momentum:
total momentum before impact – total momentum after impact
i.e. $\boldsymbol{m_1u_1+m_2u_2=m_1v_1+m_2v_2}$,
where v_1 and v_2 are the velocities of m_1 and m_2 after impact.

3 (i) **Newton's second law of motion** may be stated as:

'the rate of change of momentum is directly proportional to the applied force producing the change, and takes place in the direction of this force'.

In the SI system, the units are such that:

$$\text{the applied force} = \text{rate of change of momentum}$$

$$= \frac{\text{change of momentum}}{\text{change of time}} \qquad (1)$$

(ii) When a force is suddenly applied to a body due to either a collision with another body or being hit by an object such as a hammer, the time taken in equation (1) is very small and difficult to measure. In such cases, the total effect of the force is measured by the change of momentum it produces.
(iii) Forces which act for very short periods of time are called **impulsive forces**. The product of the impulsive force and the time during which it acts is called the **impulse** of the force and is equal to the change of momentum produced by the impulsive force,

i.e. **impulse = applied force × time = change in linear momentum**

(iv) Examples where impulsive forces occur include when a gun recoils and when a free-falling mass hits the ground. Solving problems associated with such occurrences often requires the use of the equation of motion:

$$v^2=u^2+2as.$$

4 When a pile is being hammered into the ground, the ground resists the movement of the pile and this resistance is called a **resistive force**.
Newton's third law of motion may be stated as:

'for every force there is an equal and opposite force'.

The force applied to the pile is the resistance force. The pile excerts an equal and opposite force on the ground.

5 In practise when impulsive forces occur, momentum is not entirely conserved and some energy is changed into heat, noise, and so on.

8 Linear and angular motion

1 The unit of angular displacement is the **radian**, where one radian is the angle subtended at the centre of a circle by an arc equal in length to the radius, see *Figure 8.1*.

The relationship between angle in radians (θ), are length (s) and radius of a circle (r) is:

$$s = r\theta \qquad (1)$$

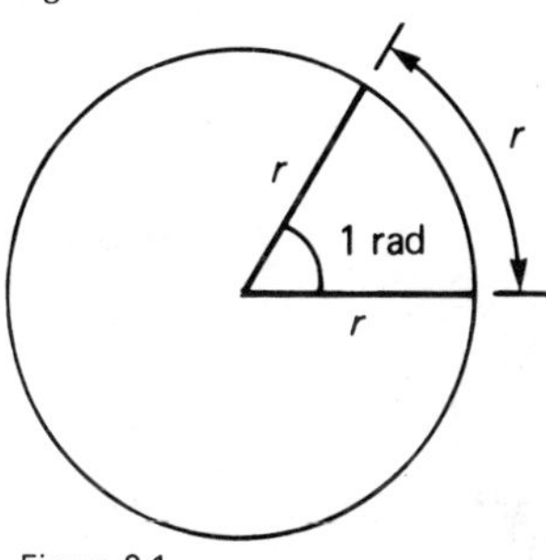

Figure 8.1

Since the arc length of a complete circle is $2\pi r$ and the angle subtended at the centre is 360°, then, from equation (1), for a complete circle,
$2\pi r = r\theta$ or $\theta = 2\pi$ radians.

Thus, 2π radians corresponds to 360° (2)

2 (i) **Linear velocity**, v, is defined as the rate of change of linear displacement, s, with respect to time, t, and for motion in a straight line:

$$\text{linear velocity} = \frac{\text{change of distance}}{\text{change of time}},$$

$$\text{i.e. } \boldsymbol{v = \frac{s}{t}} \qquad (3)$$

The unit of linear velocity is metres per second (m/s).
(ii) **Angular velocity**
The speed of revolution of a wheel or a shaft is usually measured in revolutions per minute or revolutions per second but these units do not form part of a coherent system of units. The basis used in SI units is the angle turned through in one second.

Angular velocity is defined as the rate of change of angular displacement, θ, with respect to time, t, and for an object rotating about a fixed axis at a constant speed:

$$\text{angular velocity} = \frac{\text{angle turned through}}{\text{time taken}},$$

$$\text{i.e. } \omega = \frac{\theta}{t} \quad (4)$$

The unit of angular velocity is radians per second (rad/s). An object rotating at a constant speed of n revolutions per second subtends an angle of $2\pi n$ radians in one second, that is, its angular velocity

$$\omega = 2\pi n \textbf{ rad/s} \quad (5)$$

(iii) From equation (1), $s = r\theta$ and from equation (4), $\theta = \omega t$, hence

$$s = r\omega t \text{ or } \frac{s}{t} = \omega r.$$

However, from equation (3),

$$v = \frac{s}{t}, \text{ hence } \boldsymbol{v = \omega r} \quad (6)$$

Equation (6) gives the relationship between linear velocity, v, and angular velocity, ω.

3 (i) **Linear acceleration**, a, is defined as the rate of change of linear velocity with respect to time. For an object whose linear velocity is increasing uniformly:

$$\text{linear acceleration} = \frac{\text{change of linear velocity}}{\text{time taken}}$$

$$\text{i.e. } a = \frac{v_2 - v_1}{t} \quad (7)$$

The unit of linear acceleration is metres per second squared (m/s^2). Rewriting equation (7) with v_2 as the subject of the formula, gives:

$$v_2 = v_1 + at \quad (8)$$

(ii) **Angular acceleration**, α, is defined as the rate of change of angular velocity with respect to time. For an object whose angular velocity is increasing uniformly:

$$\text{angular acceleration} = \frac{\text{change of angular velocity}}{\text{time taken}}$$

$$\text{that is, } \alpha = \frac{\omega_2 - \omega_1}{t} \quad (9)$$

The unit of angular acceleration is radians per second squared (rad/s^2). Rewriting equation (9) with ω_2 as the subject of the formula gives:

$$\omega_2=\omega_1+\alpha t \qquad (10)$$

(iii) From equation (6), $v=\omega r$. For motion in a circle having a constant radius r, v_2 in equation (7) is given by $v_2=\omega_2 r$ and $v_1=\omega_1 r$, hence equation (7) can be written as:

$$a=\frac{\omega_2 r-\omega_1 r}{t}=\frac{r(\omega_2-\omega_1)}{t}$$

But from equation (9), $\dfrac{\omega_2-\omega_1}{t}=\alpha$

$$\text{Hence } a=r\alpha \qquad (11)$$

Equation (11) gives the relationship between linear acceleration a and angular acceleration α.

4 (i) From equation (3), $s=vt$, and if the linear velocity is changing uniformly from v_1 to v_2, then
s = mean linear velocity × time,

$$\text{i.e. } s=\left(\frac{v_1+v_2}{2}\right)t \qquad (12)$$

(ii) From equation (4), $\theta=\omega t$, and if the angular velocity is changing uniformly from v_1 to v_2, then

s = mean linear velocity × time,

$$\text{i.e. } \theta=\left(\frac{\omega_1+\omega_2}{2}\right)t \qquad (13)$$

(iii) Two further equations of linear motion may be derived from equations (8) and (11):

$$s=v_1 t+\frac{1}{2}at^2 \qquad (14)$$

$$\text{and } v_2^2=v_1^2+2as \qquad (15)$$

(iv) Two further equations of angular motion may be derived from equations (10) and (12):

$$\theta=\omega_1 t+\frac{1}{2}\alpha t^2 \qquad (16)$$

$$\text{and } \omega_2^2=\omega_1^2+2\alpha\theta \qquad (17)$$

Table 8.1

s = arc length (m)	r = radius of circle (m)
t = time (s)	θ = angle (rad)
v = linear velocity (m/s)	ω = angular velocity (rad/s)
v_1 = initial linear velocity (m/s)	ω_1 = initial angular velocity (rad/s)
v_2 = final linear velocity (m/s)	ω_2 = final angular velocity (rad/s)
a = linear acceleration (m/s^2)	α = angular acceleration (rad/s^2)
n = speed of revolutions (revolutions per second)	

Equation No.	*Linear motion*	*Angular motion*
1	$s = r\theta$ m	
2		2π rad = 360°
3 and 4	$v = \frac{s}{t}$ m/s	$\omega = \frac{\theta}{t}$ rad/s
5		$\omega = 2\pi n$ rad/s
6	$v = \omega r$ m/s	
8 and 10	$v_2 = (v_1 + at)$ m/s	$\omega_2 = (\omega_1 + \alpha t)$ rad/s
11	$a = r\alpha$ m/s^2	
12 and 13	$s = \left(\frac{v_1 + v_2}{2}\right) t$ m	$\theta = \left(\frac{\omega_1 + \omega_2}{2}\right) t$ rad
14 and 16	$s = (v_1 t + \frac{1}{2}at^2)$ m	$\theta = (\omega_1 t + \frac{1}{2}\alpha t^2)$ rad
15 and 17	$v_2^2 = (v_1^2 + 2as)$ (m/s)^2	$\omega_2^2 = (\omega_1^2 + 2\alpha\theta)$ (rad/s)^2

5 *Table 8.1* summarises the principal equations of linear and angular motion for uniform changes in velocities and constant accelerations and also gives the relationship between linear and angular quantities.

6 A vector quantity is represented by a straight line lying along the line of action of the quantity and having a length which is proportional to the size of the quantity. Thus **ab** in *Figure 8.2* represents a velocity of 20 m/s, whose line of action is due west.

The bold letters, **ab**, indicate a vector quantity and the order of the letters indicate that the time of action is from a to b.

7 Consider two aircraft A and B flying at a constant altitude, A

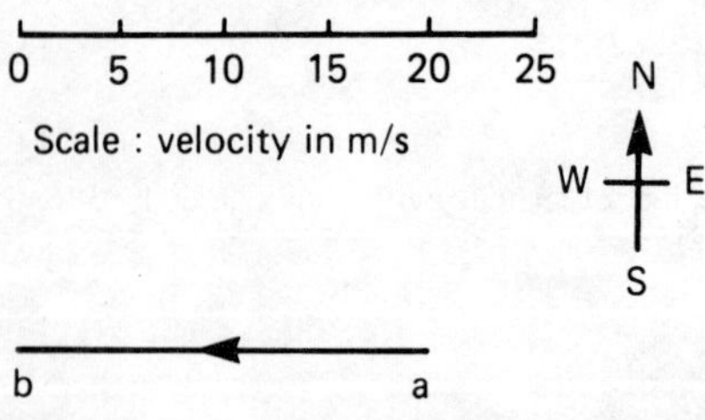

Figure 8.2

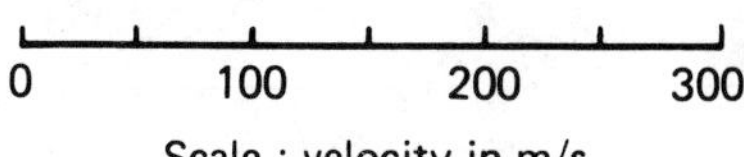

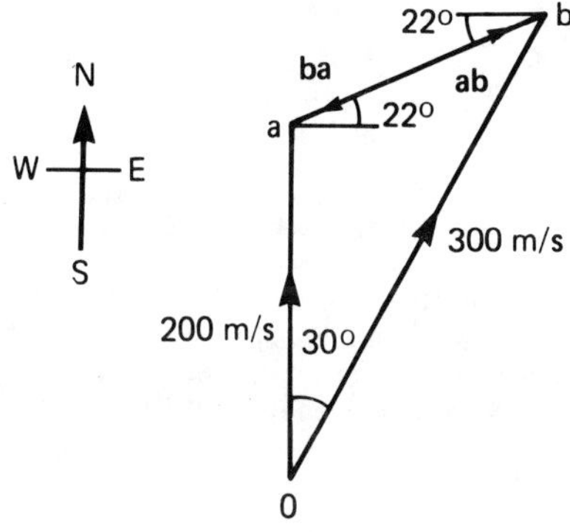

Figure 8.3

travelling due north at 200 m/s and B travelling 30° east of north, written N 30° E, at 300 m/s, as shown in *Figure 8.3.* Relative to a fixed point o, **oa** represents the velocity of A and **ob** the velocity of B. The velocity of B relative to A, that is the velocity at which B seems to be travelling to an observer on A, is given by **ab**, and by measurement is 160 m/s in a direction E 22° N.

The velocity of A relative to B, that is, the velocity at which A seems to be travelling to an observer on B, is given by **ba** and by measurement is 160 m/s in a direction W 22° S.

9 Friction

1 When an object, such as a block of wood, is placed on a floor and sufficient force is applied to the block, the force being parallel to the floor, the block slides across the floor. When the force is removed, motion of the block stops; thus there is a force which resists sliding. This force is called **dynamic** or **sliding friction**. A force may be applied to the block which is insufficient to move it. In this case, the force resisting motion is called the **static friction** or **striction**. Thus there are two categories into which a frictional force may be split:

(i) dynamic or sliding friction force which occurs when motion is taking place, and

(ii) static friction force which occurs before motion takes place.

2 There are three factors which affect the size and direction of frictional forces.

(i) The size of the frictional force depends on the type of surface (a block of wood slides more easily on a polished metal surface than on a rough concrete surface).

(ii) The size of the frictional force depends on the size of the force acting at right angles to the surfaces in contact, called the **normal force**. Thus, if the weight of a block of wood is doubled, the frictional force is doubled when it is sliding on the same surface.

(iii) The direction of the frictional force is always opposite to the direction of motion. Thus the frictional force opposes motion, as shown in *Figure 9.1*.

3 The **coefficient of friction**, μ, is a measure of the amount of friction existing between two surfaces. A low value of coefficient of friction indicates that the force required for sliding to occur is less

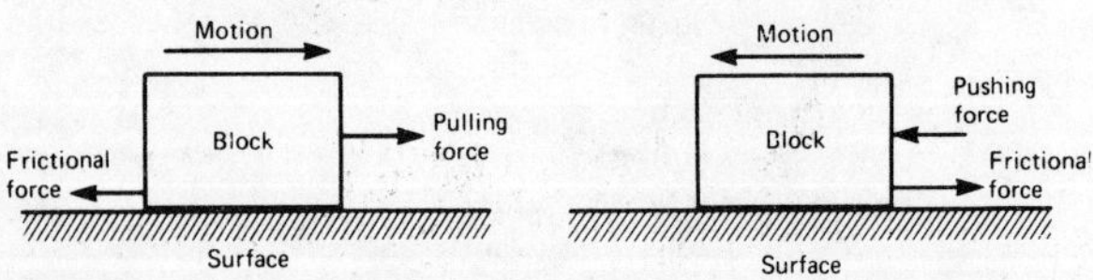

Figure 9.1

than the force required when the coefficient of friction is high. The value of the coefficient of friction is given by

$$\mu = \frac{\text{frictional force, } (F)}{\text{normal force, } (N)}$$

Transposing gives: **frictional force = μ × normal force,**

$$\boxed{F = \mu N}$$

The direction of the forces given in this equation are as shown in ***Figure 9.2*. The coefficient of frction is the ratio of a force to a** force, and hence has no units. Typical values for the coefficient of

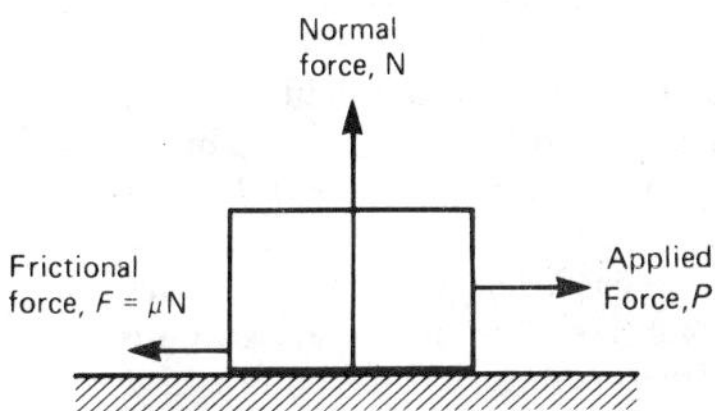

Figure 9.2

friction when sliding is occurring, i.e. the dynamic coefficient of friction are:

for polished oiled metal surfaces, less than 0.1
for glass on glass, 0.4
for rubber on tarmac, close to 1.0

4 In some applications, a low coefficient of friction is desirable, for example, in bearings, pistons moving within cylinders, on ski runs, and so on. However, for such applications as force being transmitted by belt drives and braking systems, a high value of coefficient is necessary.

Advantages and disadvantages of frictional forces

5 (a) Instances where frictional forces are an advantage include:

(i) Almost all fastening devices rely on frictional forces to keep them in place once secured, examples being screws, nails, nuts, clips and clamps.

(ii) Satisfactory operation of brakes and clutches rely on frictional forces being present.

(iii) In the absence of frictional forces, most accelerations along a horizontal surface are impossible. For example, a person's shoes just slip when walking is attempted and the types of a car just rotate with no forward motion of the car being experienced.

(b) **Disadvantages of frictional forces** include:

(i) Energy is wasted in the bearings associated with shafts, axles and gears due to heat being generated.
(ii) Wear is caused by friction, for example, in shoes, brake lining materials and bearings.
(iii) Energy is wasted when motion through air occurs (it is much easier to cycle with the wind rather than against it).

6 Two examples of **design implications** which arise due to frictional forces and how lubrication may or may not help are:
(i) Bearings are made of an alloy called white metal, which has a relatively low melting point. When the rotating shaft rubs on the white metal bearing, heat is generated by friction, often in one spot and the white metal may melt in this area, rendering the bearing useless. Adequate lubrication (oil or grease), separates the shaft from the white metal, keeps the coefficient of friction small and prevents damage to the bearing. For very large bearings, oil is pumped under pressure into the bearings and the oil is used to remove the heat generated, often passing through oil coolers before being recirculated. Designers should ensure that the heat generated by friction can be dissipated.
(ii) Wheels driving belts, to transmit force from one place to another, are used in many workshops. The coefficient of friction between the wheel and the belt must be high, and it may be increased by dressing the belt with a tar-like substance. Since frictional force is proportional to the normal force, a slipping belt is made more efficient by tightening it, thus increasing the normal and hence the frictional force. Designers should incorporate some belt tension mechanism into the design of such a system.

10 **Waves**

1 **Wave motion** is a travelling disturbance through a medium or through space, in which energy is transferred from one point to another without movement of matter.

2 **Examples where wave motion occurs include:**

(i) Water waves, such as are produced when a stone is thrown into a still pool of water;
(ii) waves on strings;
(iii) waves on stretched springs;
(iv) sound waves;
(v) light waves (see page 43);
(vi) radio waves;
(vii) infra-red waves, which are emitted by hot bodies;
(viii) ultra-violet waves, which are emitted by very hot bodies and some gas discharge lamps;
(ix) X-ray waves, which are emitted by metals when they are bombarded by high speed electrons;
(x) gamma-rays, which are emitted by radioactive elements.

Examples (i) to (iv) are **mechanical waves** and they require a medium (such as air or water) in order to move. Examples (v) to (x) are **electromagnetic waves** and do not require any medium — they can pass through a vacuum.

3 There are two types of wave, these being transverse and longitudinal waves.

(i) **Transverse waves** are where the particles of the medium move perpendicular to the direction of movement. For example, when a stone is thrown into a pool of still water, the ripple moves radially outwards but the movement of a floating object shows that the water at a particular point merely moves up and down. Light and radio waves are other examples of transverse waves.
(ii) **Longitudinal waves** are where the particles of the medium vibrate back and forth parallel to the direction of the wave travel. Examples include sound waves and waves in springs.

4 *Figure 10.1* shows a cross-section of a typical wave.

(i) **Wavelength** is the distance between two successive identical parts of a wave (for example, between two crests as shown in

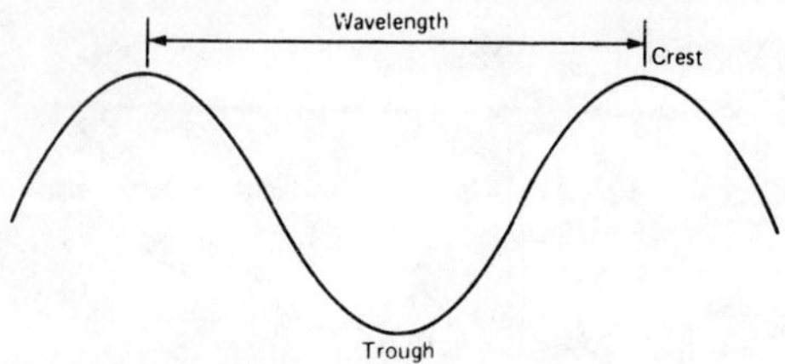

Figure 10.1

Figure 10.1). The symbol for wavelength is λ (Greek lambda) and its unit is metres.

(ii) **Frequency** is the number of complete waves (or cycles) passing a fixed point in one second. The symbol for frequency is f and its unit is the hertz, Hz.

(iii) The **velocity**, v of a wave is given by:

elocity = frequency × wavelength, i.e. $\boxed{v=f\lambda}$

The unit of velocity is metres per second. Thus, for example, if BBC radio 4 is transmitted at a frequency of 200 kHz and a wavelength of 1500 m, the velocity of the radio wave v is given by

$$v=f\lambda=(200\times 10^3)(1500)=\mathbf{3\times 10^8\ m\ s^{-1}}.$$

Fizeau's experiment to measure the velocitv of light

5 Fizeau, in 1849, was the first person to measure the velocity of light by a method not involving terrestrial observations. The principle of his experiment was as follows. Light passes between two teeth on a rotating toothed wheel which 'chops', the light into a series of flashes. Each flash then travels a large distance and is reflected back along its path. If the wheel has rotated so that a tooth has now reached the position from which the flash originated the returning flash of light is stopped by the tooth and is not observed at the eyepiece. *Figure 10.2* shows the arrangement.

The source of light is focussed by the lens L_1 at S. In the time it takes the light to travel from S, via the lenses L_2 and L_3 to the mirror M and back via L_3 and L_2 to the toothed wheel the latter has rotated so that tooth *a* now interrupts the light and the light source is eclipsed. In Fizeau's experiment the light travelled a distance of 10.72 miles. The wheel had 720 teeth and thus has to turn through 1/1440 of a revolution while the light travelled this

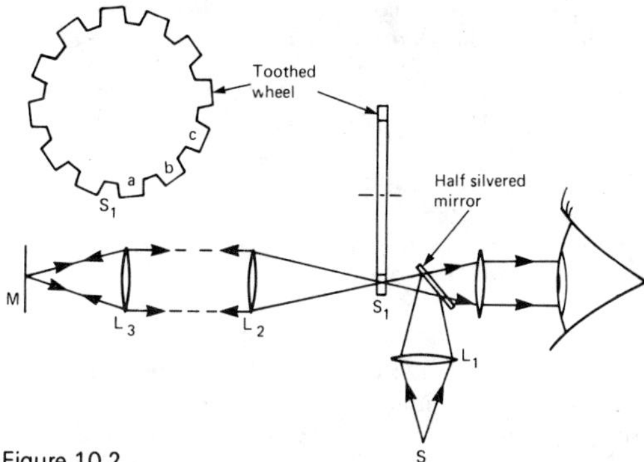

Figure 10.2

distance. Fizeau observed the first eclipse as 12.6 rev/s giving a value for the velocity of light of 3.133×10^8 m s^{-1}. Today's accepted value is $(2.997931 \pm 0.000003) \times 10^8$ m s^{-1}.

6 The **speed of waves** depends on the properties of the medium through which they travel.

(i) The speed of transverse waves in strings or springs is given by $v=\sqrt{(T/\mu)}$ where T is the tension and μ is the mass per unit length of the string or spring. For example, if a spring of mass 0.5 kg is stretched so that its length becomes 1.5 m and the tension in the spring is 48 N then:

the speed of transverse waves in the spring

$$v=\sqrt{\left(\frac{T}{\mu}\right)}=\sqrt{\left(\frac{48}{(0.5/1.5)}\right)}$$

$$=\sqrt{144}=\mathbf{12\ m\ s^{-1}}$$

(ii) The speed of longitudinal waves (sound) in a gas is given by

$$v=\sqrt{\left(\frac{\gamma p}{\rho}\right)}$$

where γ is a numerical constant,
p is the pressure of the gas
and ρ is the density of the gas.

(iii) The speed of longitudinal waves in a solid is given by

$$v=\sqrt{\left(\frac{E}{\rho}\right)}$$

where E is Young's modulus of elasticity
and ρ is the density of the solid.

Thus for steel, of density 7.8×10^3 kg m^{-3} and Young's modulus 2×10^{11} Pa, the speed of longitudinal waves

$$v = \sqrt{\left(\frac{E}{\rho}\right)}$$

$$= \sqrt{\left(\frac{2 \times 10^{11}}{7.8 \times 10^3}\right)} = \mathbf{5.1 \times 10^3\ m\ s^{-1}}.$$

7 **Electromagnetic waves** are waves which need no medium in which to propagate. We do not talk of oscillating particles in this case and the waves are produced by oscillating electric and magnetic fields perpendicular to each other and to the direction of propagation of the wave (see *Figure 10.3*). Electromagnetic waves are thus transverse waves.

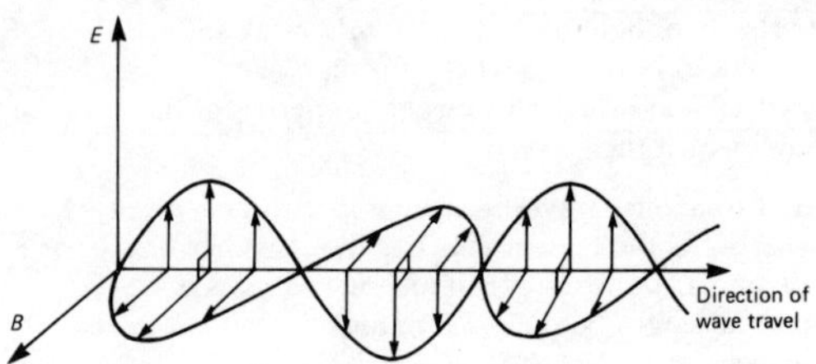

Figure 10.3

Table 10.1 gives some details of the family of electromagnetic waves. The speed of electromagnetic waves in vacuo is the same for all wavelengths. It is approximately 3×10^8 m s^{-1} and is normally given by the symbol c.

8 A transverse wave is said to be **plane polarised** if the oscillations are restricted to one plane. In the case of an electromagnetic wave it is said to be plane polarized if the electric field vector, E, is restricted to one plane. Radio waves, radar waves and microwaves are plane polarised as a result of the nature of the source. Visible light, infra red, ultra violet, X-rays and γ-rays are often unpolarised and the plane of vibration of E changes randomly.

9 If a source of sound is moving relative to an observer the observed frequency f_0 is not equal to the true frequency f. They

Table 10.1

Name of radiation		λ, *Wavelength of radiation* (m)
		10^{-14}
γ-rays	X-rays	10^{-13}
		10^{-12}
		10^{-11}
		10^{-10}
		10^{-9}
Ultra violet		10^{-8}
		10^{-7}
Visible light		10^{-6}
		10^{-5}
Infra red		10^{-4}
		10^{-3}
Microwaves		10^{-2}
		10^{-1}
Short radio waves		1
		10^{1}
Medium radio waves		10^{2}
		10^{3}
Long radio waves		10^{4}
		10^{5}

are related by the following equation

$$f_0 = f\,\frac{(v-y)}{(v-x)}$$

where v is the velocity of the sound;
x is the velocity of the source;
y is the velocity of the observer.

All three velocities are measured in the same direction. For example, a stationary observer measures the frequency of a source of sound travelling with a velocity of 130 m s^{-1} towards him as 1.5 kHz. If the velocity of sound is 330 m s^{-1} then the observed frequency is given by:

$$\text{observed frequency} = (\text{true frequency})\,\frac{(v-y)}{(v-x)}$$

$$\text{i.e. } 1.5\times 10^3 = (\text{true frequency})\left(\frac{330-0}{330-130}\right)$$

from which,

$$\text{true frequency} = \frac{1.5\times 10^3\times 200}{330} = \mathbf{909\ Hz}.$$

10 (i) White light consists of a continuous band of colours called a **continuous spectrum**. All visible wavelengths are present.
(ii) An **emission spectrum** is produced by giving extra energy to the atoms of a gas at low pressure. This energy is then radiated as electromagnetic waves of specific wavelength which are characteristic of the gas. These wavelengths may be in the infra-red, the visible and/or the ultra violet region. An emission spectrum may be a mixture of wavelengths due to the fact that two or more elements are present.
(iii) An **absorption spectrum** is produced when white light is passed through a gas which absorbs those wavelengths which it would normally emit. The emerging light is deficient in just those wavelengths. This is because, although the gas reradiates those wavelengths which it has absorbed, they are reradiated in all directions and thus the intensity is diminished in the direction of the observer. When viewed using a spectroscope the absorption spectrum is seen as a continuous spectrum crossed with black lines corresponding to the absorbed wavelengths.

11 **Reflection** is a change in direction of a wave while the wave remains in the same medium. There is no change in the speed of a reflected wave. All waves are reflected when they meet a surface through which they cannot pass. For example,

(i) light waves are reflected by mirrors;
(ii) water waves are reflected at the end of a bath or by a sea wall;
(iii) sound waves are reflected at a wall (which can produce an echo);
(iv) a wave reaching the end of a spring or string is reflected; and
(v) television waves are reflected by satellites above the Earth.

Experimentally, waves produced in an open tank of water may readily be observed to reflect off a sheet of glass placed at right angles to the surface of the water.

12 **Refraction** is a change in direction of a wave as it passes from one medium to another. All waves refract, and examples include:

(i) a light wave changing its direction at the boundary between air and glass (see *Figure 10.4*);
(ii) sea waves refracting when reaching more shallow water; and
(iii) sound waves refracting when entering air of different

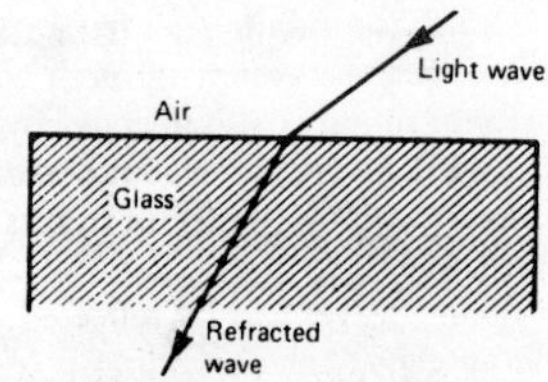

Figure 10.4

temperature (see para. 14).

Experimentally, if one end of a water tank is made shallow the waves may be observed to travel more slowly in these regions and are seen to change direction as the wave strikes the boundary of the shallow area. The greater the change of velocity the greater is the bending or refraction.

13 A **sound wave** is a series of alternate layers of air, one layer at a pressure slightly higher than atmospheric, called compressions, and the other slightly lower, called refraction. In other words, **sound is a pressure wave**. *Figure 10.5(a)* represents layers of undisturbed air. *Figure 10.5(b)* shows what happens to the air when a sound wave passes.

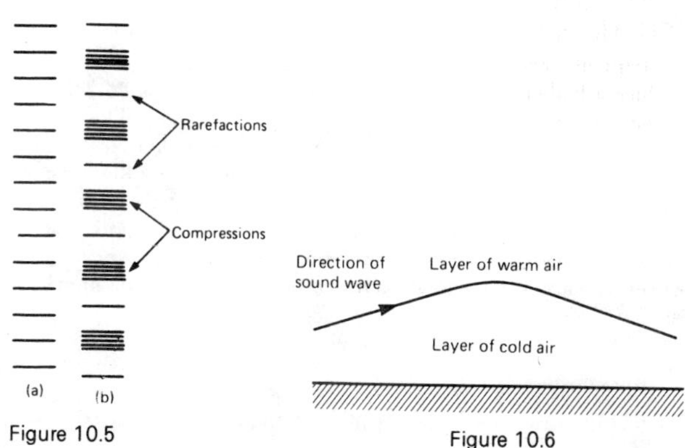

Figure 10.5

Figure 10.6

Characteristics of sound waves

14 (i) Sound waves can travel through solids, liquids and gases, but not through a vacuum.

(ii) Sound has a finite (i.e. fixed) velocity, the value of which depends on the medium through which it is travelling. The velocity of sound is also affected by temperature. Some typical values for the velocity of sound are: air 331 m s^{-1} at 0°C, and 342 m s^{-1} at 18°C, water 1410 m s^{-1} at 20°C and iron 5100 m s^{-1} at 20°C.

(iii) Sound waves can be reflected, the most common example being an echo. Echo-sounding is used for charting the depth of the sea.

(iv) Sound waves can be refracted. This occurs, for example, when sound waves meet layers of air at different temperatures. If a sound wave enters a region of higher temperature the medium has different properties and the wave is bent as shown in *Figure 10.6*, which is typical of conditions that occur at night.

15 Sound waves are produced as a result of vibrations:

(i) In brass instruments, such as trumpets and trombones, or wind instruments, such as clarinets and oboes, sound is due to the vibration of columns of air.

(ii) In stringed instruments, such as guitars and violins, sound is produced by vibrating strings causing air to vibrate. Similarly, the vibration of vocal chords produces speech.

(iii) Sound is produced by a tuning fork due to the vibration of the metal prongs.

(iv) Sound is produced in a loudspeaker due to vibrations in the cone.

16 The pitch of a sound depends on the frequency of the vibration; the higher the frequency, the higher is the pitch. The frequency of sound depends on the form of the vibrator. The valves of a trumpet or the slide of a trombone lengthen or shorten the air column and the fingers alter the length of strings on a guitar or violin. The shorter the air column or vibrating string the higher the frequency and hence pitch. Similarly, a short tuning fork will produce a higher pitch note than a long tuning fork. Frequencies between about 20 Hz and 20 kHz can be perceived by the human ear.

11 Interference and diffraction

1 At the point where two waves cross, the total displacement is the vector sum of the individual displacements due to each wave at that point. This is the **principle of superposition**. If the two waves are either both transverse or both longitudinal, **interference** effects may be observed. It is not necessary for the two waves to have the same frequencies or amplitudes for the above statements to be true, although these are the waves considered in this chapter.

2 Consider two transverse waves of the same frequency and amplitude travelling in opposite directions superimposed on one another. Interference takes place between the two waves and a **standing** or **stationary wave** is produced. The standing wave is shown in *Figure 11.1*.

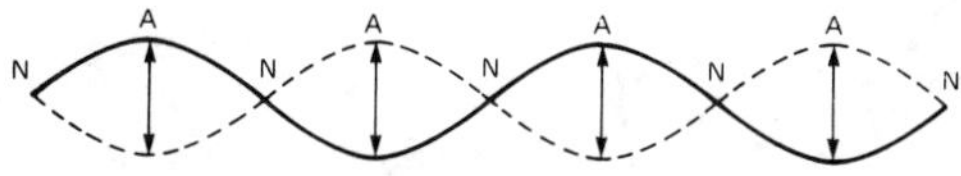

Figure 11.1

The wave does not progress to the left or right and certain parts of the wave called **nodes**, labelled N in the diagram, do not oscillate. Those positions on the wave which undergo maximum disturbance are called **antinodes**, labelled A. The distance between adjacent nodes or adjacent antinodes is $\lambda/2$, where λ is the wavelength. Standing waves may be set up in a string, for example, when a wave is reflected at the end of the string and is superimposed on the incoming wave. Under these circumstances standing waves are produced only for certain frequencies. Also the nodes may not be perfect because the reflected wave may have a slightly reduced amplitude.

3 Two sound (longitudinal) waves of the same amplitude and frequency travelling in opposite directions and superimposed on each other also produce a standing wave. In this case there are

also displacement nodes where the medium does not oscillate and displacement antinodes where the displacement is a maximum.

4 The interference effects mentioned above are not always restricted to the line between the two sources of waves. Two dimensional interference patterns are produced on the surface of water in a ripple tank for example. In this case two dippers, usually oscillating in phase and with the same frequency, produce circular ripples on the surface of the water and interference takes place where the circular ripples overlap. The resulting interference pattern is shown in *Figure 11.2.* The sources of the waves are S_1 and S_2.

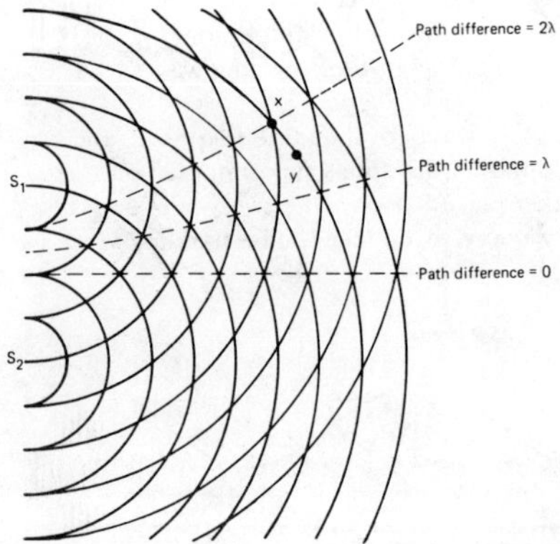

Figure 11.2

Consider a point X on the surface. (S_2X-S_1X) is called the path difference. If $(S_2X-S_1X)=n\lambda$, where n is an integer, and λ is the wavelength, the waves arriving at X from S_1 and S_2 must be in phase. (In *Figure 11.2*, $(S_2X-S_1X)=2\lambda$.) At X, constructive interference takes place and the resultant amplitude is a maximum. At a second point, Y, positioned such that $(S_2Y-S_1Y)=(n+\frac{1}{2})\lambda$, the waves arriving at Y from S_1 and S_2 are out of phase. The resultant amplitude at Y is a minimum and destructive interference has taken place. (In *Figure 11.2* $(S_2Y-S_1Y)=\lambda$, that is, $n=1$).

5 When sea waves are incident on a barrier which is parallel to them a disturbance is observed beyond the barrier in that region where it might be thought that the water would remain

undisturbed. This is because waves may spread round obstacles into regions which would be in shadow if the energy travelled exactly in straight lines. This phenomenon is called **diffraction**. All waves whether transverse or longitudinal exhibit this property. If light, for example, is incident on a narrow slit, diffraction takes place. The diffraction pattern on a screen placed beyond the slit is not perfectly sharp. The intensity of the image varies as shown in *Figure 11.3*.

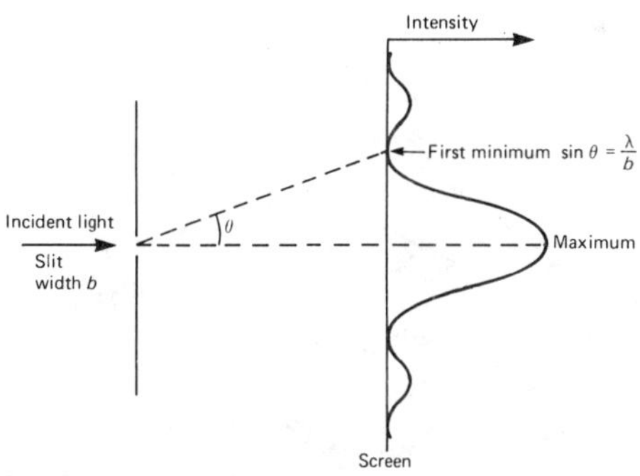

Figure 11.3

A consequence of diffraction is that if light from two sources which are close together pass through a slit or small circular aperture, the diffraction patterns of the two sources may overlap to such an extent that they appear to be one source. If they are to be distinguished as two separate sources, the angular separation, θ in radians, of the two sources, must be greater than λ/b where λ is the wavelength of the light and b is the width of the slit (see *Figure 11.4*). For a circular aperture the condition becomes $\theta > 1.22\lambda/b$. This is known as the **Rayleigh criterion**.

6 If light falls on two narrow parallel slits with a small separation light passes through both slits and because of diffraction there is an overlapping of the light and interference takes place. This is shown in *Figure 11.5*. The interference effects are similar to those described for water ripples in para 4.

Suppose light from the two slits meet at a point on a distant screen. Since the distance between the slits is much less than the slit to screen distance the two light beams will be very nearly parallel. See *Figure 11.6*. If the path difference is $n\lambda$, where n is an

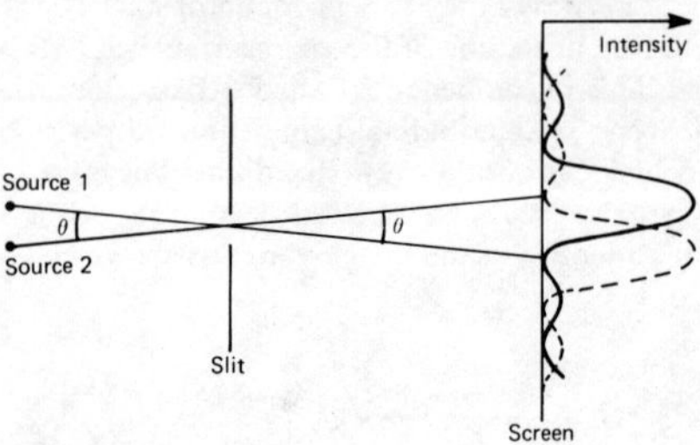

Figure 11.4

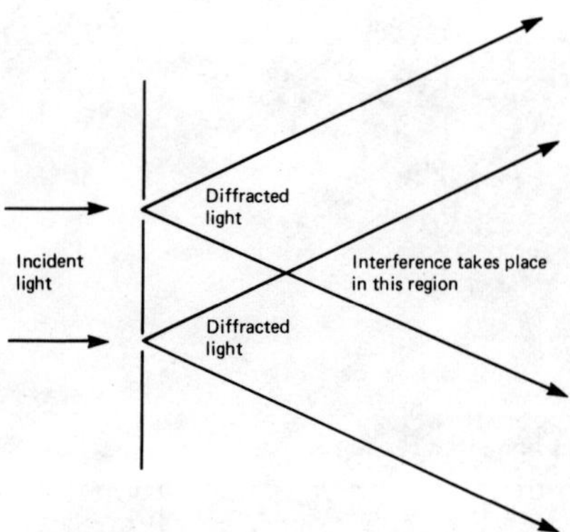

Figure 11.5

integer and λ is the wavelength there will be constructive interference and a maximum intensity occurs on the screen.

But from *Figure 11.6*, the path difference is BC, that is, $d \sin \theta$. Thus for a maximum intensity on the screen,

$$n\lambda = d \sin \theta$$

that is, $\sin \theta = \frac{n\lambda}{d}$

Thus maximum values occur where

$$\sin \theta = \frac{\lambda}{d}, \frac{2\lambda}{d}, \frac{3\lambda}{d} \text{ etc}$$

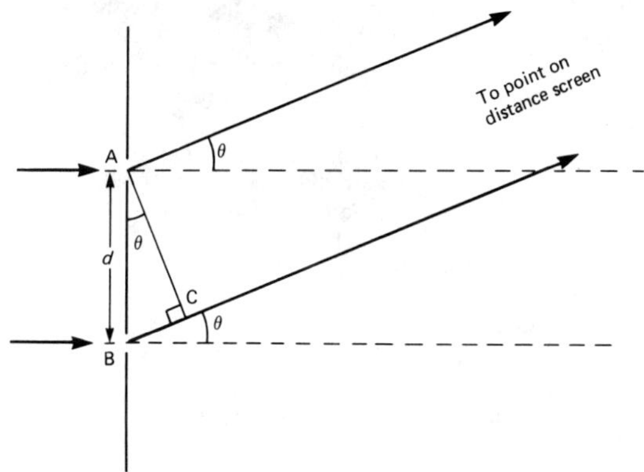

Figure 11.6

The intensity of the interference pattern on the screen at various distances from the polar axis is shown in *Figure 11.7.* The pattern is modified by the type of diffraction pattern produced by a single slit.

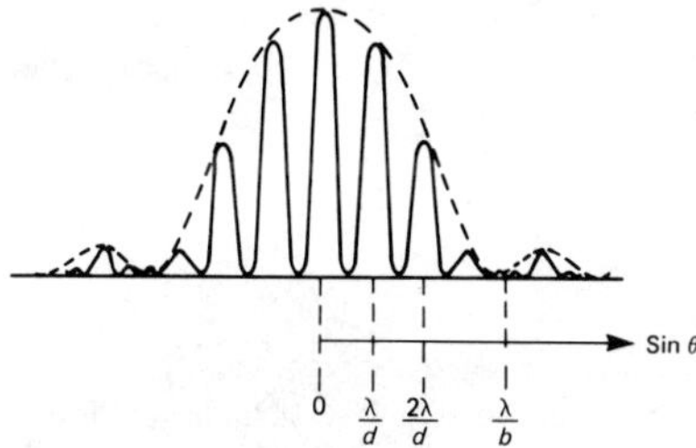

Figure 11.7

7 A diffraction grating is similar to the two slit arrangement, but with a very large number of slits. Very sharp values of maximum intensity are produced in this case. If the slit separation is d and light is incident along the normal to the grating, the condition for a maximum is $n\lambda = d \sin \theta$. When $n = 0$ then $\sin \theta = 0$,

and the light travels straight through the grating to give the **zero order** maximum (marked O in *Figure 11.7*).

If $n=1$, then $\sin\theta=\lambda/d$. This gives the direction of the first order maximum and the path difference from adjacent slits to the **first order** maximum is λ.

If $n=2$, then $\sin\theta=2\lambda/d$. This gives the direction of the **second order** maximum. The path difference is 2λ. If white light is incident on a diffraction grating a continuous spectrum is produced because the angle at which the first order emerges from the grating depends on the wavelength. Thus the diffraction grating may be used to determine the wavelengths present in a source of light.

8 Atoms in a crystal diffract X-rays which are incident upon them and information may be gained about crystal structure from the analysis of the diffraction pattern obtained. When X-rays strike atoms in a crystal, each atom scatters the X-rays in all directions. However in certain directions constructive interference takes place. In *Figure 11.8* a lattice of atoms is shown, in which X-ray strike atoms and are scattered. The X-rays emerging in a particular direction are considered.

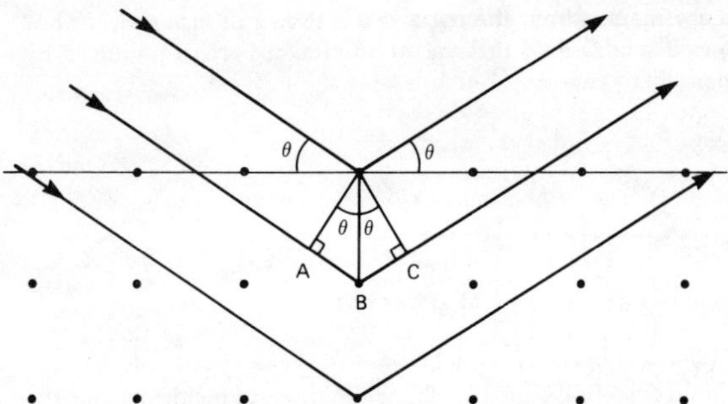

Figure 11.8

Three planes of atoms are shown. The X-rays 'reflected' from the top and middle planes (and any other pair of adjacent planes) will be in phase if their path difference is $n\lambda$, where n is an integer and λ is the wavelength of the X-rays.
The path difference $=AB+BC=d\sin\theta+d\sin\theta=2d\sin\theta$
where d is the separation between planes.
Thus the condition for constructive interference is $\boldsymbol{n\lambda=2d\sin\theta}$.
The angle θ is called the **glancing angle** and the equation is known as **Bragg's Law**.

12 Light rays

1 (i) Light is an electromagnetic wave (see page 29) and the straight line paths followed by very narrow beams of light, along which light energy travels, are called **rays**.
(ii) The behaviour of light rays may be investigated by using a **ray-box**. This consists merely of a lamp in a box containing a narrow slit which emits rays of light.
(iii) Light always travels in straight lines although its direction can be changed by reflection or refraction.

Reflection of light

2 *Figure 12.1* shows a ray of light, called the incident ray, striking a plane mirror at O, and making an angle i with the normal, which is a line drawn at right angles to the mirror at O, i is called the **angle of incidence**. The light ray reflects as shown making an angle r with the normal, r is called the **angle of reflection**.

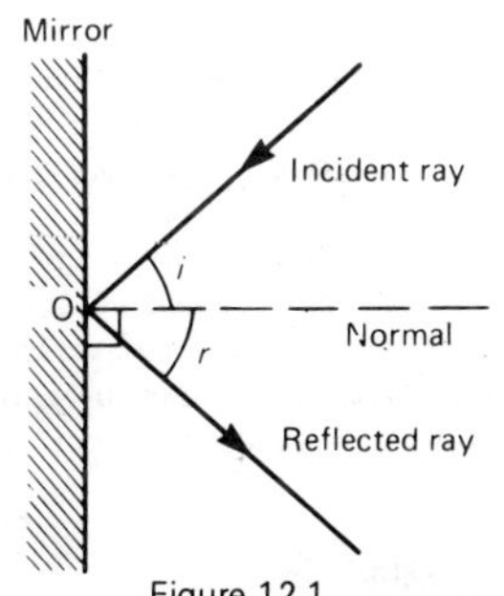

Figure 12.1

There are **two laws of reflection**:
(i) The angle of incidence is equal to the angle of reflection (i.e. $i=r$ in *Figure 12.1*).
(ii) The incident ray, the normal at the point of incidence and the reflected ray all lie in the same plane.

A **simple periscope arrangement** is shown in *Figure 12.2*. A ray of light from O strikes a plane mirror at an angle of 45° at point P. Since from the laws of reflection the angle of incidence i is equal to the angle of reflection r then $i=r=45°$. Thus angle OPQ = 90° and the light is reflected through 90°. The ray then strikes another mirror at 45° at point Q. Thus a = b = 45°, angle PQR = 90° and the light ray is again reflected through 90°. Thus the light from O finally travels in the direction QR, which is parallel to OP, but displaced by the distance PQ. The arrangement thus acts as a periscope.

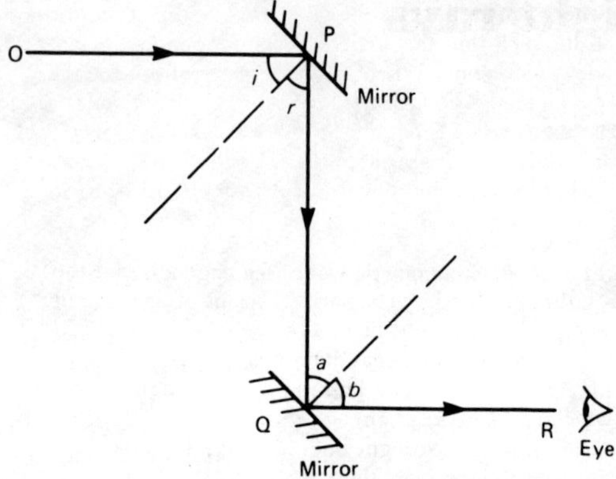

Figure 12.2

Refraction of light

3 (i) When a ray of light passes from one medium to another the light undergoes a change in direction. This displacement of light rays is called **refraction**.
(ii) *Figure 12.3* shows the path of a ray of light as it passes through a parallel sided glass block. The incident ray AB which has an

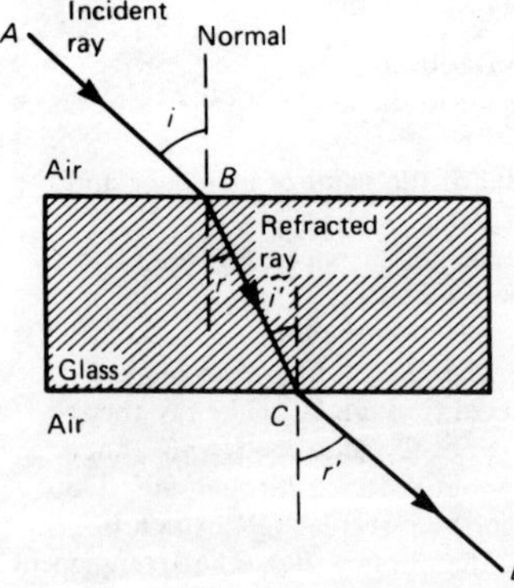

Figure 12.3

angle of incidence i enters the glass block at B. The direction of the ray changes to BC such that the angle r is less than angle i. r is called the angle of refraction. When the ray emerges from the glass at C the direction changes to CD, angle r' being greater than i'. The final emerging ray CD is parallel to the incident ray AB.
(iii) In general, when entering a more dense medium from a less dense medium, light is refracted towards the normal and when it passes from a dense to a less dense medium it is refracted away from the normal.

4 (i) **Lenses** are pieces of glass or other transparent material with a spherical surface on one or both sides. When light is passed through a lens it is refracted.
(ii) Lenses are used in spectacles, magnifying glasses and microscopes, telescopes, cameras and projectors.
(iii) There are a number of different shaped lenses and two of the most common are shown in *Figure 12.4*. *Figure 12.4(a)* shows a **bi-convex lens**, so called since both its surfaces curve outwards. *Figure 12.4(b)* shows a **bi-concave lens**, so called since both of its surfaces curve inwards. The line passing through the centre of curvature of the lens surface is called the **principal axis**.

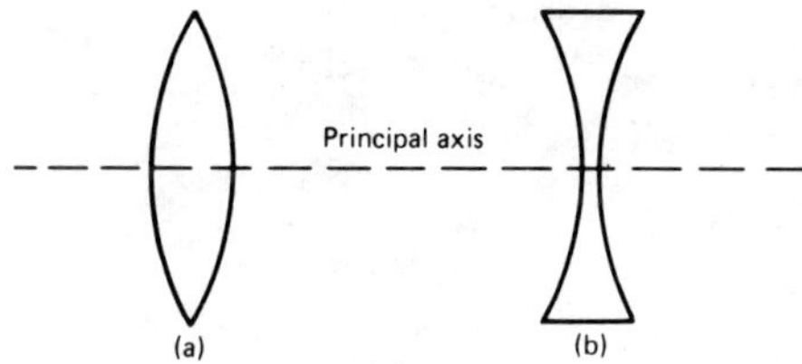

Figure 12.4

5 (i) *Figure 12.5* shows a number of parallel rays of light passing through a bi-convex lens. They are seen to converge at a point F on the principal axis.
(ii) *Figure 12.6* shows parallel rays of light passing through a bi-concave lens. They are seen to diverge such that they appear to come from a point F which lies between the source of light and the lens, on the principal axis.
(iii) In both *Figure 12.5* and *Figure 12.6*, F is called the **principal focus** or the **focal point**, and the distance from F to the centre of the lens is called the **focal length** of the lens.

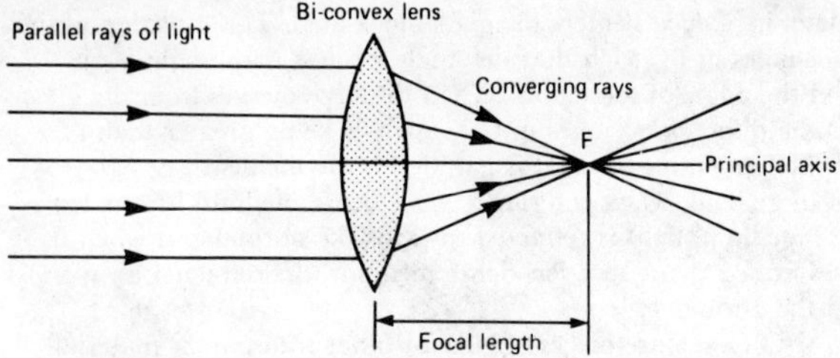

Figure 12.5

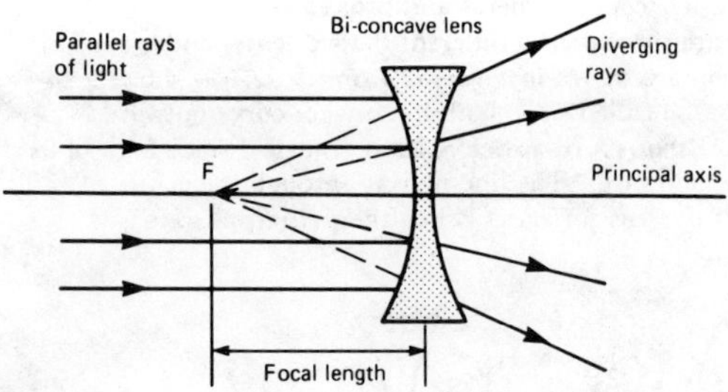

Figure 12.6

6 An **image** is the point from which reflected rays of light entering the eye appear to have originated. If the rays actually pass through the point then a **real image** is formed. Such images can be formed on a screen. *Figure 12.7* illustrates how the eye collects rays from an object after reflection from a plane mirror. To the eye, the rays appear to come from behind the mirror and the eye sees what seems to be an image of the object as far behind the mirror as the object is in front. Such an image is called a **virtual image** and this type cannot be shown on a screen.

7 Lenses are important since they form images when an object emitting light is placed at an appropriate distance from the lens.

(a) Bi-convex lenses

(i) *Figure 12.8* shows an object O (a source of light) at a distance of more than twice the focal length from the

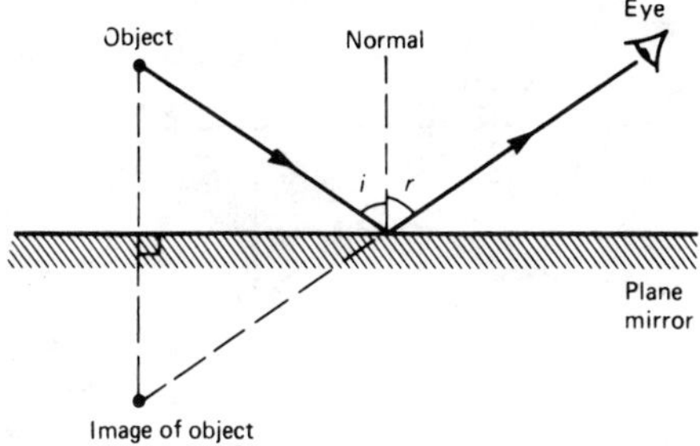

Figure 12.7

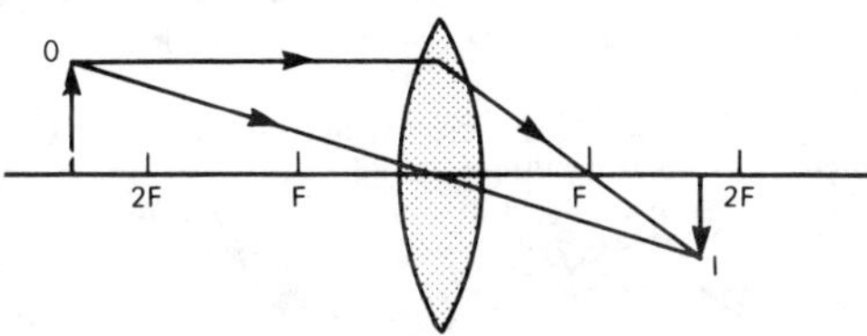

Figure 12.8

lens. To determine the position and size of the image two rays only are drawn, one parallel with the principal axis and the other passing through the centre of the lens. The image, I, produced is real, inverted (i.e. upside down), smaller than the object (i.e. diminished) and at a distance between one and two times the focal length from the lens. This arrangement is used in a **camera**.

(ii) *Figure 12.9* shows an object O at a distance of twice the focal length from the lens. This arrangement is used in a **photocopier**.

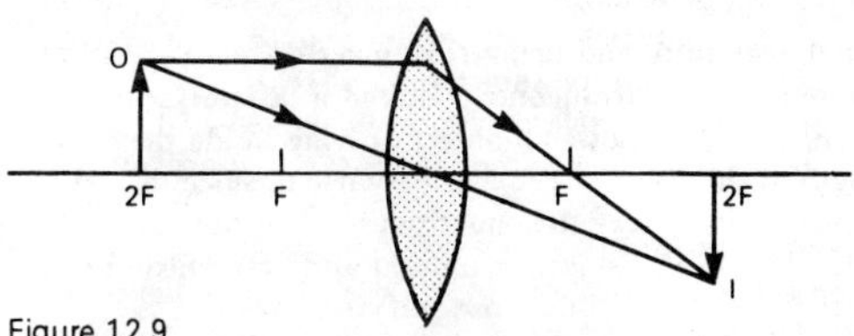

Figure 12.9

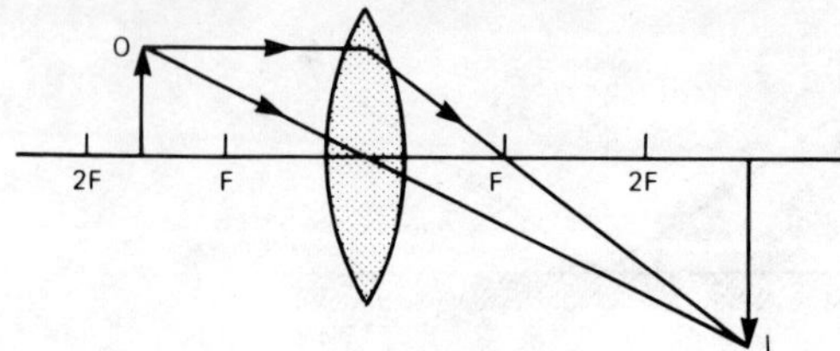

Figure 12.10

(iii) *Figure 12.10* shows an object O at a distance of between one and two focal lengths from the lens. The image I is real, inverted, magnified (i.e. greater than the object) and at a distance of more than twice the focal length from the lens. This arrangement is used in a **projector**.

(iv) *Figure 12.11* shows an object O at the focal length of the lens. After passing through the lens the rays are parallel. Thus the image I can be considered as being

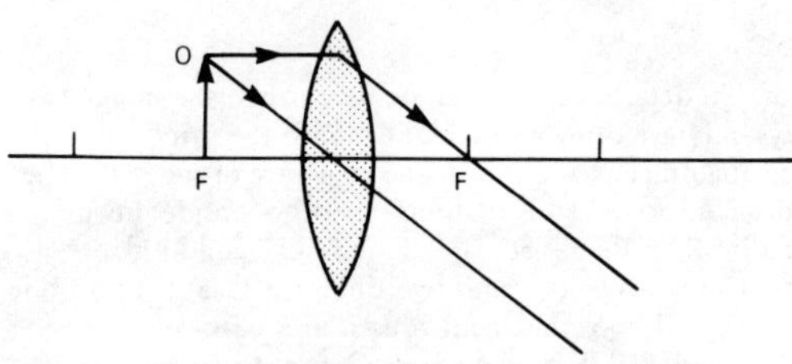

Figure 12.11

found at infinity and being real, inverted and very much magnified. This arrangement is used in a **spotlight**.

(v) *Figure 12.12* shows an object O lying inside the focal length of the lens. The image I is virtual, since the rays of light only appear to come from it, is on the same side of the lens as the object, is upright and magnified. This arrangement is used in a **magnifying glass**.

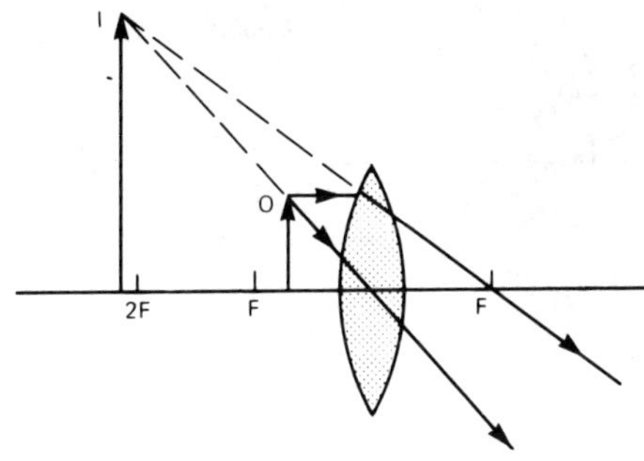

Figure 12.12

(b) **Bi-concave lenses**

For a bi-concave lens, as shown in *Figure 12.13*, the object O can be any distance from the lens and the image I formed is virtual upright, diminished and is found on the same side of the lens as the object. This arrangement is used in some types of **spectacles**.

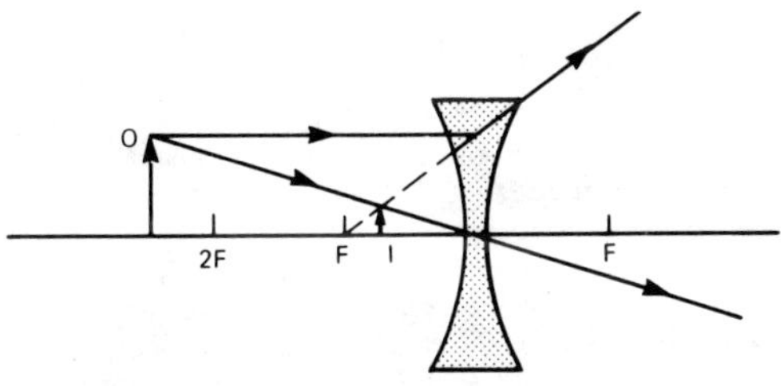

Figure 12.13

8 A **compound microscope** is able to give large magnification by the use of two (or more) lenses. An object O is placed outside the focal length F_O of a bi-convex lens, called the objective lens (since it is near to the object), as shown in *Figure 12.14*. **This produces a real, inverted, magnified image I_1. This** image then acts as the object for the eyepiece lens (i.e. the lens nearest the eye), and falls inside the focal length F_O of this lens.

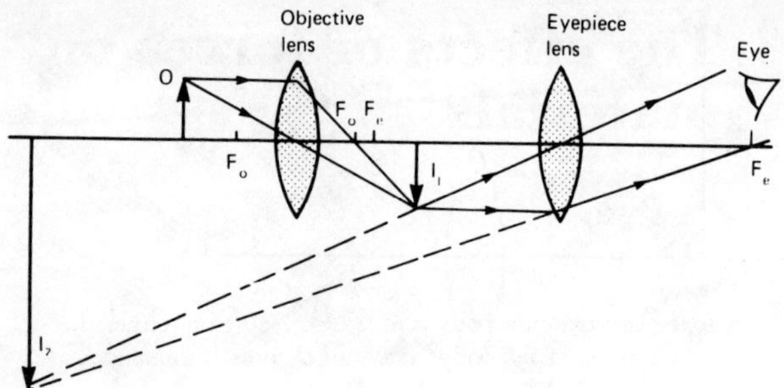

Figure 12.14

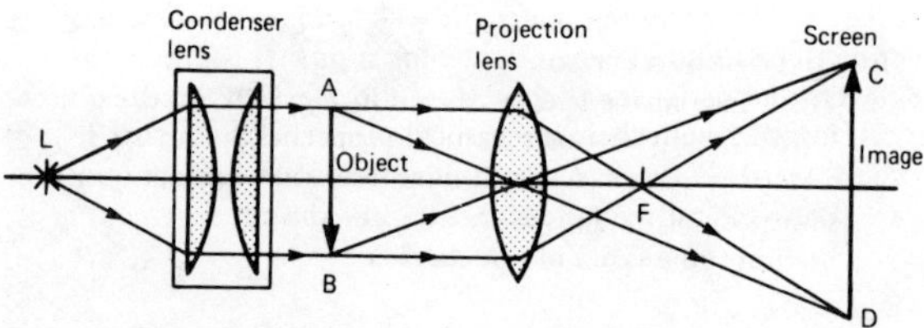

Figure 12.15

The eyepiece lens then produces a magnified, virtual, inverted image I_2 as shown in *Fiugre 12.14*.

9 A **simple projector arrangement** is shown in *Figure 12.15* and consists of a source of light and two lens systems. L is a brilliant source of light, such as a tungsten filament. One lens system called the condenser (usually consisting of two converting lenses as shown), is used to produce an intense illumination of the object AB, which is a slide transparency or film. The second lens, called the projection lens, is used to form a magnified, real, upright image of the illuminated object on a distant screen CD.

13 The effects of forces on materials

1 A **force** exerted on a body can cause a change in either the shape or the motion of the body. The unit of force is the *newton*, **N**.

2 No solid body is perfectly rigid and when forces are applied to it, changes in dimensions occur. Such changes are not always perceptible to the human eye since they are so small. For example, the span of a bridge will sag under the weight of a vehicle and a spanner will bend slightly when tightening a nut. It is important for engineers and designers to appreciate the effects of forces on materials, together with their mechanical properties.

3 The three main types of mechanical force that can act on a body are (i) tensile, (ii) compressive, and (iii) shear.

Tensile force

4 Tension is a force which tends to stretch a material, as shown in *Figure 13.1(a)*.
Examples include:
(i) the rope or cable of a crane carrying a load is in tension;
(ii) rubber bands, when stretched, are in tension;
(iii) a bolt; when a nut is tightened, a bolt is under tension.
A tensile force, i.e. one producing tension, increases the length of the material on which it acts.

Compressive force

5 Compression is a force which tends to squeeze or crush a material, as shown in *Figure 13.1(b)*. Examples include:
(i) a pillar supporting a bridge is in compression;
(ii) the sole of a shoe is in compression;
(iii) the job of a crane is in compression.
A compressive force, i.e. one producing compression, will decrease the length of the material on which it acts.

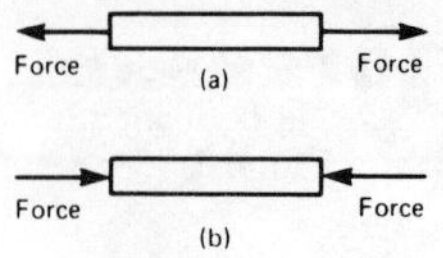

Figure 13.1

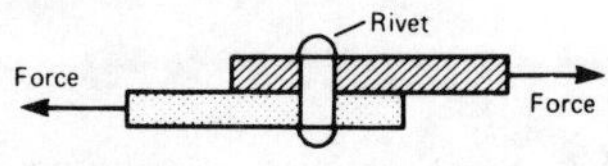

Figure 13.2

Shear force

6 Shear is a force which tends to slide one face of the material over an adjacent face. Examples include:

(i) a rivet holding two plates together is in shear if a tensile force is applied between the plates, as shown in *Figure 13.2*.
(ii) a guillotine cutting sheet metal, or garden shears each provide a shear force,
(iii) a horizontal beam is subject to shear force,
(iv) transmission joints on cars are subject to shear forces.

A shear force can cause a material to bend, slide or twist.

7 Forces acting on a material cause a change in dimensions and the material is said to be in a state of **stress**. Stress is the ratio of the applied force F to cross-sectional area A of the material. The symbol used for tensile and compressive stress is σ (Greek letter sigma). The unit of stress is the Pascal Pa, where 1 Pa = 1 N/m^2.

Hence $$\boldsymbol{\sigma = \frac{F}{A}} \textbf{ Pa,}$$

where F is the force in newtons and A is the cross-sectional area in square metres.

For tensile and compressive forces, the cross-sectional area is that which is at right angles to the direction of the force.

8 The fractional change in a dimension of a material produced by a force is called the **strain**.

For a tensile or compressive force, strain is the ratio of the change of length to the original length. The symbol used for strain is ε (Greek epsilon). For a material of length l metres which changes in length by an amount x metres when subjected to stress,

$$\boldsymbol{\varepsilon = \frac{x}{l}}$$

Strain is dimensionless and is often expressed as a percentage,

i.e. $$\textbf{Percentage strain} = \frac{x}{l} \times \textbf{100\%}$$

9 (i) **Elasticity** is the ability of a material to return to its original shape and size on the removal of external forces.
(ii) **Plasticity** is the property of a material of being permanently deformed by a force without breaking. Thus if a material does not return to the original shape, it is said to be plastic.
(iii) Within certain load limits, mild steel, copper, polythene and rubber are examples of elastic materials; lead and plasticine are examples of plastic materials.

10 If a tensile force applied to a uniform bar of mild steel is gradually increased and the corresponding extension of the bar is measured, then provided the applied force is not too large, a graph depicting these results is likely to be **as shown in *Figure 13.3*.** Since the graph is a straight line,

extension is directly proportional to the applied force.

11 If the applied force is large, it is found that the material no longer returns to its original length when the force is removed. The

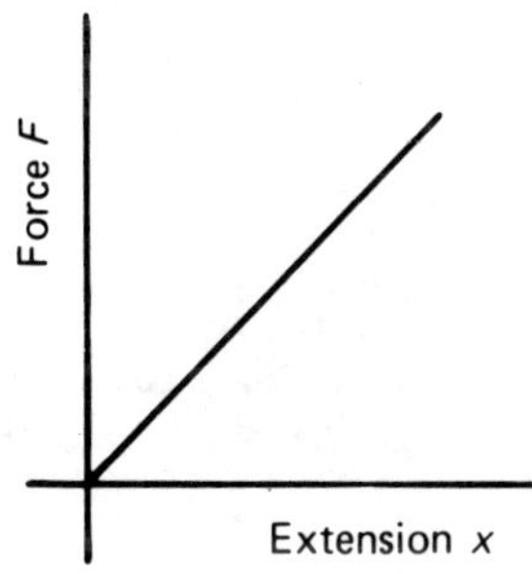

Figure 13.3

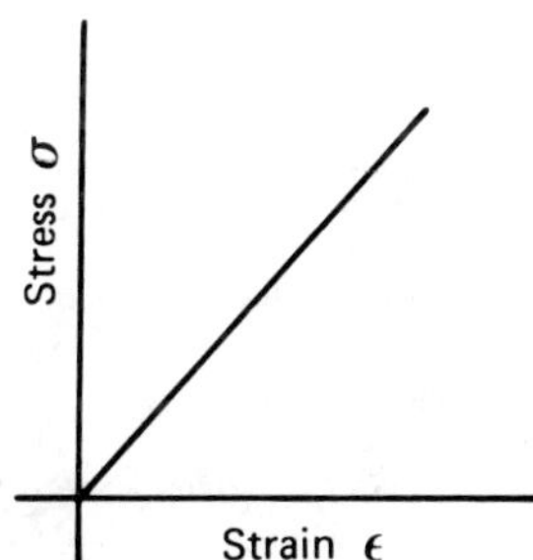

Figure 13.4

Stress $\sigma = \frac{F}{A}$, from para. 7,

and since, for a particular bar, A can be considered as constant

then, $F \alpha \sigma$.

Strain $\varepsilon = \frac{x}{l}$, from para. 8,

and since for a particular bar l is constant,

then, $x \alpha \varepsilon$.

Hence for stress applied to a material below the elastic limit a graph of stress/strain will be as shown in *Figure 13.4*, and is a similar shape to the force/extension graph of *Figure 13.3*.

12 **Hooke's law** states:

'Within the elastic limit, the extension of a material is proportional to the applied force.'

It follows, from para. 11, that:

'Within the elastic limit of a material, the strain produced is directly proportional to the stress producing it.'

13 Within the elastic limit, stress $\propto$ strain

Hence, stress= (a constant) × strain

This constant of proportionality is called **Young's Modulus of Elasticity** and is given by the symbol E. The value of E may be determined from the gradient of the straight line portion of the stress/strain graph. The dimensions of E are pascals (the same as for stress, since strain is dimensionless).

$$\boxed{E=\frac{\sigma}{\varepsilon}\ \text{Pa}}$$

Some typical values for Young's modulus of elasticity, E, include: aluminium 70 GPa (i.e. 70×10^9 Pa), brass 100 GPa, copper 110 GPa, diamond 1200 GPa, mile steel 210 GPa, lead 18 GPa, tungsten 410 GPa, cast iron 110 GPa, zinc 110 GPa.

14 A material having a large value of Young's modulus is said to have a high value of stiffness, where stiffness is defined as:

$$\textbf{Stiffness}=\frac{\textbf{force } F}{\textbf{extension } x}$$

For example, mild steel is much stiffer than lead.

Since $E=\dfrac{\sigma}{\varepsilon}$ and $\sigma=\dfrac{F}{A}$ and $\varepsilon=\dfrac{x}{l}$

then $E=\dfrac{F/A}{x/l}$, i.e. $E=\dfrac{Fl}{Ax}=\left(\dfrac{F}{x}\right)\left(\dfrac{l}{A}\right)$

i.e. $E=(\textbf{stiffness})\times\left(\dfrac{l}{A}\right)$

Stiffness $(=F/x)$ is also the gradient of the force/extension graph

$$\text{Hence: } E = (\textbf{gradient of force/extension graph})\left(\frac{l}{A}\right).$$

Since l and A for a particular specimen are constant, the greater Young's modulus the greater the stiffness.

15 A **tensile test** is one in which a force is applied to a specimen of a material in increments and the corresponding extension of the specimen noted. The process may be continued until the specimen breaks into two parts and this is called testing to destruction. The testing is usually carried out using a universal testing machine which can apply either tensile or compressive forces to a specimen in small, accurately measured steps. BS 18 gives the standard procedure for such a test. Test specimens of a material are made to standard shapes and sizes and two typical test pieces are shown in *Figure 13.5*. The results of a tensile test may be

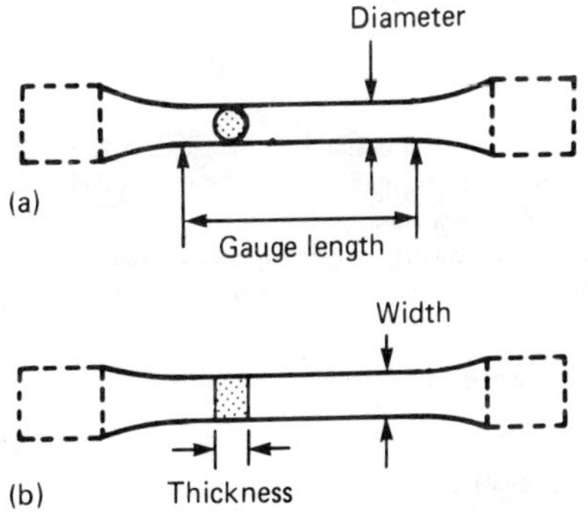

Figure 13.5

plotted on a load/extension graph and a typical graph for a mild steel specimen is shown in *Figure 13.6*

(i) Between A and B is the region in which Hooke's law applies and stress is directly proportional to strain. The gradient of AB is used when determining Young's modulus of elasticity (see para. 14).

(ii) Point B is the **limit of proportionality** and is the point at which stress is no longer proportional to strain when a further load is applied.

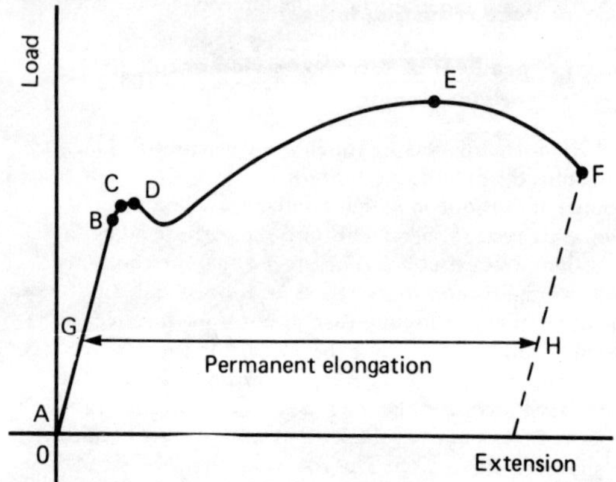

Figure 13.6

(iii) Point C is the **elastic limit** and a specimen loaded to this point will effectively return to its original length when the load is removed, i.e. there is negligible permanent extension.
(iv) Point D is called the **yield point** and at this point there is a sudden extension with no increase in load. The yield stress of the material is given by:

$$\textbf{yield stress} = \frac{\textbf{load where yield begins to take place}}{\textbf{original cross-sectional area}}$$

The yield stress gives an indication of the ductility of the material (see para. 16).
(v) Between points D and E extension takes place over the whole gauge length of the specimen.
(vi) Point E gives the maximum load which can be applied to the specimen and is used to determine the ultimate tensile strength (UTS) of the specimen (often just called the tensile strength).

$$\textbf{UTS} = \frac{\textbf{maximum load}}{\textbf{original cross-sectional area}}$$

(vi) Between points E and F the cross-sectional area of the specimen decreases, usually about half way between the ends, and a **waist** or **neck** is formed before fracture.

The **percentage reduction in area**

$$= \frac{\textbf{increase in length during test to destruction}}{\textbf{original length}} \times \mathbf{100\%}$$

The percentage reduction in area provides information about the malleability of the material (see para. 16).

The value of stress at point F is greater than at point E since although the load on the specimen is decreasing as the extension increases, the cross-sectional area is also reducing.

(viii) At point F the specimen fractures.

(ix) Distance GH is called the **permanent elongation** and

Percentage elongation

$$= \frac{\textbf{increase in length during test to destruction}}{\textbf{original length}} \times \mathbf{100\%}$$

16 (i) **Ductility** is the ability of a material to be plastically deformed by elongation, without fracture. This is a property which enables a material to be drawn out into wires. For ductile materials such as mild steel, copper and gold, large extensions can result before fracture occurs with increasing tensile force. Ductile materials usually have a percentage elongation value of about 15% or more.

(ii) **Brittleness** is the property of a material manifested by fracture without appreciable prior plastic deformation. Brittleness is a lack of ductility and brittle materials, such as cast iron, glass, concrete, brick and ceramics, have virtually no plastic stage, the elastic stage being followed by immediate fracture. Little or no 'waist' occurs before fracture in a brittle material undergoing a tensile test, and there is no noticeable yield point.

(iii) **Malleability** is the property of a material whereby it can be shaped when cold by hammering or rolling. A malleable material is capable of undergoing plastic deformation without fracture.

Shear stress and strain

17 For a shear force the shear stress is equal to force/area, where the area is that which is parallel to the direction of the force. The symbol for shear stress is the Greek letter tau, τ.

Hence from *Figure 13.7*, **shear stress**, $\boldsymbol{\tau = \frac{F}{bd}}$

Shear strain is denoted by the Greek letter gamma, γ, and with reference to *Figure 13.7*,

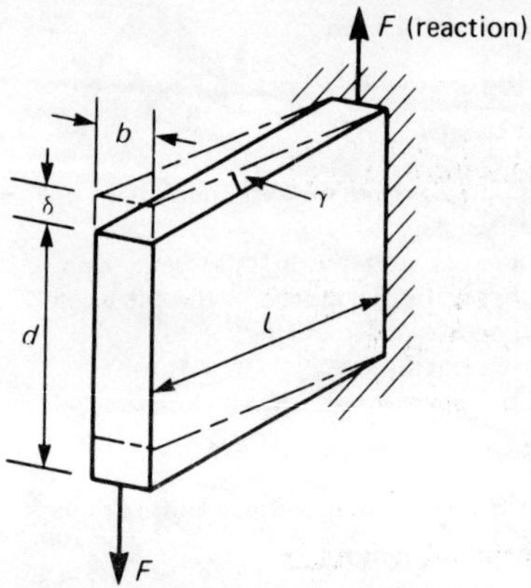

Figure 13.7

shear strain $\gamma = \dfrac{\delta}{l}$.

Modulus of rigidity, $G = \dfrac{\text{shear stress}}{\text{shear stress}}$,

i.e. $\boldsymbol{G = \dfrac{\tau}{\gamma}}$.

For any metal the modulus of rigidity G is approximately 0.4 of the modulus of elasticity, E.

Torsional stress and strain

18 With reference to *Figure 13.8*

$$\frac{\tau}{\gamma} = \frac{T}{J} = \frac{G\theta}{l}$$

where
τ = shear stress at radius r
T = torque on shaft
J = polar second moment of area of section of shaft
G = modulus of rigidity
θ = angle of twist (radians) in a length l of shaft.

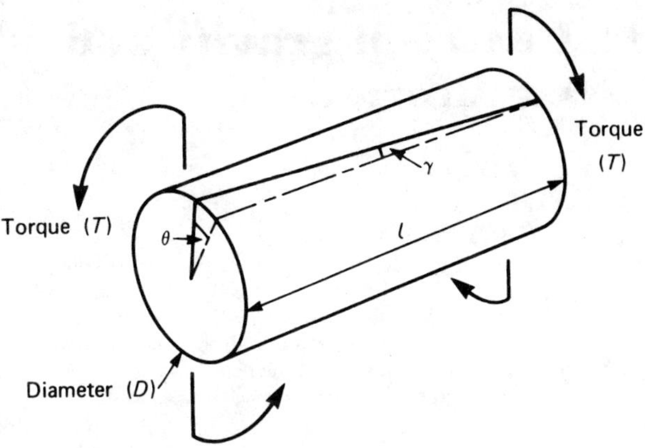

Figure 13.8

The polar second moment of area of a solid shaft is:

$$J = \frac{\pi D^4}{32}$$

where D is the diameter.

The polar second moment of area of a hollow shaft is:

$$J = \frac{\pi}{32}(D^4 - d^4)$$

where D = external diameter and d = internal diameter.

14 Centre of gravity and equilibrium

1 The **centre of gravity** of an object is a point where the resultant gravitational force acting on the body may be taken to act. For objects of uniform thickness lying in a horizontal plane, the centre of gravity is vertically in line with the point of balance of the object. For a thin uniform rod the point of balance and hence the centre of gravity is halfway along the rod, as shown in *Figure 14.1(a)*.

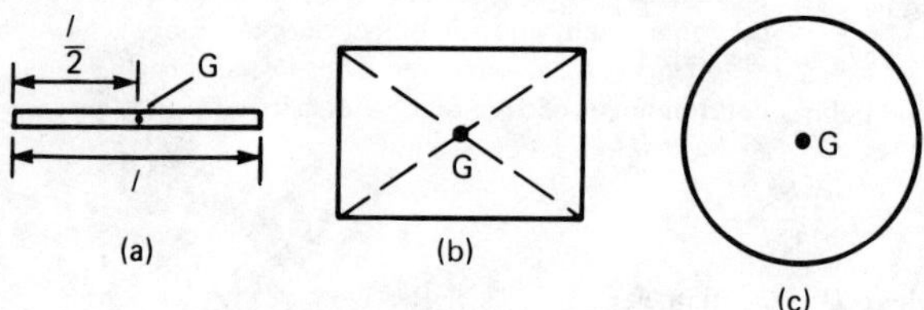

Figure 14.1

A thin flat sheet of a material of uniform thickness is called a **lamina** and the centre of gravity of a rectangular lamina lies at the point of intersection of its diagonals, as shown in *Figure 14.1(b)*. The centre of gravity of a circular lamina is at the centre of the circle, as shown in *Figure 14.1(c)*.

2 An object is in **equilibrium** when the forces acting on the object are such that there is no tendency for the object to move. The state of equilibrium of an object can be divided into three groups.

(a) If an object is in **stable equilibrium** and it is slightly disturbed by pushing or pulling (i.e. a disturbing force is applied), the centre of gravity is raised and when the disturbing force is removed, the object returns to its original position. Thus a ball bearing in a hemispherical cup is in stable equilibrium, as shown in *Figure 14.2(a)*.

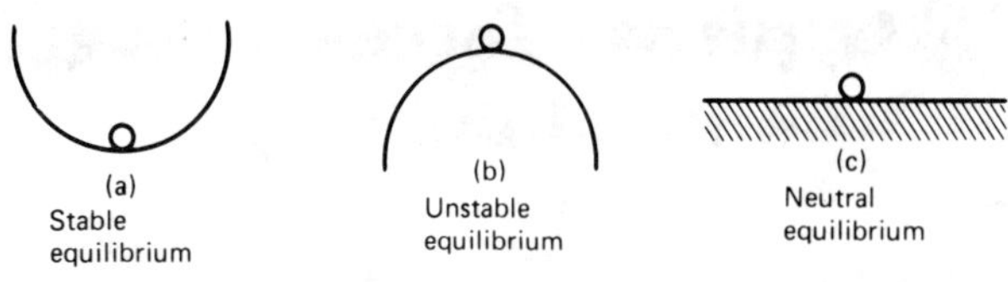

Figure 14.2

(b) An object is in **unstable equilibrium** if, when a disturbing force is applied, the centre of gravity is lowered and the object moves away from its original position. Thus, a ball bearing balanced on top of a hemispherical cup is in unstable equilibrium, as shown in *Figure 14.2(b)*.

(c) When an object in **neutral equilibrium** has a disturbing force applied, the centre of gravity remains at the same height and the object does not move when the disturbing force is removed. Thus, a ball bearing on a flat horizontal surface is in neutral equilibrium, as shown in *Figure 14.2(c)*.

15 Coplanar forces acting at a point

1 When forces are all acting in the same plane, they are called **coplanar**. When forces act at the same time and at the same point, they are called **concurrent forces**.

2 Force is a vector quantity and thus has both a magnitude and a direction. A vector can be represented graphically by a line drawn to scale in the direction of the line of action of the force. Vector quantities may be shown by using bold, lower case letters, thus **ab** in *Figure 15.1* represents a force of 5 newtons acting in a direction due east.

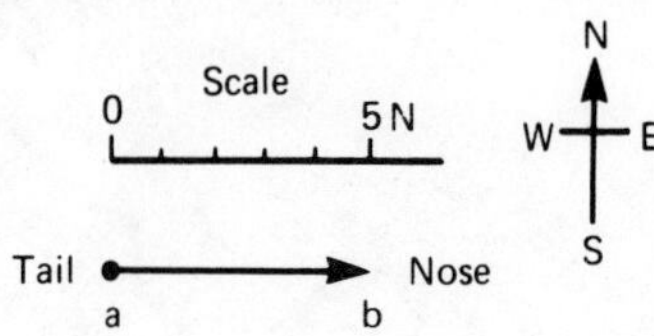

Figure 15.1

The resultant of two coplanar forces

3 For two forces acting at a point, there are three possibilities.

(a) For forces acting in the same direction and having the same line of action, the single force having the same effect as both of the forces, called the **resultant force** or just the **resultant**, is the arithmetic sum of the separate forces. Forces of F_1 and F_2 acting at point P, as shown in *Figure 15.2(a)* have exactly the same effect on point P as force F shown in *Figure 15.2(b)*, where $F = F_1 + F_2$ and acts in the same direction as F_1 and F_2. Thus, F is the resultant of F_1 and F_2.

(b) For forces acting in opposite directions along the same line of action, the resultant force is the arithmetic difference between the two forces. Forces of F_1 and F_2

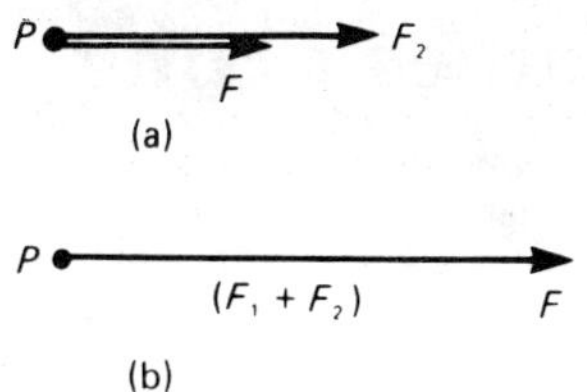

Figure 15.2

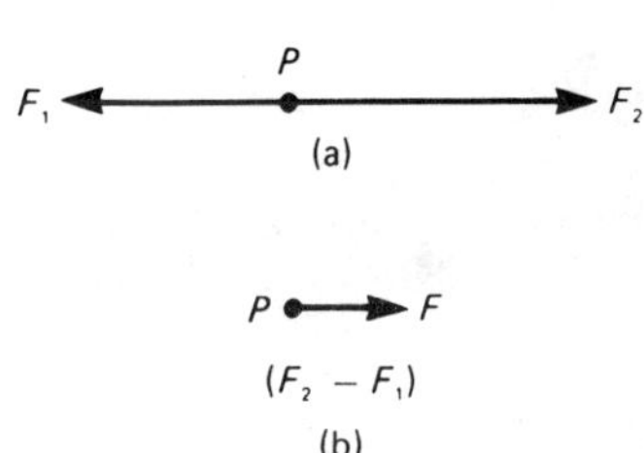

Figure 15.3

acting at point P as shown in *Figure 15.3(a)*, have exactly the same effect on point P as force F shown in *Figure 15.3(b)*, where $F = F_2 - F_1$ and acts in the direction of F_2, since F_2 is greater than F_1. Thus F is the resultant of F_1 and F_2.

(c) When two forces do not have the same line of action, the magnitude and direction of the resultant force may be found by a procedure called vector addition of forces. There are two graphical methods of performing **vector addition**, known as the triangle of forces method and the parallelogram of forces method.

The triangle of forces method

4 (i) Draw a vector representing one of the forces, using an appropriate scale and in the direction of its line of action.
(ii) From the **nose** of this vector and using the same scale, draw a vector representing the second force in the direction of its line of action.
(iii) The resultant vector is represented in both magnitude

and direction by the vector drawn from the **tail** of the first vector to the nose of the second vector.

Thus, for example, to determine the magnitude and direction of the resultant of a force of 15 N acting horizontally to the right and a force of 20 N, inclined at an angle of 60° to the 15 N force, using the triangle of forces method: With reference to *Figure 15.4* and using the above procedure:

(i) **ab** is drawn 15 units long horizontally;

(ii) from b, **bc** is drawn 20 units long, inclined at an angle of 60° to ab. (Note, in angular measure, an angle of 60° from ab means 60° in an anticlockwise direction.)

(iii) By measurement, the resultant **ac** is 30.5 units long inclined at an angle of 35° to ab.

Hence the resultant force is **30.5 N** inclined at an angle of **35**° to the 15 N force.

The parallelogram of forces method:

5 (i) Draw a vector representing one of the forces, using an appropriate scale and in the direction of its line of action.

(ii) From the **tail** of this vector and using the same scale draw a vector representing the second force in the direction of its line of action.

(iii) Complete the parallelogram using the two vectors drawn in (i) and (ii) as two sides of the parallelogram.

(iv) The resultant force is represented in both magnitude and direction by the vector corresponding to the diagonal of the parallelogram drawn from the tail of the vectors in (i) and (ii).

Thus, for example, to determine the magnitude and direction of the resultant of a 250 N force acting at an angle of 135° and a force of 400 N acting at an angle of −120°, using the parallelogram of force method: With reference to *Figure 15.5* and using the above procedure:

(i) **ab** is drawn at an angle of 135° and 250 units in length;

(ii) **ac** is drawn at an angle of −120° and 400 units in length;

(iii) bc and cd are drawn to complete the parallelogram;

(iv) **ad** is drawn. By measurement **ad** is 413 units long at an angle of −156°.

Hence the resultant force is **413 N** at an angle of **−156°**.

6 An alternative to the graphical methods of determining the resultant of two coplanar forces is by **calculation**. This can be achieved by trigonometry using the cosine rule and the sine rule, or by resolution of forces (see para. 9).

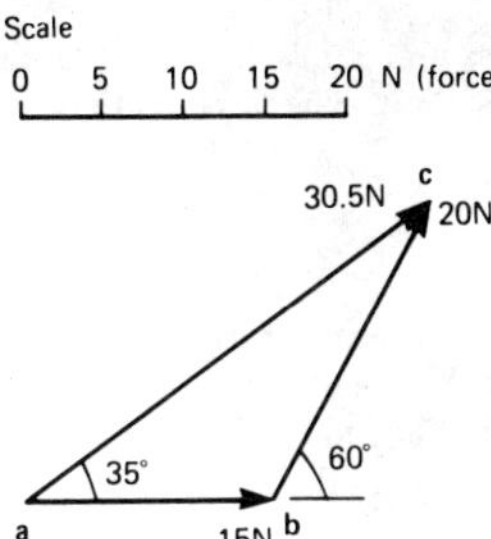

Figure 15.4

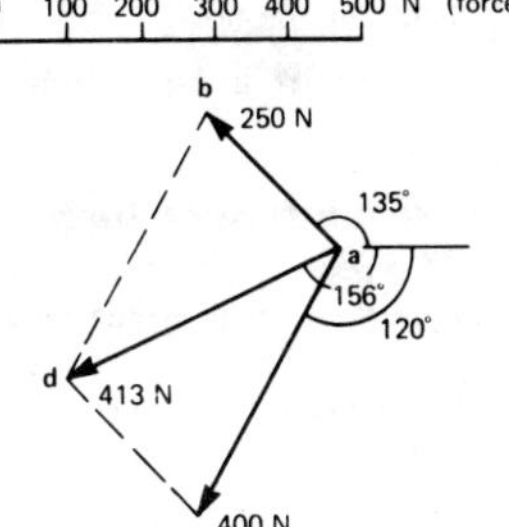

Figure 15.5

The resultant of more than two coplanar forces

7 For the three coplanar forces F_1, F_2 and F_3 acting at a point as shown in *Figure 15.6*, the vector diagram is drawn using the nose to tail method. The procedure is:

(i) Draw **oa** to scale to represent force F_1 in both magnitude and direction (see *Figure 15.7*).
(ii) From the nose of **oa**, draw **ab** to represent force F_2.
(iii) From the nose of **ab**, draw **bc** to represent force F_3.
(iv) The resultant vector is given by length **oc** in *Figure 15.7*. The direction of resultant **oc** is from where we started, i.e. point o, to where we finished, i.e. point c. When acting by itself, the resultant force, given by **oc**, has the same effect on the point as forces F_1, F_2 and F_3 have when acting together. The resulting vector diagram of *Figure 15.7* is called the **polygon of forces**.

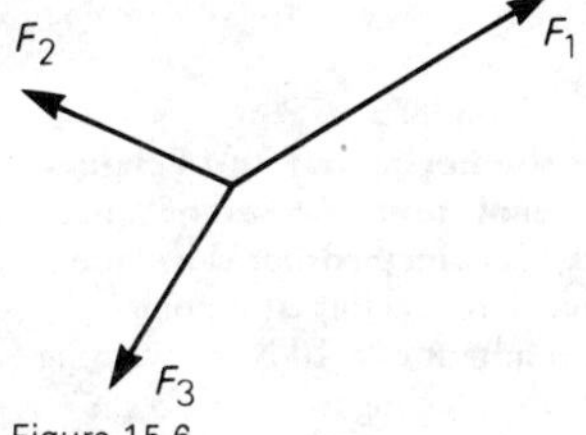

Figure 15.6

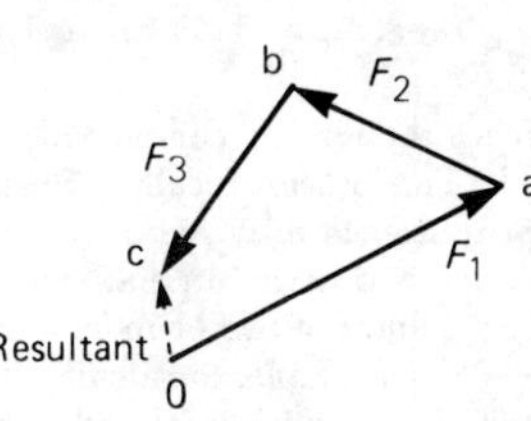

Figure 15.7

8 When three or more coplanar forces are acting at a point and the vector diagram closes, there is no resultant. The forces acting at the point are in **equilibrium**.

Resolution of forces

9 A vector quantity may be expressed in terms of its **horizontal and vertical components**. For example, a vector representing a force of 10 N at an angle of 60° to the horizontal is shown in *Figure 15.8*. If the horizontal line oa and the vertical line

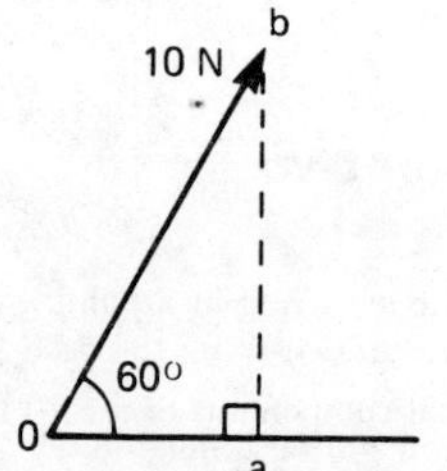

Figure 15.8

ab are constructed as shown, then oa is called the horizontal component of the 10 N force and ab the vertical component of the 10 N force.
From trigonometry,

$$\cos 60° = \frac{oa}{ob}.$$

Hence the vertical component, ab = 10 sin 60°.

$$\sin 60° = \frac{ab}{ob}.$$

Hence the vertical component, ab = 10 sin 60°.

This process is called '**finding the horizontal and vertical components of a vector**' or '**the resolution of a vector**', and can be used as an alternative to graphical methods for calculating the resultant of two or more coplanar forces acting at a point.

For example, to calculate the resultant of a 10 N force acting at 60° to the horizontal and a 20 N force acting at −30° to the horizontal (see *Figure 15.9*) the procedure is as follows:

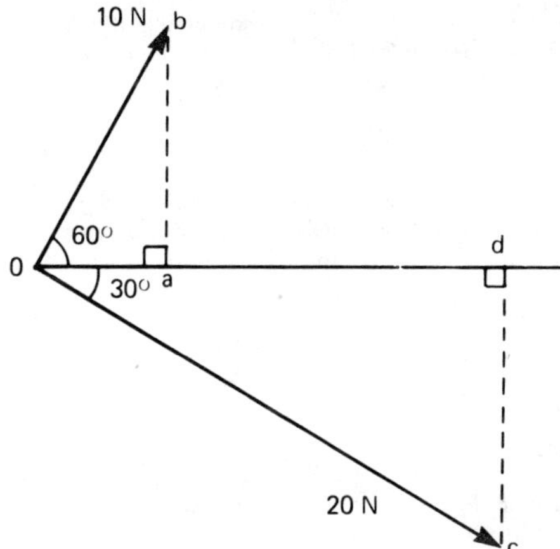

Figure 15.9

(i) Determine the horizontal and vertical components of the 10 N force, i.e.

horizontal component, oa = 10 cos 60° = 5.0 N
vertical component, ab = 10 sin 60° = 8.66 N

(ii) Determine the horizontal and vertical components of the 20 N force, i.e.

horizontal component, od = 20 cos (= 30°) = 17.32 N
vertical component, cd = 20 sin (= 30°) = – 10.0 N.

(iii) Determine the total horizontal component, i.e.

oa + od = 5.0 + 17.32 = 22.32 N

(iv) Determine the total vertical component, i.e.

ab + cd = 8.66 + (– 10.0) – 1.34 N

(v) Sketch the total horizontal and vertical components as shown in ***Figure 15.10*. The resultant of the two components is given by**

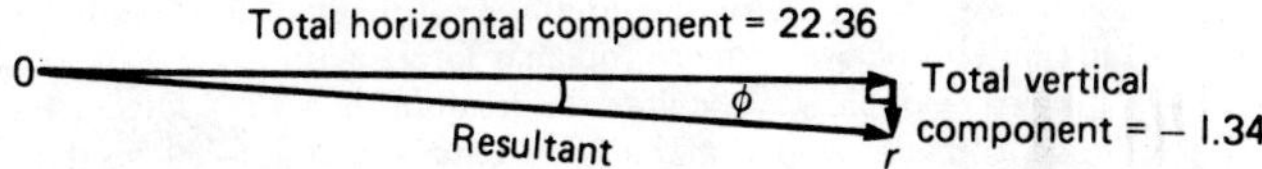

Figure 15.10

length **or** and, by Pythagoras' theorem,

$$\text{or} = \sqrt{[(22.32)^2 + (1.34)^2]} = 22.36 \text{ N}$$

and using trigonometry,

$$\text{angle } \phi = \arctan\left(\frac{1.34}{2.32}\right) = 3° \ 26'.$$

Hence the resultant of the 10 N and 20 N forces shown in *Figure 15.9* is 22.36 N at an angle of $-3°26'$ to the horizontal.

The above example demonstrates the use of resolution of forces for calculating the resultant of two coplanar forces acting at a point. However, the method may be used for more than two forces acting at a point.

Summary

10 (a) To determine the resultant of two coplanar forces acting at a point, there are four methods commonly used. These are, by drawing:
1 triangle of forces method,
2 parallelogram of forces method,
and by calculation:
3 use of cosine and sine rules,
4 resolution of forces.

(b) To determine the resultant of more than two coplanar forces acting at a point, there are two methods commonly used. These are, by drawing:
1 polygon of forces method,
and by calculation:
2 resolution of forces.

16 Simply supported beams

1 When using a spanner to tighten a nut, a force tends to turn the nut in a clockwise direction. This turning effect of a force is called the **moment of a force** or more briefly, just a **moment**. The size of the moment acting on the nut depends on two factors:

(a) the size of the force acting at right angles to the shank of the spanner, and

(b) the perpendicular distance between the point of application of the force and the centre of the nut.

In general, with reference to *Figure 16.1*, the moment M of a force acting at a point P = force × perpendicular distance between the line of action of the force and P.

i.e. $M = F \times d$

The unit of a moment is the newton metre (Nm). Thus, if force F in *Figure 16.1* is 7 N and distance d is 3 m, then the moment at P is 7(N) × 3(m), i.e. 21 Nm.

2 If more than one force is acting on an object and the forces do not act at a point, then the turning effect of the forces, that is, the moment of the forces, must be considered.

Figure 16.2 shows a beam with its support (known as its pivot or fulcrum), at P, acting vertically upwards, and forces F_1 and F_2 acting vertically downwards at distances a and b respectively from the fulcrum.

A beam is said to be in **equilibrium** when there is no tendency for it to move.

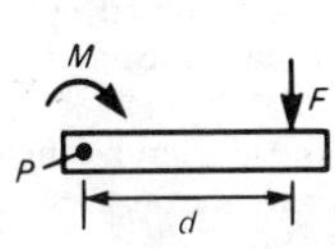

Figure 16.1

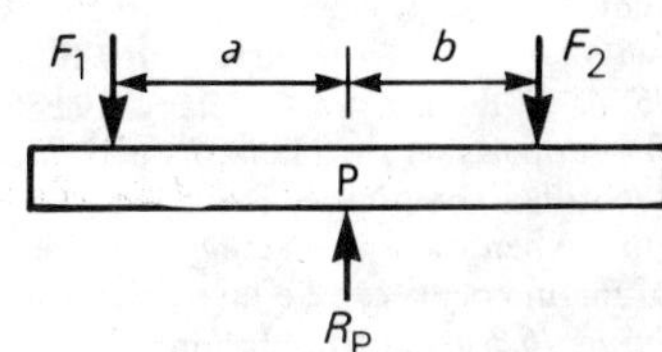

Figure 16.2

There are two conditions for equilibrium:

(i) the sum of the forces acting vertically downwards must be equal to the sum of the forces acting vertically upwards, i.e. for *Figure 16.2*,

$$R_p = F_1 + F_2, \text{ and}$$

(ii) the total moment of the forces acting on a beam must be zero; for the total moment to be zero:

> *'the sum of the clockwise moments about any point must be equal to the sum of the anticlockwise moments about that point'.*

This statement is known as the **principle of moments**. Hence, taking moments about P in *Figure 16.2*, $F_2 \times b$ = the clockwise moment, and $F_1 \times a$ = the anticlockwise moment. Thus for equilibrium:

$$F_1 a = F_2 b.$$

3 (i) A **simply supported beam** is one which rests on two supports and is free to move horizontally.

(ii) Two typical simply supported beams having loads acting at given points on the beam (called **point loading**), as shown in *Figure 16.3.*

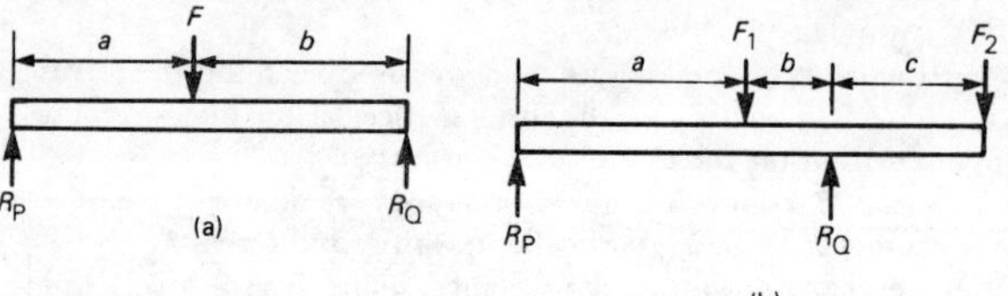

Figure 16.3

A man whose mass exerts a force F vertically downwards, standing on a wooden plank which is simply supported at its ends, may, for example, be represented by the beam diagram of *Figure 16.3(a)* if the mass of the plank is neglected. The forces exerted by the supports on the plank, R_P and R_Q, act vertically upwards, and are called **reactions**.

(iii) When the forces acting are all in one plane, the algebraic sum of the moments can be taken about **any** point. For the beam in *Figure 16.3(a)*, at equilibrium:

(i) $R_P + R_Q = F$, and

(ii) taking moments about R_p, $F_a = R_Q b$.
(Alternatively, taking moments about F, $R_P a = R_Q b$).
For the beam in *Figure 16.3(b)*, at equilibrium:
(i) $R_P + R_Q = F_1 + F_2$, and
(ii) taking moments about R_Q, $R_P(a+b) + F_2 c = F_1 b$.
(iv) Typical practical applications of simply supported beams witl point loadings include bridges, beams in buildings and beds of machine tools.

For example, for the beam shown in *Figure 16.4*, the force acting on support A, R_A, and distance d are calculated as follows:

(Forces acting in an upward direction)
= (Forces acting in a downward direction)

i.e. $R_A + 40 = 10 + 15 + 30$

and $R_A = 10 + 15 + 30 - 40 = \mathbf{15\ N}$

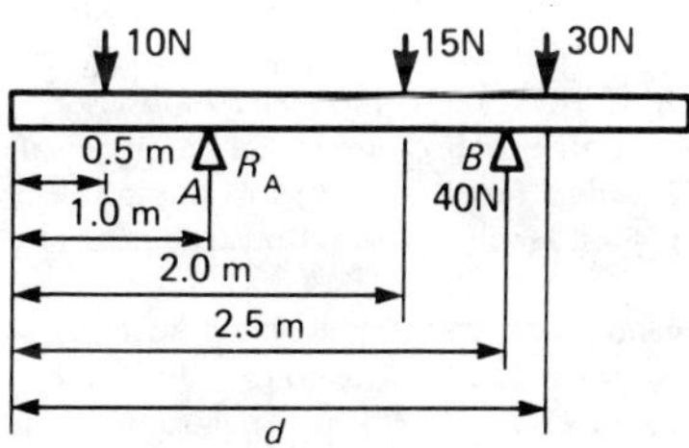

Figure 16.4

Taking moments about the left hand end of the beam and applying the principle of moments gives:

clockwise moments = anticlockwise moments

i.e. $(10 \times 0.5) + (15 \times 2.0) + (30d) - (15 \times 1.0) + (40 \times 2.5)$

i.e. $35 + 30d = 15$

from which, distance $d = \dfrac{115 - 35}{30} = \mathbf{2\frac{2}{3}\ m}$

Shearing force and bending moments

4 As stated in para. 3, for equilibrium of a beam, the forces to the left of any section such as X in *Figure 16.5*, must balance the forces to the right. Also the moment about X of the forces to the left must balance the moment about X of the forces to the right.

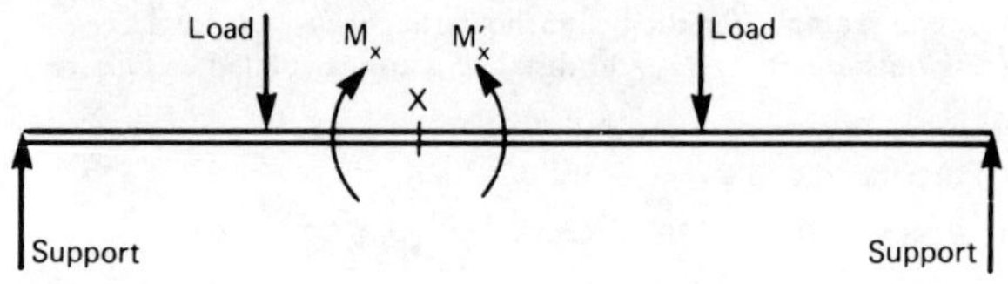

Figure 16.5

Although for equilibrium the forces and moments cancel, the magnitude and nature of these forces and moments are important as they determine both the stresses at X and the beam curvature and deflection. The resultant force to the left of X and the resultant force to the right of X (forces or components of forces transverse to the beam), constitute a pair of forces tending to shear the beam at this section. **Shearing force** is defined as the force transverse to the beam at a given section tending to cause it to shear at that section.

By convention, if the tendency is to shear as shown in *Figure 16.6(a)*, the shearing force is regarded as positive, i.e. $+F$; if the tendency to shear is as shown in *Figure 16.6(b)*, it is regarded as negative, i.e. $-F$.

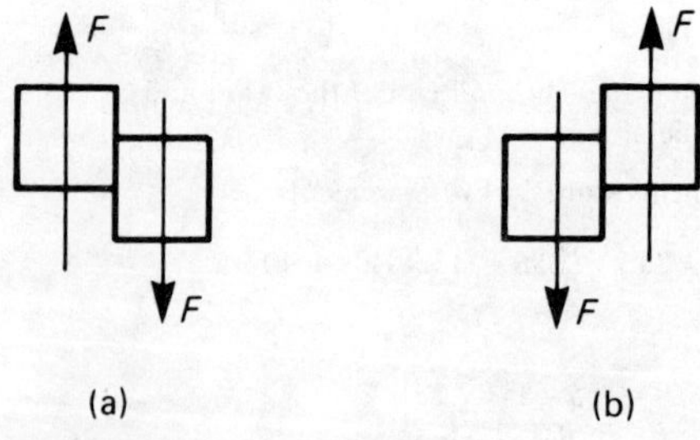

Figure 16.6

5 The **bending moment** at a given section of a beam is defined as the resultant moment about that section of either all of the forces to its left — or of all of the forces to its right. In *Figure 16.5* it is M_X or M'_X. These moments, clockwise to the left and anticlockwise to the right, will cause the beam to bend concave upwards, called '**sagging**'. By convention this is regarded as positive bending (i.e. the bending moment is a positive bending moment). Where the curvature produced is concave downwards (called '**hogging**'), the bending moment is regarded as negative.

The values of shearing force and bending moment will usually vary along a beam. Diagrams showing the shearing force and bending moment for all sections of a beam are called shearing force and bending moment diagrams respectively.

Shearing forces and shearing force diagrams are less important than bending moments, but can be very useful in giving pointers to the more important bending moment diagrams. For example, wherever the shearing force is zero, the bending moment

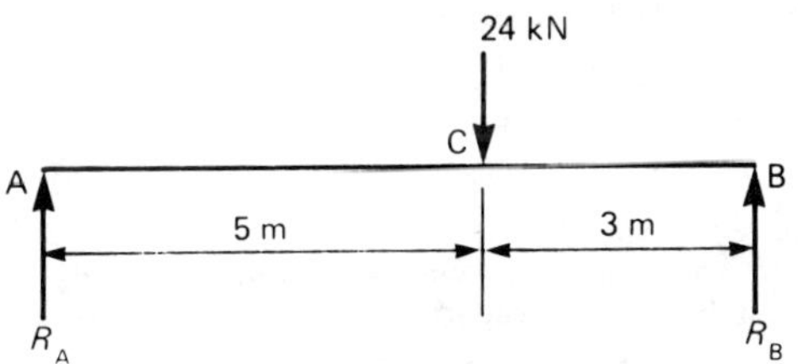

Figure 16.7

will be a maximum or a minimum. The shearing force and bending moment diagrams for the beam shown in *Figure 16.7* are obtained as follows:

It is first necessary to calculate the reactions at A and B. The beam is simply-supported at A and B which means that it rests on supports at these points giving vertical reactions. The general **conditions for equilibrium require that the resultant moment about** any point must be zero, and total upward force must equal total downward force. Therefore, taking moments about A, the moment of R_B must balance the moment of the load at C:

$$R_B \times 8\ \text{m} = 24\ \text{kN} \times 5\ \text{m} = 120\ \text{kNm}$$

$$R_B = \frac{120\ \text{kNm}}{8\ \text{m}} = 15\ \text{kN, and}$$

$$R_A = 24\ \text{kN} - 15\ \text{kN} = 9\ \text{kN.}$$

Immediately to the right of A the shearing force is due to R_A and is therefore 9 kN. As this force to the left of the section considered, is upwards, the shearing force is positive. The shearing force is the same for all points between A and C as no other forces come on the beam between these points.

When a point to the right of C is considered, the load at C as well as R_A must be considered, or alternatively, R_B on its own. The shearing force is 15 kN, either obtained from $R_B = 15$ kN, or from load at $C - R_A = 15$ kN. For any point between C and B the force to the right is upwards and the shearing force is therefore negative. It should be noted that the shearing force changes suddenly at C.

The bending moment at A is zero, as there are no forces to its left. At a point 1 m to the right of A the moment of the only force R_A to the left of the point is $R_A \times 1$ m$=9$ kNm. As this moment to the left is clockwise the bending moment is positive, i.e. it is $+9$ kNm. At points 2 m, 3 m, 4 m and 5 m to the right of A the bending moments are respectively:

$R_A \times 2$ m $= 9$ kN $\times 2$ m $= 18$ kNm
$R_A \times 3$ m $= 9$ kN $\times 3$ m $= 27$ kNm
$R_A \times 4$ m $= 9$ kN $\times 4$ m $= 36$ kNm
$R_A \times 5$ m $= 9$ kN $\times 5$ m $= 45$ kNm
All are positive bending moments.

For points to the right of C, the load at C as well as R_A must be considered or, more simply, R_B alone can be used. At points 5 m, 6 m and 7 m from A the bending moments are respectively:

$R_B \times 3$ m $= 15$ kN $\times 3$ m $= 45$ kNm
$R_B \times 2$ m $= 15$ kN $\times 2$ m $= 30$ kNm
$R_B \times 1$ m $= 15$ kN $\times 1$ m $= 15$ kNm

As these moments to the right of the points considered are anticlockwise they are all positive bending moments. At B the bending moment is zero as there is no force to its right. The results are summarised in the table below.

Distance from A (m)	0	1	2	3	4	5	6	7	8
Shearing force (kN)	+9	+9	+9	+9	+9	+9 −15	−15	−15	−15
Bending moment (kNm)	0	+9	+18	+27	+36	+45	+30	+15	0

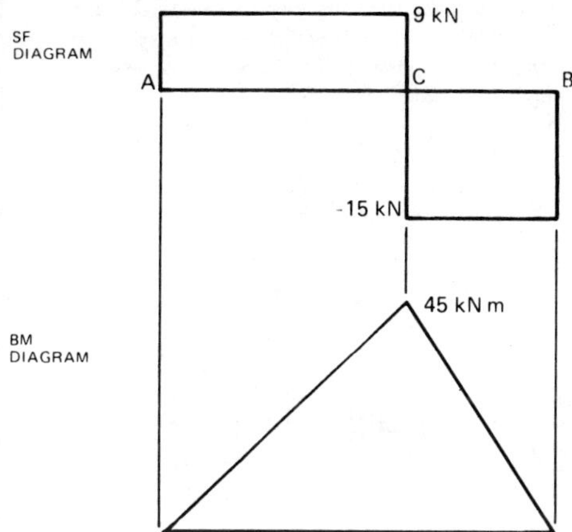

Figure 16.8

Making use of the above values, the diagrams are as shown (in the usual manner) in *Figure 16.8*: **A stepped shearing force diagram,** with horizontal and vertical lines only, is always obtained when the beam carries concentrated loads only. A sudden change in shearing force occurs where the concentrated loads, including the reactions at supports, occur. For this type of simple loading the bending moment diagram always consists of straight lines, usually sloping. Sudden changes of bending moment cannot occur except in the unusual circumstances of a moment being applied to a beam as distinct from a load.

6 Bending stress

$$\frac{\sigma}{y}=\frac{M}{I}\left(=\frac{E}{R}\right)$$

where

σ = stress due to bending at distance y from the neutral axis;
M = bending moment;
I = second moment of area of section of beam about its neutral axis;
E = modulus of elasticity;
R = radius of curvature.

Section modulus $Z = I/y_{max}$.

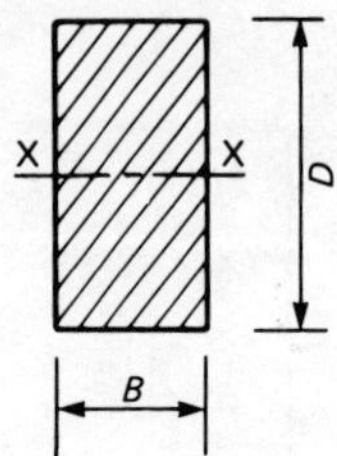

Figure 16.9

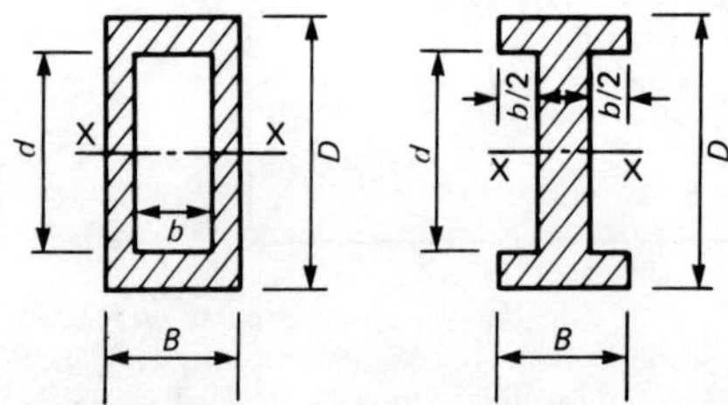

Figure 16.10

The second moments of area of the beam sections most commonly met with are (about the central axis XX):

(a) **Solid rectangle** (*Figure 16.9*).

$$I=\frac{BD^3}{12}$$

(b) **Symmetrical hollow rectangle or *I* section** (*Figure 16.10*)

$$I=\frac{BD^3-bd^3}{12}$$

(c) **Solid rod** (*Figure 16.11*)

$$I=\frac{\pi D^4}{64}$$

(d) **Tube** (*Figure 16.12*)

$$I=\frac{\pi(D^4-d^4)}{64}$$

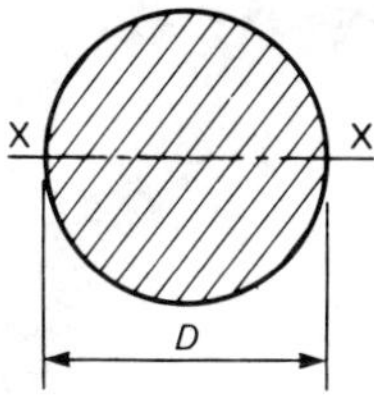

Figure 16.11

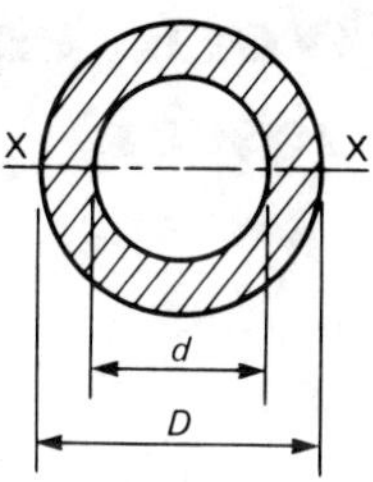

Figure 16.12

The neutral axis of any section, where bending produces no strain and therefore no stress, always passes through the centroid of the section. For the symmetrical sections listed above this means that for vertical loading the neutral axis is the horizontal axis of symmetry.

For example, let the maximum bending moment on a beam be 120 Nm. If the beam section is rectangular 18 mm wide and 36 mm deep, the maximum bending stress is calculated as follows:

Second moment of area of section about the neutral axis,

$$I=\frac{bd^3}{12}=\frac{(18)(36)^3}{12}=6.9984\times 10^4\ \text{mm}^4$$

Maximum distance from neutral axis,

$$y=\frac{36}{2}=18\ \text{mm.}$$

Since $\sigma/y=M/I$ then the maximum bending stress σ will occur where M and y have their maximum values, i.e.

$$\sigma=\frac{My}{I}=\frac{120\ \text{Nm}\times 18\ \text{mm}}{6.9984\times 10^4\ \text{mm}^4}=120\ \text{Nm}\times\frac{18\times 10^{-3}\ \text{m}}{6.9984\times 10^4}\times 10^{-12}\ \text{m}^4$$

$$=30.86\ \text{MN/m}^2$$

$$=\mathbf{30.86\ MPa}.$$

17 Work, energy and power

1 Fuel, such as oil, coal, gas or petrol, when burnt, produces heat. Heat is a form of energy and may be used, for example, to boil water or to raise steam. Thus fuel is useful since it is a convenient method of storing energy, that is, **fuel is a source of energy**.

2 (i) If a body moves as a result of a force being applied to it, the force is said to do work on the body. The amount of work done is the product of the applied force and the distance, i.e.

Work done = force × distance moved in the direction of the force

(ii) The unit of work is the **joule**, **J**, which is defined as the amount of work done when a force of 1 newton acts for a distance of 1 metre in the direction of the force.

Thus, 1 J = 1 Nm.

3 If a graph is plotted of experimental values of force (on the vertical axis) against distance moved (on the horizontal axis) a force-distance graph or work diagram is produced. **The area under the graph represents the work done**.

For example, a constant force of 20 N used to raise a load a height of 8 m may be represented on a force-distance graph as shown in *Figure 17.1(a)*. The area under the graph shown shaded, represents the work done.

Hence, work done = 20 N × 8 m = **160 J**

Similarly, a spring extended by 20 mm by a force of 500 N may be represented by the work diagram shown in *Figure 17.1(b)*.

$$\text{Work done} = \text{shaded area} = \tfrac{1}{2} \times \text{base} \times \text{height}$$
$$= \tfrac{1}{2} \times (20 \times 10^{-3})\ \text{m} \times 500\ \text{N} = \mathbf{5\ J}$$

4 **Energy** is the capacity, or ability, to do work. The unit of energy is the joule, the same as for work. Energy is expended when work is done.

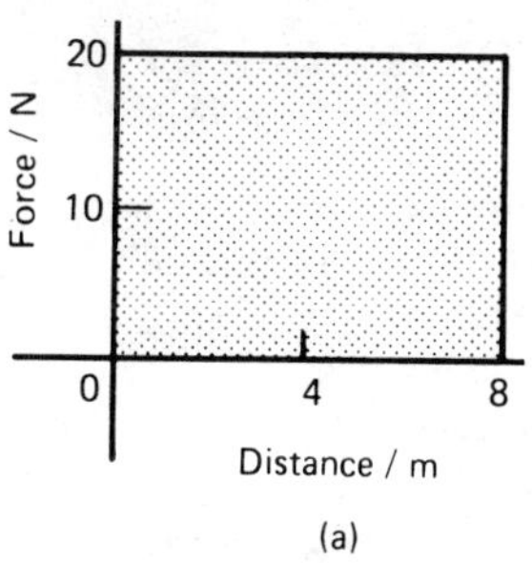

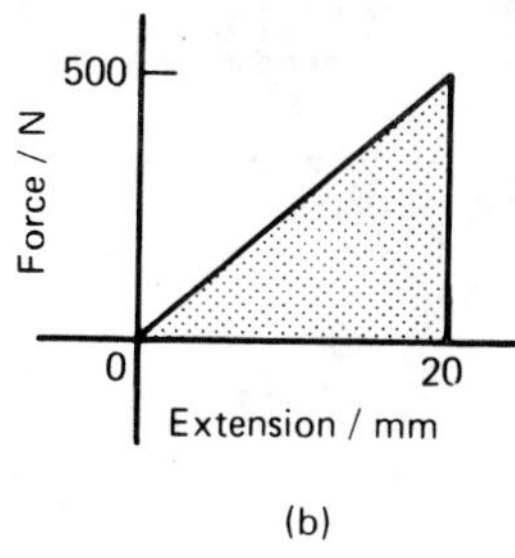

Figure 17.1

5 There are several **forms of energy** and these include:

(i) Mechanical energy;
(ii) Heat or thermal energy;
(iii) Electrical energy;
(iv) Chemical energy;
(v) Nuclear energy;
(vi) Light energy;
(vii) Sound energy.

6 Energy may be converted from one form to another. **The principle of conservation of energy** states that the total amount of energy remains the same in such conversions, i.e. energy cannot be created or destroyed. Some examples of energy conversions include:

(i) Mechanical energy is converted to electrical energy by a generator.
(ii) Electrical energy is converted to mechanical energy by a motor.
(iii) Heat energy is converted to mechanical energy by a steam engine.
(iv) Mechanical energy is converted to heat energy by friction.
(v) Heat energy is converted to electrical energy by a solar cell.
(vi) Electrical energy is converted to heat energy by an electric fire.
(vii) Heat energy is converted to chemical energy by living plants.
(viii) Chemical energy is converted to heat energy by chemical reactions.
(ix) Heat energy is converted to electrical energy by a thermocouple.
(x) Chemical energy is converted to electrical energy by batteries.
(xi) Electrical energy is converted to light energy by a light bulb.

(xii) Sound energy is converted to electrical energy by a microphone.

(xiii) Electrical energy is converted to chemical energy by electrolysis.

7 **Efficiency** is defined as the ratio of the useful output energy to the input energy. The symbol for efficiency is η (Greek letter eta).

Hence, **efficiency**, $\eta = \dfrac{\textbf{useful output energy}}{\textbf{input energy}}$

Efficiency has no units and is often stated as a percentage. A perfect machine would have an efficiency of 100%. However, all machines have an efficiency lower than this due to friction and other losses. Thus, if the input energy to a motor is 1000 J and the output energy is 800 J then the efficiency is

$$\frac{800}{1000} \times 100\%, \text{ i.e. } 80\%.$$

8 **Power** is a measure of the rate at which work is done or at which energy is converted from one form to another.

$$\textbf{Power } P = \frac{\textbf{energy used}}{\textbf{time taken}} \left(\text{or } P = \frac{\text{work done}}{\text{time taken}}\right)$$

The unit of power is the **watt**, **W**, where 1 watt is equal to 1 joule per second. The watt is a small unit for many purposes and a larger unit called the kilowatt, kW, is used, where 1 kW = 1000 W. The power output of a motor which does 120 kJ of work is 30 s is thus given by

$$P = \frac{120 \text{ kJ}}{30 \text{ s}} = 4 \text{ kW}.$$

(For electrical power, see page 158)

9 Since, word done = force × distance

$$\text{Then, power} = \frac{\text{work done}}{\text{time taken}} = \frac{\text{force} \times \text{distance}}{\text{time taken}}$$

$$= \text{force} \times \frac{\text{distance}}{\text{time taken}}$$

$$\text{However, } \frac{\text{distance}}{\text{time taken}} \doteq \text{velocity}.$$

Hence, **power = force × velocity**.

Thus, for example, if a lorry is travelling at a constant speed of 72 km/h and the force resisting motion is 800 N, then the tractive power necessary to keep the lorry moving at this speed is given by:

$$\text{power} = \text{force} \times \text{velocity} = (800\ \text{N})\left(\frac{72}{3.6}\ \text{m/s}\right) = 16\,000\ \frac{\text{Nm}}{\text{s}}$$

$$= 16\,000\ \text{J/s} = \mathbf{16\ kW}.$$

10 (i) Mechanical engineering is concerned principally with two kinds of energy, these being potential energy and kinetic energy.
(ii) **Potential energy** is energy due to the position of a body. The force exerted on a mass of m kg is mg N (where $g = 9.81$ N/kg, the earth's gravitational field). When the mass is lifted vertically through a height h m above some datum level, the work done is given by: force × distance = $(mg)(h)$ J. This work done is stored as potential energy in the mass.

Hence **potential energy** = mgh **joules** (the potential energy at the datum level being taken as zero).
(iii) **Kinetic energy** is the energy due to the motion of a body. Suppose a resultant force F acts on an object of mass m originally at rest and accelerates it to a velocity v in a distance s.

$$\text{Work done} = \text{force} \times \text{distance} = Fs = (ma)(s),$$

where a is the acceleration.

However, $v^2 = 2as$, from which $a = \frac{v^2}{2s}$

Hence, work done $= m\left(\frac{v^2}{2s}\right)s = \frac{1}{2}mv^2$

This energy is called the kinetic energy of the mass m,

i.e. **kinetic energy** $= \frac{1}{2}mv^2$ **joules.**

For example, at the instant of striking, a hammer of mass 30 kg has a velocity of 15 m/s. The kinetic energy in the hammer is given by:

$$\text{Kinetic energy} = \frac{1}{2}mv^2 = \frac{1}{2}(30\ \text{kg})(15\ \text{m/s})^2 = \mathbf{3375\ J}.$$

11 (i) Energy may be converted from one form to another. The principle of conservation of energy states that the total amount of energy remains the same in such conversions, i.e. energy cannot be created or destroyed.

(ii) In mechanics, the potential energy possessed by a body is frequently converted into kinetic energy, and vice versa. When a mass is falling freely, its potential energy decreases as it loses height, and its kinetic energy increases as its velocity increases. Ignoring air frictional losses, at all times:

potential energy + kinetic energy = a constant.

(iii) If friction is present, then work is done overcoming the resistance due to friction and this is dissipated as heat. Then,

p.e. + k.e. = final energy + work done overcoming frictional resistance.

(iv) Kinetic energy is not always conserved in collisions. Collisions in which kinetic energy is conserved (i.e. stays the same) are called **elastic collisions**, and those in which it is not conserved are termed **inelastic collisions**.

Kinetic energy of rotation

12 (i) The tangential velocity v of a particle of mass m moving at an angular velocity ω rad/s at a radius r metres (see *Figure 17.2*) is given by $v = \omega r$ **m/s**.

(ii) The kinetic energy of a particle of mass m is given by:

$$\textbf{kinetic energy} = \tfrac{1}{2} mv^2 = \tfrac{1}{2} m(\omega r)^2 = \tfrac{1}{2} m\omega^2 r^2 \textbf{ joules.}$$

(iii) The total kinetic energy of a system of masses rotating at different radii about a fixed axis but with the same angular velocity ω, as shown in *Figure 17.3*, is given by:

$$\text{total kinetic energy} = \frac{1}{2} m_1\omega^2 r_1^2 + \frac{1}{2} m_2\omega^2 r_2^2 + \frac{1}{2} m_3\omega^2 r_3^2$$

$$= (m_1 r_1^2 + m_2 r_2^2 + m_3 r_3^2)\,\frac{\omega^2}{2}$$

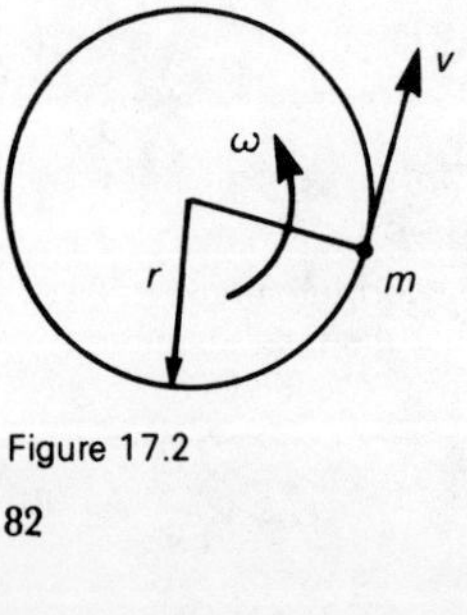

Figure 17.2

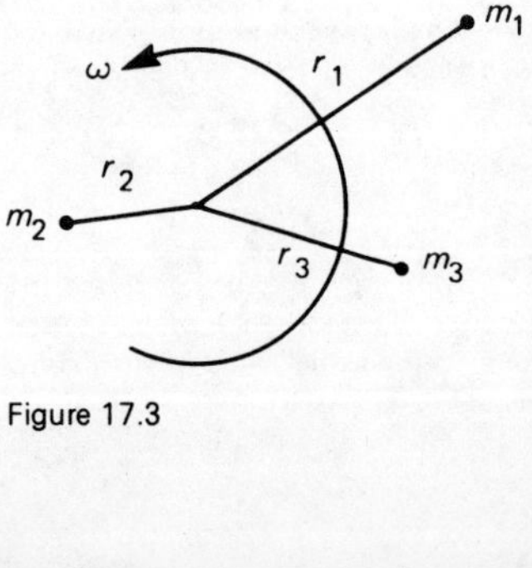

Figure 17.3

In general, this may be written as:

$$\textbf{total kinetic energy} = (\sum mr^2)\frac{\omega^2}{2} = I\frac{\omega^2}{2},$$

where $I\ (=\sum mr^2)$ is called the **moment of inertia** of the system about the axis of rotation.

The moment of inertia of a system is a measure of the amount of work done to give the system an angular velocity of ω rad/s, or the amount of work which can be done by a system turning at ω rad/s.

In general, total kinetic energy $= I\dfrac{\omega^2}{2} = Mk^2\dfrac{\omega^2}{2}$,

where $M\ (=\sum m)$ is the total mass and k is called the **radius of gyration** of the system for the given axis.

If all of the mass were concentrated at the **radius of gyration** it would give the same moment of inertia as the actual system.

Flywheels

13 The function of a flywheel is to restrict fluctuations of speed by absorbing and releasing large quantities of kinetic energy for small speed variations.

To do this they require large moments of inertia and to avoid excessive mass they need to have radii of gyration as large as possible. Most of the mass of a flywheel is usually in its rim.

18 Simple machines

1 A machine is a device which can change the magnitude or line of action, or both magnitude and line of action of a force. A simple machine usually amplifies an input force, called the **effort**, to give a larger output force, called the **load**. Some typical examples of simple machines include pulley systems, screw-jacks, gear systems and lever systems.

2 The **force ratio** or **mechanical advantage** is defined as the ratio of load to effort, i.e.

$$\textbf{force ratio} = \frac{\textbf{load}}{\textbf{effort}} \quad (1)$$

Since both load and effort are measured in newtons, force ratio is a ratio of the same units and thus is a dimensionless quantity.

3 The **movement ratio** or **velocity ratio** is defined as the ratio of the distance moved by the effort to the distance moved by the load, i.e.

$$\textbf{movement ratio} = \frac{\textbf{distance moved by the effort}}{\textbf{distance moved by the load}} \quad (2)$$

Since the numerator and denominator are both measured in metres, movement ratio is a ratio of the same units and thus is a dimensionless quantity.

4 (i) The **efficiency of a simple machine** is defined as the ratio of the force ratio to the movement ratio, i.e.

$$\text{efficiency} = \frac{\text{force ratio}}{\text{movement ratio}}$$

Since the numerator and denominator are both dimensionless quantities, efficiency is a dimensionless quantity. It is usually expressed as a percentage, thus:

$$\textbf{efficiency} = \frac{\textbf{force ratio}}{\textbf{movement ratio}} \times \textbf{100\%} \quad (3)$$

(ii) Due to the effects of friction and inertia associated with the movement of any object, some of the input energy to a machine is converted into heat and losses occur. Since losses occur, the energy output of a machine is less than the energy input, thus the mechanical effieiency of any machine cannot reach 100%. For example, a simple machine raises a load of 160 kg through a distance of 1.6 m. The effort applied to the machine is 200 N and moves through a distance of 16 m.

$$\text{Thus, the force ratio} = \frac{\text{load}}{\text{effort}} = \frac{160\ \text{kg}}{200\ \text{N}} = \frac{160 \times 9.81\ \text{N}}{200\ \text{N}} = \mathbf{7.85},$$

$$\text{the movement ratio} = \frac{\text{distance moved by effort}}{\text{distance moved by load}}$$

$$= \frac{16\ \text{m}}{1.6\ \text{m}} = \mathbf{10}$$

$$\text{and, the effieiency} = \frac{\text{force ratio}}{\text{movement ratio}} \times 100\%$$

$$\frac{7.85}{10} \times 100\% = \mathbf{78.5\%}.$$

(iii) For simple machines, the relationship between effort and load is of the form: $F_e = aF_l + b$, where F_e is the effort, F_l is the load and a and b are constants. From equation (1)

$$\text{force ratio} = \frac{\text{load}}{\text{effort}} = \frac{F_l}{F_e} = \frac{F_e}{aF_l + b}$$

Dividing both numerator and denominator by F_l gives:

$$\frac{F_l}{aF_l + b} = \frac{1}{a + \dfrac{b}{F_l}}$$

When the load is large, F_l is large and b/F_l is small compared with a. The force ratio then becomes approximately equal to $1/a$ and is called the **limiting force ratio**.

The **limiting efficiency** of a simple machine is defined as the ratio of the limiting force ratio to the movement ratio, i.e.

$$\text{limiting efficiency} = \frac{1}{a}\ (\text{movement ratio})$$

$$= \frac{1}{a \times \text{movement ratio}} \times 100\%,$$

where a is the constant for the law of the machine: $F_e = aF_l + b$.

Due to friction and inertia, the limiting effieiency of simple machines is usually well below 100%. For example, in a test on a simple machine, the effort-load graph was a straight line of the form $F_e = aF_l + b$. Two values lying on the graph were at $F_e = 10$ N, $F_l = 30$ N and at $F_e = 74$ N, $F_l = 350$ N. The movement ratio of the machine was 17.

The equation $F_e = aF_l + b$ is of the form $y = mx + c$, where m is the gradient of the graph. The slope of the line passing through points (x_1, y_1) and (x_2, y_2) of the graph $y = mx + c$ is given by:

$$m = \frac{y_2 - y_1}{x_2 - x_1}$$

Thus for $F_e = aF_l + b$, the slope a is given by:

$$a = \frac{74 - 10}{350 - 30} = \frac{64}{320} = 0.2$$

Hence the limiting force ratio $= \frac{1}{a} = \frac{1}{0.2} = \mathbf{5}$

$$\text{The limiting efficiency} = \frac{1}{a \times \text{movement ratio}} \times 100\%$$

$$= \frac{1}{0.2 \times 17} \times 100\% = \mathbf{29.4\%}$$

5 A pulley system is a simple machine. A single-pulley system, shown in *Figure 18.1(a)*, changes the line of action of the effort, but does not change the magnitude of the force.

A two-pulley system, shown in *Figure 18.1(b)*, changes both the line of action and the magnitude of the force. Theoretically, each of the ropes marked (i) and (ii) share the load equally, thus the theoretical effort is only half of the load, i.e. the theoretical force ratio is 2. In practice the actual force ratio is less than 2 due to losses.

A three-pulley system is shown in *Figure 18.1(c)*. Each of the ropes marked (i), (ii) and (iii) carry one-third of the load, thus the theoretical force ratio is 3. In general, for a multiple pulley system having a total of n pulleys, the theoretical force ratio is n. Since the theoretical efficiency of a pulley system (neglecting losses) is 100% and since from equation (3):

$$\text{efficiency} = \frac{\text{force ratio}}{\text{movement ratio}} \times 100\%,$$

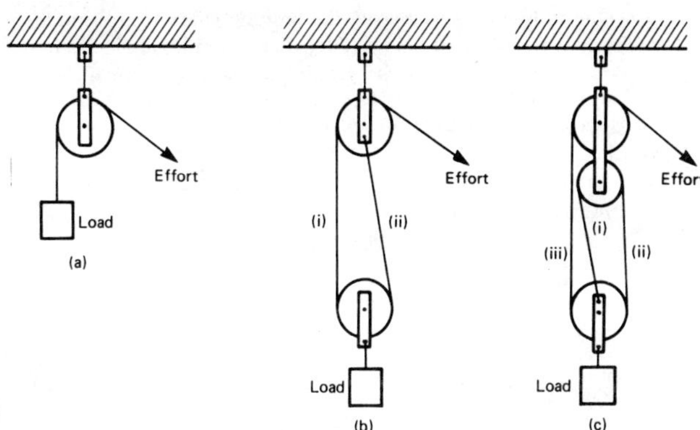

Figure 18.1

it follows that when the force ratio is n,

$$100=\frac{n}{\text{movement ratio}}\times 100,$$

that is the movement ratio is also n.

For example, a load of 80 kg is lifted by a three-pulley system and the applied effort is 392 N.

$$\text{The force ratio}=\frac{\text{load}}{\text{effort}}=\frac{80\times 9.81}{392}=\mathbf{2}$$

The movement ratio = 3 (since it is a three-pulley system)

$$\text{Thus the efficiency}=\frac{\text{force ratio}}{\text{movement ratio}}\times 100\%$$

$$=\frac{2}{3}\times 100\%=\mathbf{66.67\%}$$

6 A **simple screw-jack** is shown in *Figure 18.2* and is a simple machine since it changes both the magnitude and the line of action of a force.

The screw of the table of the jack is located in a fixed nut in the body of the jack. As the table is rotated by means of a bar, it raises or lowers a load placed on the table. For a single-start thread, as shown, for one complete revolution of the table, the

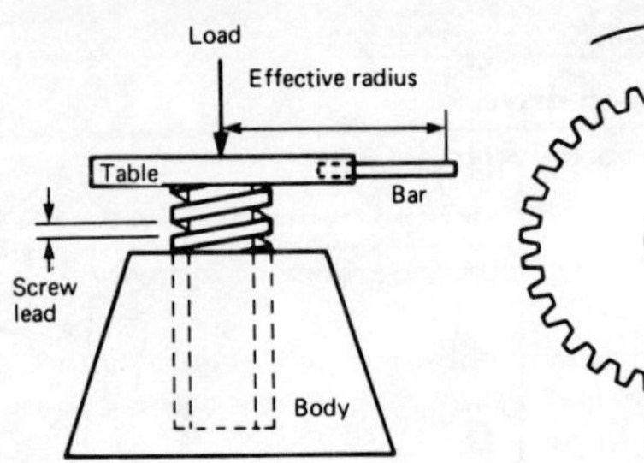

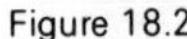
Figure 18.2

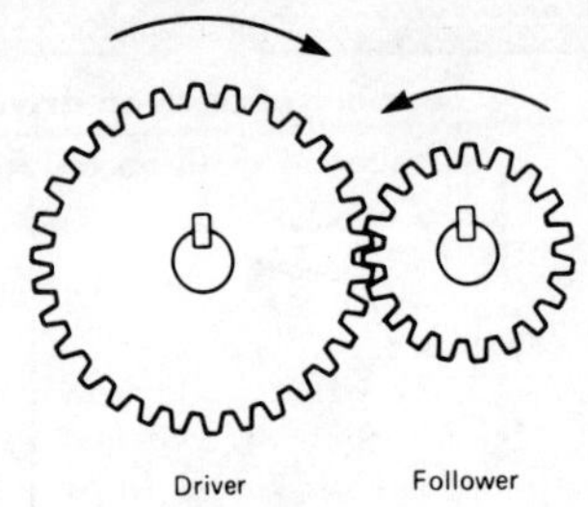

Figure 18.3

effort moves through a distance $2\pi r$ and the load moves through a **distance equal to the lead of the screw, say** l. **Thus:**

$$\text{the movement ratio} = \frac{2\pi r}{l} \qquad (4)$$

For example, a screw jack is used to support the axle of a car, the load on it being 2.4 kN. The screw jack has an effort arm of effective radius 200 mm and a single-start square thread having a lead of 5 mm. If an effort of 60 N is **required to raise the car axle:**

$$\text{Force ratio} = \frac{\text{load}}{\text{effort}} = \frac{2400\text{ N}}{60\text{ N}} = \mathbf{40}$$

$$\text{Movement ratio} = \frac{2\pi r}{l} = \frac{2\pi\ 200\text{ mm}}{5\text{ mm}} = \mathbf{251.3}$$

$$\text{Hence, efficiency} = \frac{40}{251.3} \times 100\% = \mathbf{15.9\%}.$$

7 (i) A **simple gear train** is used to transmit rotary motion and can change both the magnitude and the line of action of a force, hence is a simple machine. The gear train shown in *Figure 18.3* consists of **spur gears** and has an effort applied to one gear, called the **driver** and a load applied to the other gear, called the **follower**.

(ii) In such a system, the teeth on the wheels are so spaced that they exactly fill the circumference with a whole number of identical teeth, and the teeth on the driver and follower mesh without interference. Under these conditions, the number of teeth on the driver and follower are in direct proportion to the circumference of these wheels, i.e.

$$\frac{\textbf{number of teeth on driver}}{\textbf{number of teeth on follower}} = \frac{\textbf{circumference of driver}}{\textbf{circumference of follower}} \quad (5)$$

(iii) If there are, say, 40 teeth on the driver and 20 teeth on the follower then the follower makes two revolutions for each revolution of the driver. In general

$$\frac{\text{the number of revolutions made by the driver}}{\text{the number of revolutions made by the follower}} = \frac{\text{the number of teeth on the follower}}{\text{the number of teeth on the driver}} \quad (6)$$

It follows from equation (6) that the speeds of the wheels in a gear train are inversely proportional to the number of teeth.

(iv) The ratio of the speed of the driver wheel to that of the follower is the movement ratio, i.e.

$$\textbf{movement ratio} = \frac{\textbf{speed of driver}}{\textbf{speed of follower}} = \frac{\textbf{teeth on follower}}{\textbf{teeth on driver}} \quad (7)$$

(v) When the same direction of rotation is required on both the driver and the follower an **idler wheel** is used as shown in *Figure 18.4*. Let the driver, idler and follower be A, B and C respectively, and let N be the speed of rotation and T be the number of teeth. Then from equation (7),

$$\frac{N_B}{N_A} = \frac{T_A}{T_B} \text{ and } \frac{N_C}{N_B} = \frac{T_B}{T_C}$$

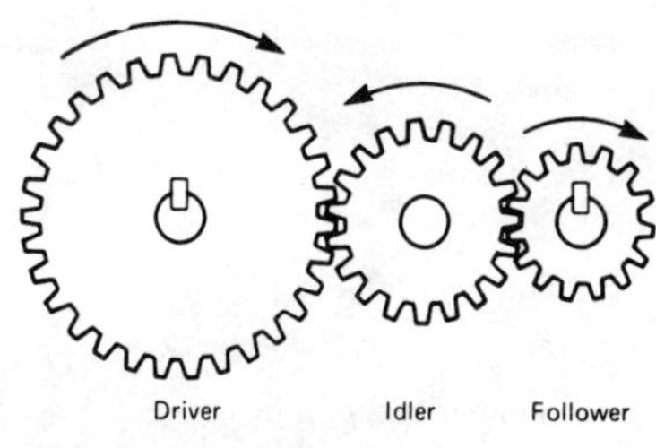

Figure 18.4

$$\text{Thus } \frac{\text{speed of A}}{\text{speed of C}} = \frac{N_A}{N_C} = \frac{N_B \dfrac{T_B}{T_A}}{N_B \dfrac{T_B}{T_C}} = \frac{T_B}{T_A} \times \frac{T_C}{T_B} = \frac{T_C}{T_A}$$

This shows that the movement ratio is independent of the idler, only the direction of the follower being altered.

(vi) **A compound gear train is shown in *Figure 18.5*, in which gear**

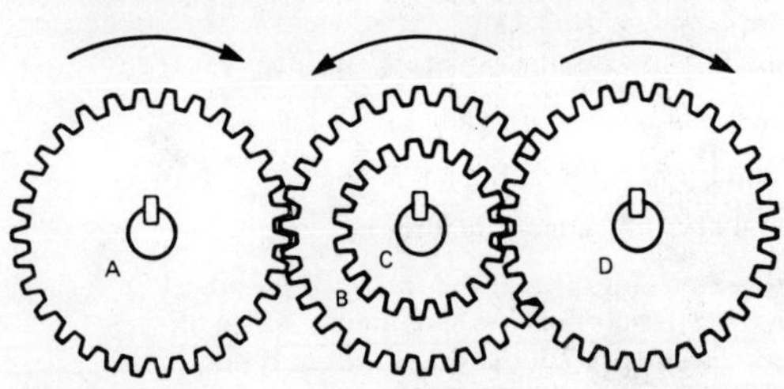

Figure 18.5

wheels B and C are fixed to the same shaft and hence $N_B = N_C$. From equation (7),

$$\frac{N_A}{N_B} = \frac{T_B}{T_A} \text{ i.e. } N_B = N_A \times \frac{T_A}{T_B}$$

Also, $\dfrac{N_D}{N_C} = \dfrac{T_C}{T_D}$, i.e. $N_D = N_C \times \dfrac{T_C}{T_D}$. But $N_B = N_C$,

$$\text{Hence, } N_D = N_A \times \frac{T_A}{T_B} \times \frac{T_C}{T_D} \qquad (8)$$

For compound gear **trains having say *P* gear wheels,**

$$N_P = N_A \times \frac{T_A}{T_B} \times \frac{T_C}{T_D} \times \frac{T_E}{T_F} \ldots \times \frac{T_O}{T_P} \qquad (9)$$

from which

$$\textbf{movement ratio} = \frac{N_A}{N_P} = \frac{T_B}{T_A} \times \frac{T_D}{T_C} \times \ldots \times \frac{T_P}{T_O}$$

For example, a compound gear train consists of a driver gear A having 40 teeth, engaging with gear B, having 160 teeth. Attached

to the same shaft as B, gear C has 48 teeth and meshes with gear D on the output shaft having 96 teeth.
Thus from equation (8),

$$\text{movement ratio} = \frac{\text{speed of A}}{\text{speed of D}} = \frac{T_B}{T_A} \times \frac{T_D}{T_C}$$

$$= \frac{160}{40} \times \frac{96}{48} = \mathbf{8}$$

If the force ratio is, say, 6, then the efficiency is $\frac{6}{8} \times 100\% = \mathbf{75\%}$.

8 A **lever** can alter both the magnitude and the line of action of a force and is thus classed as a simple machine. There are three types or orders of levers, as shown in *Figure 18.6*.

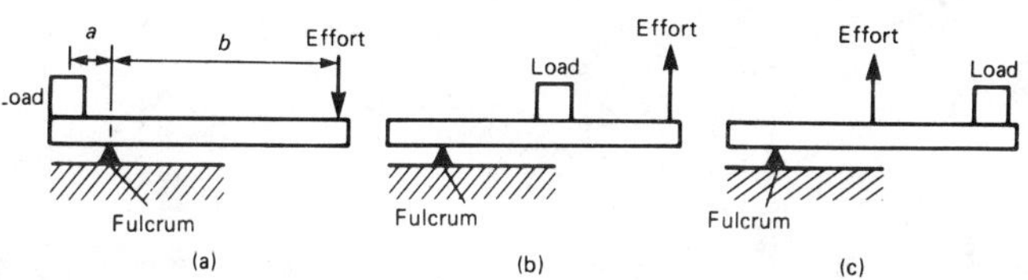

Figure 18.6

(i) A lever of the first order has the fulcrum placed between the effort and the load, as shown in *Figure 18.6(a)*.
(ii) A lever of the second order has the load placed between the effort and fulcrum, as shown in *Figure 18.6(b)*.
(iii) A lever of the third order has the effort applied between the load and the fulcrum, as shown in *Figure 18.6(c)*.
Problems on levers can largely be solved by applying the principle of moments (see page 70). Thus for the lever shown in *Figure 18.6(a)*, when the lever is in equilibrium

anticlockwise moment = clockwise moment

i.e. $a \times F_l = b \times F_e$

Thus **force ratio** $= \frac{F_l}{F_C} = \frac{b}{a}$

$$= \frac{\textbf{distance of effort from fulcrum}}{\textbf{distance of load from fulcrum}} \qquad (10)$$

For example, the load on a first-order lever is 1.2 kN, the distance between the fulcrum and load is 0.5 m and the distance between the fulcrum and effort is 1.5 m. Thus, at equilibrium,

anticlockwise moment = clockwise moment

i.e. $1200 \times 0.5 = \text{effort} \times 1.5$

from which, $\text{effort} = \dfrac{1200 \times 0.5}{1.5} = \mathbf{400\ N}$

$$\text{Force ratio} = \frac{F_l}{F_e} = \frac{1200}{400} = \mathbf{3} \left(\text{or force ratio} = \frac{b}{a} = \frac{1.5}{0.5} = 3 \right)$$

This result shows that to lift a load of say 300 N, an effort of 100 N is required.

19 Heat energy

1 (i) **Heat** is a form of energy and is measured in joules.
(ii) **Temperature** is the degree of hotness or coldness of a substance.
Heat and temperature are thus **not** the same thing. For example, twice the heat energy is needed to boil a full container of water than half a container — that is, different amounts of heat energy are needed to cause an equal rise in the temperature of different amounts of the same substance.

2 Temperature is measured either (i) on the **Celsius (°C) scale** (formerly Centigrade), where the temperature at which ice melts, i.e. the freezing point of water, is taken as 0°C and the point at which water boils under normal atmospheric pressure is taken as 100°C, or (ii) on the **thermodynamic scale**, in which the unit of temperature is the kelvin (K). The kelvin scale uses the same temperature interval as the Celsius scale but as its zero takes the 'absolute zero of temperature' which is at about −273°C.

Hence kelvin temperature = degree Celsius + 273

i.e. **K = (°C) + 273**

Thus, for example,

0°C = 273 K, 25°C = 298 K and 100°C = 373 K.

3 A **thermometer** is an instrument which measures temperature. Any substance which possesses one or more properties which vary with temperature can be used to measure temperature. These properties include changes in length, area or volume, electrical resistance or in colour. Examples of temperature measuring devices include:
(i) **liquid-in-glass thermometer**, which uses the expansion of a liquid with increase in temperature as its principle of operation,
(ii) **thermocouples**, which use the e.m.f. set up when the junction of two dissimilar metals is heated,
(iii) **resistance thermometer**, which uses the change in electrical resistance caused by temperature change, and
(iv) **pyrometers**, which are devices for measuring very high

temperatures, using the principle that all substances emit radiant energy when hot, the rate of emission depending on their temperature.
(See Chapter 21, page 104).

4 (i) The **specific heat capacity** of a substance is the quantity of heat energy required to raise the temperature of 1 kg of the substance by 1°C.

(ii) The symbol used for specific heat capacity is c and the units are J/(kg °C) or J/(kg K). (Note that these units may also be written as J kg^{-1} $°C^{-1}$ or J kg^{-1} K^{-1}.)

(iii) Some typical values of specific heat capacity for the range of temperature 0°C to 100°C include:

water	4190 J/(kg °C),	ice	2100 J/(kg °C)
aluminium	950 J/kg °C),	copper	390 J/(kg °C)
iron	500 J(kg °C)	lead	130 J/(kg °C)

Hence

to raise the temperature of 1 kg of iron by 1°C requires 500 J of energy,
to raise the temperature of 5 kg of iron by 1°C requires (500 × 5) J of energy, and to raise the temperature of 5kg of iron by 40°C requires (500 × 5 × 40) J of energy, i.e. 100 kJ.

In general, the quantity of heat energy, Q, required to raise a mass m kg of a substance with a specific heat capacity c J/(kg °C) from temperature t_1 °C to t_2 °C is given by:

$$\boxed{\boldsymbol{Q = mc(t_2 - t_1)} \textbf{ joules.}}$$

5 A material may exist in any one of three states — solid, liquid or gas. If heat is supplied at a constant rate to some ice initially at, say, −30°C, its temperature rises as shown in *Figure 19.1*. Initially the temperature increases from −30°C to 0°C as shown by the line AB. It then remains constant at 0°C for the time BC required for the ice to melt into water. The energy gained by continual heating causes an increase in the vibrational energy of the water molecules in the solid ice structure. The magnitude of the vibrations increases until the attractive forces between the molecules are sufficiently reduced to allow the water molecules to exist completely in the liquid state. The constant temperature is observed because the heat energy is absorbed in this way. When the ice is completely melted to water, continual heating raises the temperature to 100°C, as shown by CD in *Figure 19.1*. The temperature then remains constant at 100°C along DE. The energy provided during this time is being used to increase the energy of the water molecules until they are all moving rapidly enough to overcome the forces of attraction in the

liquid phase and move freely in the gas phase. Continual heating raises the temperature of the steam as shown by EF in the region where the steam is termed superheated. Changes of state from solid to liquid or liquid to gas occur without change of temperature and such changes are reversible process. When heat energy flows to or from a substance and causes a change of temperature, such as between A and B, between C and D and between E and F in *Figure 19.1*, it is called **sensible heat** (since it can be 'sensed' by a thermometer).

Heat energy which flows to or from a substance while the temperature remains constant, such as between B and C and between D and E in *Figure 19.1*, is called **latent heat** (latent means concealed or hidden).

6 (i) The **specific latent heat of fusion** is the heat required to change 1 kg of a substance from the solid state to the liquid state (or vice versa) at constant temperature.

(ii) The **specific latent heat of vaporisation** is the heat required to change 1 kg of a substance from a liquid to a gaseous state (or vice versa) at constant temperature.

(iii) The units of the specific latent heats of fusion and vaporisatioı are J/kg, or more often, kJ/kg, and some typical values are shown below.

	Latent heat of fusion (kJ/kg)	*Melting point (°C)*
Mercury	11.8	−39
Lead	22	327
Silver	100	957
Ice	335	0
Aluminium	387	660
	Latent heat of vaporisation (kJ/kg)	*Boiling point (°C)*
Oxygen	214	−183
Mercury	286	357
Ethyl alcohol	857	79
Water	2257	100

(iv) The quantity of heat Q supplied or given out during a change of state is given by:

$$\boxed{Q = mL}$$

where m is the mass in kilograms and L is the specific latent heat.

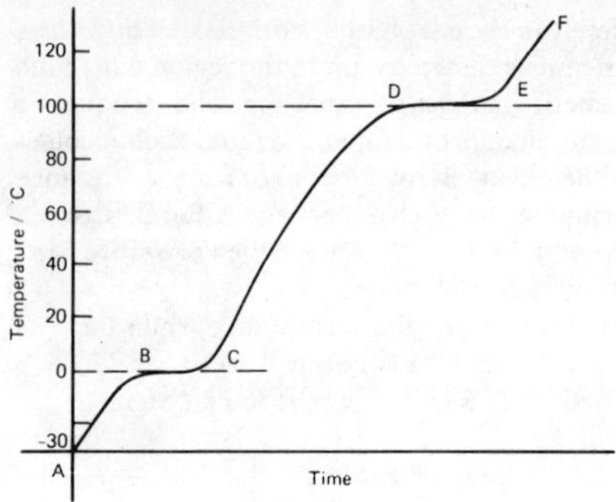

Figure 19.1

Thus, for example, the heat required to convert 10 kg of ice at 0°C to water at 0°C is given by 10 kg × 335 kJ/kg, i.e. 3350 kJ or 3.35 MJ.

Principle of operation of a refrigerator

7 The boiling point of most liquids may be lowered if the pressure is lowered. In a simple refrigerator a working fluid, such as ammonia or freon, has the pressure acting on it reduced. The resulting lowering of the boiling point causes the liquid to vaporise.

In vaporising, the liquid takes in the necessary latent heat from its surroundings, i.e. the freezer, which thus becomes cooled. The vapour is immediately removed by a pump to a condenser which is outside of the cabinet, where it is compressed and changed back into a liquid, giving out latent heat. The cycle is repeated when the liquid is pumped back to the freezer to be vaporised.

Conduction, connection and radiation

8 Heat may be **transferred** from a hot body to a cooler body by one or more of three methods, these being: (a) by **conduction**, (b) by **convection**, or (c) by **radiation**.

9 **Conduction** is the transfer of heat energy from one part of a body to another (or from one body to another) without the particles of the body moving.

Conduction is associated with solids. For example, if one end of a metal bar is heated, the other end will become hot by conduction. Metals and metallic alloys are good conductors of heat whereas air, wood, plastic, cork, glass and gases are examples of poor conductors (i.e. heat insulators).

Practical applications of conduction:

(i) A domestic saucepan or dish conducts heat from the source to the contents. Also, since wood and plastic are poor conductors of heat they are used for saucepan handles.
(ii) The metal of a radiator of a central heating system conducts heat from the hot water inside to the air outside.
10 **Convection** is the transfer of heat energy through a substance by the actual movement of the substance itself. Convection occurs in liquids and gases, but not in solids. When heated, a liquid or gas becomes less dense. It then rises and is replaced by a colder liquid or gas and the process repeats. For example, electric kettles and central heating radiators always heat up at the top first.

Examples of convection are:

(i) Natural circulation hot water heating systems depend on the hot water rising by convection to the top of a house and then falling back to the bottom of the house as it cools, releasing the heat energy to warm the house as it does so.
(ii) Convection currents cause air to move and therefore affect climate.
(iii) When a radiator heats the air around it, the hot air rises by convection and cold air moves in to take its place.
(iv) A cooling system in a car radiator relies on convection.
(v) Large electrical transformers dissipate waste heat to an oil tank. The heated oil rises by convection to the top, then sinks through cooling fins, losing heat as it does so.
(vi) In a refrigerator, the cooling unit is situated near the top. The air surrounding the cold pipes becomes heavier as it contracts and sinks towards the bottom. Warmer, less dense air is pushed upwards and in turn is cooled. A cold convection current is thus created.
11 **Radiation** is the transfer of heat energy from a hot body to a cooler one by electromagnetic waves. Heat radiation is similar in character to light waves (see Chapter 12) — it travels at the same speed and can pass through a vacuum — except that the frequency of the waves are different. Waves are emitted by a hot body, are transmitted through space (even a vacuum), and are not detected until they fall on to another body. Radiation is reflected from shining, polished surfaces but absorbed by dull, black, surfaces.

Practical applications of radiation include:

(i) heat from the sun reaching earth;
(ii) heat felt by a flame;
(iii) cooker grills;
(iv) industrial furnaces;
(v) infra-red space heaters.

Vacuum flask

12 A cross-section of a typical vacuum flask is shown in *Figure 19.2* and is seen to be a double-walled bottle with a vacuum space between them, the whole supported in a protective outer case.

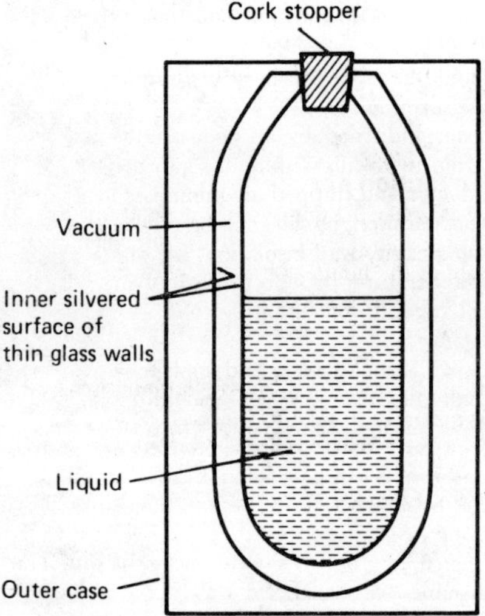

Figure 19.2

Very little heat can be transferred by conduction because of the vacuum space and the cork stopper (cork is a bad conductor of heat). Also, because of the vacuum space, no convection is possible. Radiation is minimised by silvering the two glass surfaces (radiation is reflected off of shining surfaces).

Thus a vacuum flask is an example of prevention of all three types of heat transfer and is therefore able to keep hot liquids hot and cold liquids cold.

Use of insulation in conserving fuel

13 Fuel used for heating a building is becoming increasingly expensive. By the careful use of insulation, heat can be retained in a building for longer periods and the cost of heating thus minimised.

(i) Since convection causes hot air to rise it is important to insulate the roof space, which is probably the greatest source of heat loss in the home. This can be achieved by laying fibre-glass between the wooden joists in the roof space.

(ii) Glass is a poor conductor of heat. However large losses can occur through thin panes of glass and such losses can be reduced by using double-glazing. Two sheets of glass, separated by air, are used. Air is a good insulator but the air space must not be too large otherwise convection currents can occur which would carry heat across the space.

(iii) Hot water tanks should be lagged to prevent conduction and convection of heat to the surrounding air.

(iv) Brick, concrete, plaster and wood are all poor conductors of heat. A house is made from two walls with an air gap between them. Air is a poor conductor and trapped air minimises losses through the wall. Heat losses through walls can be prevented almost completely by using cavity wall insulation, i.e. plastic-foam.

14 Besides changing temperature, the effects of supplying heat to a material can involve changes in dimensions, as well as in colour, state and electrical resistance.

Most substances expand when heated and contract when cooled, and there are many practical applications and design implications of thermal movement. (See Chapter 20).

20 Thermal expansion

1 When heat is applied to most materials, **expansion** occurs in all directions. Conversely, if heat energy is removed from a material (i.e. the material is cooled) **contraction** occurs in all directions. The effects of expansion and contraction each depend on the **change of temperature** of the material.

2 Some practical applications where expansion and contraction of solid materials must be allowed for include:

(i) Overhead electrical transmission lines are hung so that they are slack in summer, otherwise their contraction in winter might snap the conductors or bring down pylons.

(ii) Gaps need to be left in lengths of railway lines to prevent buckling in hot weather.

(iii) Ends of large bridges are often supported on rollers to allow them to expand and contract freely.

(iv) Fitting a metal collar to a shaft or a steel tyre to a wheel is often achieved by first heating them so that they expand, fitting them in position, and then cooling them so that the contraction holds them firmly in place. This is known as a 'shrink-fit'. By a similar method hot rivets are used for joining metal sheets.

(v) The amount of expansion varies with different materials. *Figure 20.1(a)* shows a bimetallic strip at room temperature (i.e. two different strips of metal riveted together).

When heated, brass expands more than steel, and since the two metals are riveted together the bimetallic strip is forced into an arc as shown in *Figure 20.1(b)*. Such a movement can be arranged to make or break an electric circuit and bimetallic strips are used, in particular, in thermostats (which are temperature operated switches) used to control central heating systems, cookers, refrigerators, toasters, irons, hot-water and alarm systems.

(vi) Motor engines use the rapid expansion of heated gases to force a piston to move.

(vii) Designers must predict, and allow for, the expansion of steel pipes in a steam-raising plant so as to avoid damage and consequent danger to health.

3 (i) Water is a liquid which at low temperatures displays an unusual effect. If cooled, contraction occurs until, at about 4°C,

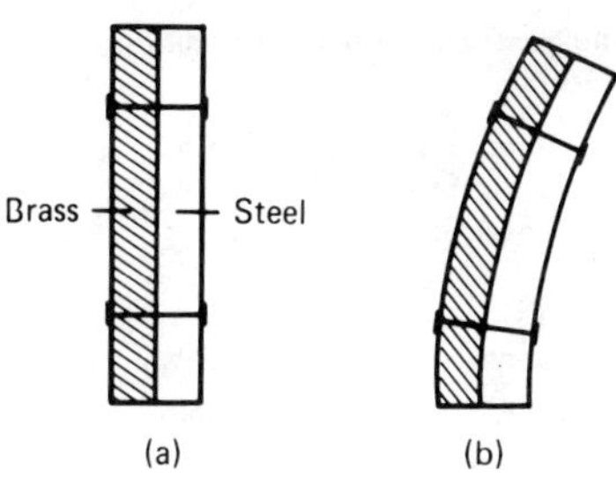

Figure 20.1

the volume is at a minimum. As the temperature is further decreased from 4°C to 0°C expansion occurs, i.e. the volume increases. When ice is formed considerable expansion occurs and it is this expansion which often causes frozen water pipes to burst.
(ii) A practical application of the expansion of a liquid is with thermometers, where the expansion of a liquid, such as mercury or alcohol, is used to measure temperature.

4 (i) The amount by which unit length of a material expands when the temperature is raised one degree is called the **coefficient of linear expansion** of the material and is represented by α (Greek alpha).
(ii) The units of the coefficient of linear expansion are m/(m K), although it is usually quoted as just /K or K^{-1}. For example, **copper has a coefficient of linear expansion value of 17×10^{-6} K^{-1}, which means that a 1 m long bar of copper expands by** 0.000017 m if its temperature is increased by 1 K (or 1°C). If a 6 m long bar of copper is subjected to a temperature rise of 25 K **then the bar will expand by** $(6 \times 0.000017 \times 25)$ m, i.e. 0.00255 m or 2.55 mm. (Since the kelvin scale uses the same temperature interval as the Celsius scale, a **change** of temperature of, say, 50°C, is the same as a **change** of temperature of 50 K.)
(iii) If a material, initially of length l_1 and at a temperature t_1 and having a coefficient of linear expansion α, has its temperature increased to t_2, then the new length l_2 of the material is given by:

new length = original length + expansion

$$l_2 = l_1 + l_1\alpha(l_2 - l_1)$$

i.e. $$\boldsymbol{l_2 = l_1[1 + \alpha(t_2 - t_1)]} \qquad (1)$$

(iv) Some typical values for the coefficient of linear expansion include:

aluminium	$23\times10^{-6}\ K^{-1}$,	brass	$18\times10^{-6}\ K^{-1}$
concrete	$12\times10^{-6}\ K^{-1}$,	copper	$17\times10^{-6}\ K^{-1}$
gold	$14\times10^{-6}\ K^{-1}$,	invar (nickel-steel alloy)	
iron	$11–12\times10^{-6}\ K^{-1}$,		$0.9\times10^{-6}\ K^{-1}$
steel	$15–16\times10^{-6}\ K^{-1}$,	nylon	$100\times10^{-6}\ K^{-1}$
zinc	$31\times10^{-6}\ K^{-1}$,	tungsten	$4.5\times10^{-6}\ K^{-1}$.

For example, the copper tubes in a boiler are 4.20 m long at a temperature of 20°C. Then, when surrounded only by feed water at 10°C, the final length of the tubes l_2 is given by:

$$l_2=l_1[1+\alpha(t_2-t_1)]=4.20[1+(17\times10^{-6})(10-20)]$$
$$=\mathbf{4.1993\ m},$$

i.e. the tube contracts by 0.7 mm when the temperature decreases from 20°C to 10°C.

When the boiler is operating and the mean temperature of the tubes is 320°C, then the length of the tube l_2 is given by:

$$l_2=l_1[1+\alpha(t_2-t_1)]=4.20[1+(17\times10^{-6})(320-20)]$$
$$=4.20[1+0.0051]=\mathbf{4.2214\ m},$$

i.e. the tube extends by 21.4 mm when the temperature rises from 20°C to 320°C.

5 (i) The amount by which unit area of a material increases when the temperature is raised by one degree is called the **coefficient of superficial (i.e. area) expansion** and is represented by β (Greek beta).

(ii) If a material having an initial surface area A_1 at temperature t_1 and having a coefficient of superficial expansion β, has its temperature increased to t_2, then the new surface area A_2 of the material is given by:

new surface area = original surface area + increase in area

i.e. $A_2=A_1+A_1\beta(t_2-t_1)$

i.e. $$\boxed{A_2=A_1[1+\beta(t_2-t_1)]} \quad (2)$$

(iii) It may be shown that the coefficient of superficial expansion is twice the coefficient of linear expansion, i.e. $\beta=2\alpha$, to a very close approximation.

6 (i) The amount by which unit volume of a material increases for a one degree rise of temperature is called the **coefficient of cubic (or volumetric) expansion** and is represented by γ (Greek gamma).

(ii) If a material having an initial volume V_1 at temperature t_1 and having a coefficient of cubic expansion γ, has its temperature raised to t_2, then the new volume V_2 of the material is given by:

new volume = initial volume + increase in volume

i.e. $V_2 = V_1 + V_1\gamma(t_2 - t_1)$

i.e. $V_2 = V_1[1 + \gamma(t_2 - t_1)]$ (3)

(iii) It may be shown that the coefficient of cubic expansion is three times the coefficient of linear expansion, i.e. $\gamma = 3\alpha$, to a very close approximation. A liquid has no definite shape and only its cubic or volumetric expansion need be considered. Thus with expansions in liquids, equation (3) is used.
(iv) Some typical values for the coefficient of cubic expansion measured at 20°C (i.e. 293 K) include:

ethyl alcohol	1.1×10^{-3} **K**$^{-1}$	mercury	1.82×10^{-4} K^{-1},
paraffin oil	9×10^{-2} K^{-1},	water	2.1×10^{-4} K^{-1}.

The coefficient of cubic expansion γ is only constant over a limited range of temperature.

For example, mercury contained in a thermometer has a volume of 476 mm^3 at 15°C. When the volume is, say, 478 mm^3, then:

$$V_2 = V_1[1 + \gamma(t_2 - t_1)]$$

$$478 = 476[1 + 1.82 \times 10^{-4}(t_2 - 15)]$$

from which, $478 = 476 + (476)(1.82 \times 10^{-4})(t_2 - 15)$

$$\frac{478 - 476}{(476)(1.82 \times 10^{-1})} = t_2 - 15$$

$$23.09 = t_2 - 15$$

Hence when the volume of mercury is 478 mm^3 the temperature is $23.09 + 15 =$ **38.09°C**.

21 Measurement of temperature

1 A change in temperature of a substance can often result in a change in one or more of its physical properties. Thus, although temperature cannot be measured directly, its effects can be measured. Some properties of substances used to determine changes in temperature include changes in dimensions, electrical resistance, state, type and volume of radiation and colour. Temperature measuring devices available are many and varied. The devices described in the following paragraphs are most often used in science and industry.

Liquid-in-glass thermometer

Construction

2 (a) A typical liquid-in-glass thermometer is shown in *Figure 21.1* and consists of a sealed stem of uniform small bore tubing, called a capillary tube, made of glass, with a cylindrical glass bulb formed at one end. The bulb and

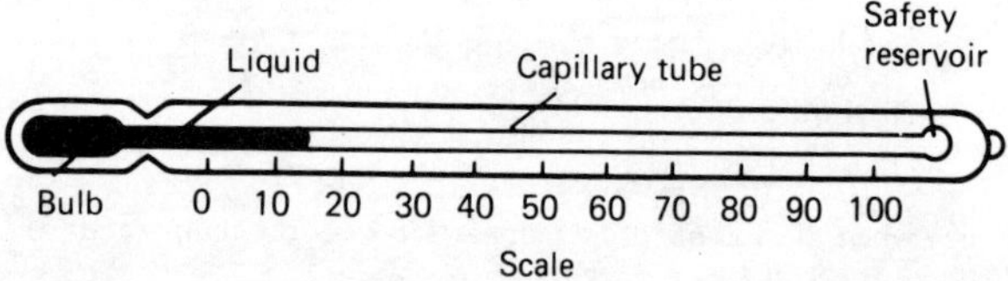

Figure 21.1

part of the stem are filled with a liquid such as mercury or alcohol and the remaining part of the tube is evacuated. A temperature scale is formed by etching graduations on the stem (see (b)). A safety reservoir is usually provided into which the liquid can expand without bursting the glass if the temperature is raised beyond the upper limit of the scale.

Principle of operation

(b) The operation of a liquid-in-glass thermometer depends on the liquid expanding with increase in temperature and contracting with decrease in temperature. The position of the end of the column of liquid in the tube is a measure of the temperature of the liquid in the bulb — shown as 15°C in *Figure 21.1*, which is about room temperature. Two fixed points are needed to calibrate the thermometer, with the interval between these points beting divided into 'degrees'. In the first thermometer, made by Celsius, the fixed points chosen were the temperature of melting ice (0°C), and that of boiling water at standard atmospheric pressure (100°C), in each case the blank stem being marked at the liquid level. The distance between these two points, called the fundamental interval, was divided into 100 equal parts, each equivalent to 1°C, thus forming the scale.

The cylindrical thermometer, with a limited scale around body temperature, the maximum and/or minimum thermometer, recording the maximum day temperature and minimum night temperature, and the Beckman thermometer, which is used only in accurate measurement of temperature change and has no fixed points, are particular types of liquid in glass thermometer which all operate on the same principle.

Advantages

(c) The liquid in glass thermometer is simple in construction, relatively inexpensive, easy to use and portable, and is the most widely used method of temperature measurement having industrial, chemical, clinical and meteorological applications.

Disadvantages

(d) Liquid in glass thermometers tend to be fragile and hence easily broken, can only be used where the liquid column is visible, cannot be used for surface temperature measurements, cannot be read from a distance and are unsuitable for high temperature measurements.

(e) **The use of mercury in a thermometer has many advantages**, for mercury:

(i) is clearly visible;

(ii) has a fairly uniform rate of expansion;

(iii) is readily obtainable in the pure state;

(iv) does not 'wet' the glass; and
(v) is a good conductor of heat.

Mercury has a freezing point of −39°C and cannot be used in a thermometer below this temperature. Its boiling point is 357°C but before this temperature is reached some distillation of the mercury occurs if the space above the mercury is a vacuum. To prevent this, and to extend the upper temperature limits to over 500°C, an inert gas such as nitrogen under pressure is used to fill the remainder of the capillary tube. Alcohol, often dyed red to be seen in the capillary tube, is considerably cheaper than mercury and has a freezing point of −113°C, which is considerably lower than for mercury. However it has a low boiling point at about 79°C.

(f) **Typical errors in liquid in glass thermometers may occur due to:**
(i) the slow cooling rate of glass,
(ii) incorrect positioning of the thermometer,
(iii) a delay in the thermometer becoming steady (i.e. slow response time), and
(iv) non-uniformity of the bore of the capillary tube, which means that equal intervals marked on the stem do not correspond to equal temperature intervals.

Thermocouple

Principle of operation

(a) At the junction between two different metals, say, copper and constantan, there exists a difference in electrical potential, which varies with the temperature of the junction. This is known as the 'thermoelectric effect'. If the circuit is completed with a second junction at a different temperature, a current will flow round the circuit. This principle is used in the thermocouple. Two different metal conductors having their ends twisted together are shown in *Figure 21.2*. If the two junctions are at different temperatures, a current I flows round the circuit.

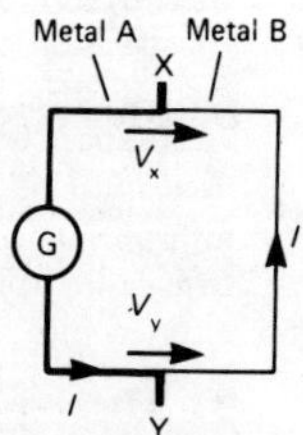

Figure 21.2

The deflection on the galvanometer G depends on the difference in temperature between junction X and Y and is caused by the difference between voltages V_X and V_Y. The higher temperature junction is usually called the 'hot junction' and the lower temperature junction the 'cold junction'. If the cold junction is kept at a constant known temperature, the galvanometer can be calibrated to indicate the temperature of the hot junction directly. The cold junction is then known as the reference junction.

In many instrumentation situations, the measuring instrument needs to be located far from the point at which the measurements are to be made. Extension leads are then used, usually made of the same material as the thermocouple but of smaller gauge. The reference junction is then effectively moved to their ends.

The thermocouple is used by positioning the hot junction where the temperature is required. The meter will indicate the temperature of the hot junction only if the reference junction is at 0°C for

(temperature of hot junction)
= (temperature of the cold junction + (temperature difference).

In a laboratory the reference junction is often placed in melting ice, but in industry it is often positioned in a thermostatically controlled oven or buried underground where the temperature is constant.

Construction and typical applications

(b) Thermocouple junctions are made by twisting together two wires of dissimilar metals before welding them. The construction of a typical copper-constantan thermocouple for industrial use is shown in *Figure 21.3.*

Apart from the actual junction the two conductors used must be electrically insulated from each other with appropriate insulation and is shown in *Figure 21.3* as twin-holed tubing. The wires and insulation are usually inserted into a sheath for protection from environments in which it might be damaged or corroded. A copper-constantan thermocouple can measure temperature from −250°C up to about 400°C, and is used typically with boiler flue gases, food processing and with sub-zero mperature measurement.

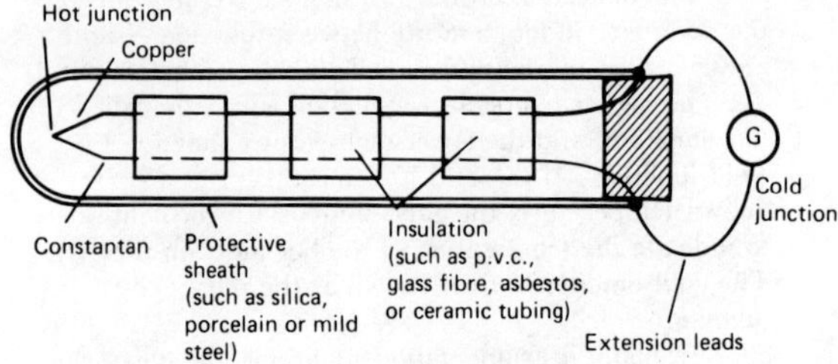

Figure 21.3

An iron-constantan thermocouple can measure temperature from −200°C to about 850°C, and is used typically in paper and pulp mills, reheat and annealing furnaces and in chemical reactors. A chromalalumel thermocouple can measure temperatures from −200°C to about 1100°C and is used typically with blast furnace gases, brick kilns and in glass manufacture.

For the measurement of temperatures above 1100°C radiation pyrometers are usually used. However, thermocouples are available made of platinum-platinum/rhodium capable of measuring temperatures up to 1400°C or tungsten-molybedenum which can measure up to 2600°C.

Advantages of thermocouples over other devices

(c) A thermocouple

(i) has a very simple relatively inexpensive construction;
(ii) can be made very small and compact;
(iii) is robust;
(iv) is easily replaced if damaged;
(v) has a small response time;
(vi) can be used at a distance from the actual measuring instrument and is thus ideal for use with automatic and remote-control systems.

(d) **Sources of error in the thermocouple which are difficult to overcome include:**

(i) voltage drops in leads and junctions;
(ii) possible variations in the temperature of the cold junction; and
(iii) stray thermoelectric effects, which are caused by the addition of further metals into the 'ideal' two metal

thermocouple circuit. Additional leads are frequently necessary for extension leads or voltmeter terminal connections.

A thermocouple may be used with a battery or mains operated electronic thermometer instead of a millivoltmeter. These devices amplify the small e.m.f.'s from the thermocouple before feeding them to a multi-range voltmeter calibrated directly with temperature scales. These devices have great accuracy and are almost unaffected by voltage drops in the leads and junctions.

Resistance thermometer

Construction

4 (a) Resistance thermometers are made in a variety of sizes, shapes and forms depending on the application for which they are designed.

A typical resistance thermometer is shown diagrammatically in *Figure 21.4*. The most common metal used for the coil in such thermometers is platinum even

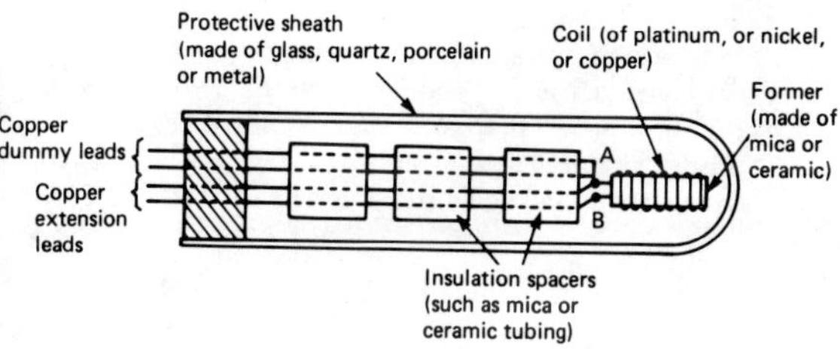

Figure 21.4

though its sensitivity is not as high as other metals such as copper and nickel. However, platinum is a very stable metal and provides reproducible results in a resistance thermometer. A platinum resistance thermometer is often used as a calibrating device. Since platinum is expensive connecting leads of another metal, usually copper, are used with the thermometer to connect it to a measuring circuit. The platinum and the connecting leads are shown joined at A and B in *Figure 21.4*, although sometimes this junction may be made outside of the sheath. However, these leads often come into close contact with the heat source which can introduce errors into the measurements. This may be eliminated by including a pair of identical leads, called dummy leads, which experience the same temperature change as the extension leads.

Principle of operation

(b) With most metals a rise in temperature causes an increase in electrical resistance, and since resistance can be measured accurately this property can be used to measure temperature. If the resistance of a length of wire at 0°C is R_o, and its resistance at θ°C is R_θ,

Then, $R_\theta = R_o(1+\alpha\theta)$,

where α is the temperature coefficient of resiatance of the material.

Rearranging gives: **temperature**, $\boldsymbol{\theta = \dfrac{R_\theta - R_o}{\alpha R_o}}$

Values of R_o and α may be determined experimentally or obtained from existing data. Thus, if R_θ can be measured, temperature θ can be calculated. This is the principle of operation of a resistance thermometer. Although a sensitive ohmmeter can be used to measure R_θ, for more accurate determinations a Wheatstone bridge circuit is used as shown in *Figure 21.5*. This circuit compares an unknown resistance R_θ with others of known

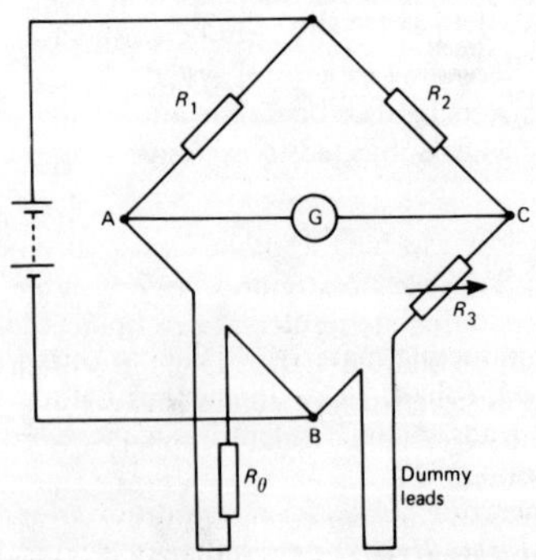

Figure 21.5

values, R_1 and R_2 being fixed values and R_3 being variable. Galvanometer G is a sensitive centre-zero microammeter. R_3 is varied until zero deflection is obtained on the galvanometer, i.e. no current flows through G and the bridge is said to be 'balanced'.

At balance: $R_2 R_\theta = R_1 R_3$, from which

$$R_\theta = \frac{R_1 R_3}{R_2},$$

and if R_1 and R_2 are of equal value, then $R_\theta = R_3$. A resistance thermometer may be connected between points A and B in *Figure 21.5* and its resistance R_θ at any temperature θ accurately measured. Dummy leads are included in arm BC to help eliminate errors caused by the extension leads which are normally necessary in such a thermometer.

(c) Resistance thermometers using a nickel coil are used mainly in the range −100°C to 300°C, whereas platinum resistance thermometers are capable of measuring with great accuracy temperatures in the range −200°C to about 800°C. This upper range may be extended up to about 1500°C if high melting point materials are used for the sheath and coil construction.

(d) Platinum is commonly used in resistance thermometers since it is chemically inert, i.e. unreactive, resists corrosion and oxidation and has a high melting point of 1769°C. A disadvantage of platinum is its slow response to temperature variation.

(e) Platinum resistance thermometers may be used as a calibrating device or in such applications as heat treating and annealing processes and it can easily be adopted for use with automatic recording or control systems. Resistance thermometers tend to be fragile and easily damaged especially when subjected to excessive vibration or shock.

Thermistors

5 A thermistor is a semiconducting material — such as mixtures of oxides of copper, mangenese, cobalt, etc. — in the form of a fused bead connected to two leads. As its temperature is increased its resistance rapidly decreases.

Typical resistance/temperature curves for a thermistor and common metals is shown in *Figure 21.6*. The resistance of a typical

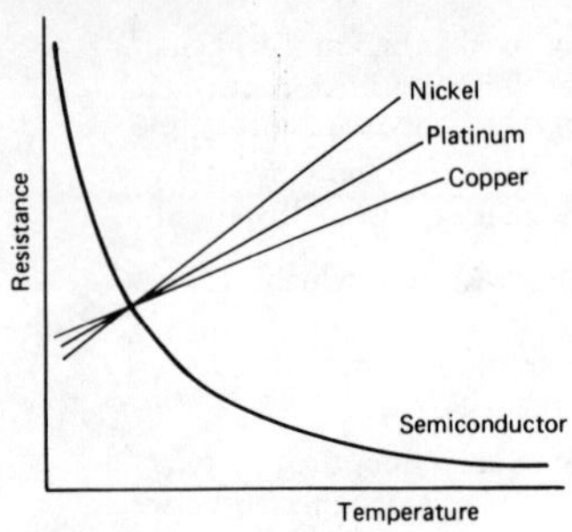

Figure 21.6

thermistor can vary from 400 Ω at 0°C to 100 Ω at 140°C.

The main advantages of a thermistor are its high sensitivity and small size. It provides an inexpensive method of measuring and detecting small changes in temperature.

Pyrometers

6 (a) A pyrometer is a device for measuring very high temperatures and uses the principle that all substances emit radiant energy when hot, the rate of emission depending on their temperature. The measurement of thermal radiation is therefore a convenient method of determining the temperature of hot sources and is particularly useful in industrial processes. There are two main types of pyrometers, these being the total radiation pyrometer and the optical pyrometer.

Pyrometers are very convenient instruments since they can be used at a safe and comfortable distance from the hot source. Thus applications of pyrometers are found in measuring the temperature of molten metals, the interiors of furnaces or the interiors of volcanos. Total radiation pyrometers can also be used in conjunction with devices which record and control temperature continuously.

Total radiation pyrometer

(b) A typical arrangement of a total radiation pyrometer is shown in *Figure 21.7*. Radiant energy from a hot source, such as a furnace, is focussed on to the hot junction of a

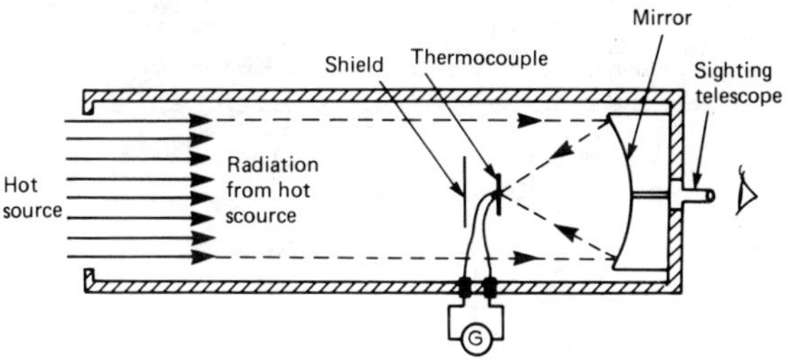

Figure 21.7

thermocouple after reflection from a concave mirror. The temperature rise recorded by the thermocouple depends on the amount of radiant energy received, which in turn depends on the temperature of the hot source. The galvanometer G shown connected to the thermocouple records the current which results from the e.m.f. developed and may be calibrated to give a direct reading of the temperature of the hot source. The thermocouple is protected from direct radiation by a shield as shown and the hot source may be viewed through the sighting telescope. For greater sensitivity, a thermopile may be used, a thermopile being a number of thermocouples connected in series. Total radiation pyrometers are used to measure temperature in the range 700°C to 2000°C.

Optical pyrometer

(c) When the temperature of an object is raised sufficiently two visual effects occur. These are that the object appears brighter and that there is a change in colour of the light emitted. These effects are used in the optical pyrometer where a comparison or matching is made between the brightness of the glowing hot source and the light from a filament of known temperature.

The most frequently used optical pyrometer is the disappearing filament pyrometer and a typical arrangement is shown in *Figure 21.8*. A filament lamp is built into a telescope arrangement which receives radiation from a hot source, an image of which is seen through an eyepiece.

A red filter is incorporated as a protection to the eye. The current flowing through the lamp is controlled

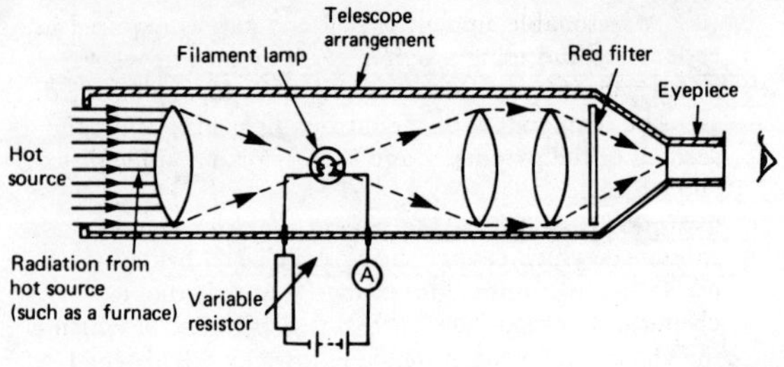

Figure 21.8

by a variable resistor. As the current is increased the temperature of the filament increases and its colour changes. When viewed through the eyepiece the filament of the lamp appears superimposed on the image of the radiant energy from the hot source. The current is varied until the filament glows as brightly as the background. It will then merge into the background and seem to disappear. The current required to achieve this is a measure of the temperature of the hot source and the ammeter can be calibrated to read the temperature directly. Optical pyrometers may be used to measure temperatures up to, and even in excess of, 3000°C.

Advantages of pyrometers

(d) (i) There is no practical limit to the temperature that a pyrometer can measure.
(ii) A pyrometer need not be brought directly into the hot zone and so is free from the effect of heat and chemical attack that can often cause other measuring devices to deteriorate in use.
(iii) Very fast rates of changes of temperature can be followed by a pyrometer.
(iv) The temperature of moving bodies can be measured.
(v) The lens system makes the pyrometer virtually independent of its distance from the source.

Disadvantages of pyrometers

(e) (i) A pyrometer is often more expensive than other temperaturing measuring devices.
(ii) A direct view of the heat process is necessary.
(iii) Manual adjustment is necessary.

(iv) A reasonable amount of skill and care is required in calibrating and using a pyrometer. For each new measuring situation the pyrometer must be re-calibrated.
(v) The temperature of the surroundings may affect the reading of the pyrometer and such errors are difficult to eliminate.

7 (a) **Temperature indicating paints** contain coloured substances which change their colour when heated to certain temperatures. This change is usually due to chemical decomposition, such as loss of water, in which the change in colour of the paint after having reached the particular temperature will be a permanent one. However in some types, the original colour returns after cooling. Temperature indicating paints are used where the temperature of inaccessible parts of apparatus and machines is required. They are particularly useful in heat-treatment processes where the temperature of the component needs to be known before a quenching operation. There are several such paints available and most have only a small temperature range so that different paints have to be used for different temperatures. The usual range of temperatures covered by these paints is from about 30°C to 700°C.

(b) **Temperature sensitive crayons** consist of fusible solids compressed into the form of a stick. The melting point of such crayons is used to determine when a given temperature has been reached. The crayons are simple to use but indicate a single temperature only, i.e. its melting point temperature. There are over a hundred different crayons available, each covering a particular range of temperature. Crayons are available for temperatures within the range of 50°C to 1400°C. Such crayons are used in metallurgical applications such as preheating before welding, hardening, annealing, tempering, or in monitoring the temperature of critical parts of machines or for checking mould temperatures in the rubber and plastics industry.

8 **Bimetallic thermometers** depend on the expansion of metal strips which operate an indicating pointer. Two thin metal strips of differing thermal expansion are welded or riveted together and the curvature of the bimetallic strip changes with temperature change. For greater sensitivity the strips may be coiled into a flat spiral or helix, one end being fixed and the other being made to rotate a pointer over a scale. Bimetallic thermometers are useful for alarm and overtemperature applications where extreme accuracy is

not essential. If the whole is placed in a sheath, protection from corrosive environments is achieved but with a reduction in response characteristics. The normal upper limit of temperature measurement by this thermometer is about 200°C, although with special metals the range can be extended to about 400°C.

9 The **mercury in steel thermometer** is an extension of the principle of the mercury in glass thermometer. Mercury in a steel bulb expands via a small bore capillary tube into a pressure indicating device, say, a Bourdon gauge, the position of the pointer indicating the amount of expansion and thus the temperature. The advantages of this instrument are that it is robust, and, by increasing the length of the capillary tube, the gauge can be placed some distance from the bulb and can thus be used to monitor temperatures in positions which are inaccessible to the liquid in glass thermometer. Such thermometers may be used to measure temperatures up to 600°C.

10 The **gas thermometer** consists of a flexible U-tube of mercury connected by a capillary tube to a vessel containing gas. The change in the volume of a fixed mass of gas at constant pressure, or the change in pressure of a fixed mass of gas at constant volume, may be used to measure temperature. This thermometer is cumbersome and rarely used to measure temperature directly. It is often used as a standard with which to calibrate other types of thermometer. With pure hydrogen the range of the instrument extends from −240°C to 1500°C and measurements can be made with extreme accuracy.

22 Solutions

1 **A solution** is formed when one substance, **the solute**, is dispersed in a different substance, **the solvent**, which is present in the largest amount. The substances can be gases, liquids and solids and solutions can be composed of any combination of these as shown in *Table 22.1*.

Table 22.1

Solution	*Examples*
Gas in gas	Air
Gas in liquid	Lemonade
Liquid in liquid	Whisky
Solid in liquid	Salt solution
Solid in solid	Metal alloys

2 A **true solution** is one which is completely **homogeneous** (i.e. the same throughout) and which can be separated perfectly into the constituent parts of the solution. The relative amounts of solute and solvent present in a solution can be expressed in a number of ways:

(a) **Percentage by mass**. This is the **number of grams of solute in 100 grams of solvent**. For example, a 1% sugar solution contains 1 g of sugar in 100 g of water, or alternatively, 1 kg of sugar in 100 kg of water.

(b) **Grams per cubic decimetre**. The cubic decimetre is one of the traditional volumes of scientific measurement, **the litre**. The composition of the solution in these terms is the **number of grams of solute in 1 cubic decimetre of solvent, i.e. g dm^{-3}**. Alternatively, the units used could be the number of kilograms in 1 cubic metre, i.e. kg m^{-3}.

(c) **Moles per cubic decimetre**. This is the most popular method of expressing the composition in terms of the

number of moles of solute in 1 cubic decimetre of solution. This means that the amount of solvent used depends upon the mass of solute used. In chemical terms it means that the actual number of molecules contained in any given volume of the solution is known. For example, a 0.1 M solution (one tenth molar) contains 0.1 moles of solute in 1 cubic decimetre of solution, or **6.023×10^{23} molecules of solute in 1 dm^3** of solution.

3 When a solution is formed, the components must be compatible with each other. Many examples are found in chemistry in which substances do not dissolve in each other; for example, sodium chloride (salt) is not soluble in propanone (acetone). As a general rule substances can be broadly classified as **polar** and **non-polar** substances. The polar substances are found in inorganic chemistry and the non-polar substances in organic chemistry. The implication is that **polar solutes** like sodium chloride **dissolve in polar solvents** like water, and **non-polar solutes** like naphthalene **dissolve in non-polar solvents** like petroleum ether (a hydrocarbon).

4 **Saturated and super-saturated solutions**. A **saturated solution** is formed when no more solute will dissolve in a fixed volume of solvent at a constant temperature. The easiest way to obtain a saturated solution is to dissolve the solute in the solvent until excess solute is present and then to filter the solution. A **supersaturated solution** is one which contains more dissolved solute than is needed to form a saturated solution at the same temperature; these solutions are usually made from hydrated salts like sodium thisulphate, $Na_2S_2O_3 . 5H_2O$ or sodium carbonate $Na_2CO_3 . 10H_2O$.

5 **Solubility**. The **solubility** of a solute in a solvent is defined **as the maximum number of kilograms of the solute which can be dissolved in 100 kilograms of solvent at a fixed temperature** in the presence of undissolved excess solid.

6 **Solubility curves**. When the solubility of ionic solids in water is investigated at different temperature the solubilities are found to vary. By finding the solubilities of a solute at different temperatures the results can be displayed graphically as the **solubility curves** shown in *Figure 22.1*.

Most substances behave like potassium nitrate and sodium nitrate in that **solubility usually increases with temperature**.

7 A **true solution** is one in which the **size of the particles** of the solute are so small that they **cannot be observed with the eye or even by using a microscope**. However, as the size of the solute particles becomes larger then the solution changes from a true solution to a **colloidal solution**.

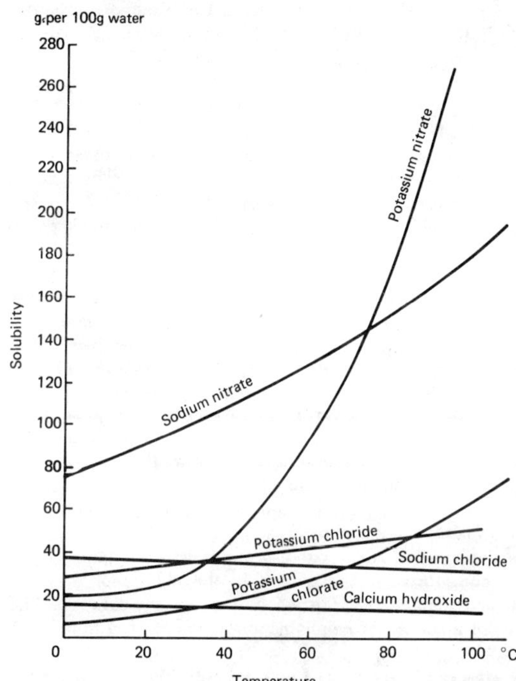

Figure 22.1

8 **Colloidal solutions**. A **colloidal solution** is formed when a **disperse phase** of particles which are between **1 nm and 100 nm** in diameter are distributed in a **dispersion medium**. The disperse phase and the dispersion medium or **continuous phase** can be gases, liquids or solids. A colloidal solution can be composed of any combination of these phases except for a mixture of gases. A variety of colloid solutions are shown in *Table 22.2*.

9 (i) A colloidal solution, or **sol**, can be classified into two main types, **lyophobic** and **lyophilic sols**.

(ii) In lyophobic sols there is **little or no interaction** between the disperse phase and the dispersion medium, and the particles are kept dispersed by electrical charges.

(iii) Lyophilic sols form a **strong interaction** between the

Table 22.2 Some examples of colloidal systems

Disperse phase	*Dispersion medium*	*Type of colloid*	*Example*
Gas	Liquid	Foam	Beer froth, whipped cream
Gas	Solid	Solid foam	Pumice, merangue
Liquid	Gas	Aerosol	Mist, cloud
Liquid	Liquid	Emulsion	Milk, mayonnaise, rubber latex
Liquid	Solid	Solid emulsion	Butter, margarine
Solid	Gas	Aerosol	Dust, smoke
Solid	Liquid	Sol, gel or paste	Paint, jelly, starch
Solid	Solid	Solid sol or solid gel	Coloured glass, pearl

disperse phase and the continous phase such that the disperse phase is strongly absorbed by the continuous phase, and it is this absorption which keeps the particles dispersed.

(iv) Lyophilic sols can form **semi-solid masses** called **gels** or pastes when the disperse phase is present in high concentration.

10 The properties of colloidal solutions are that they:

(i) do not separate easily on standing;

(ii) undergo **electrophoresis** if electrically charged (see para 11);

(iii) undergo **dialysis** (see para 12);

(iv) show the **Tyndall** effect with a beam of light (see para 13);

(v) can be **coagulated** or precipitated by the addition of electrolyte according to the Hardy-Schultze rule (see para 14). The effects are not identical for lyophobic sols as can be seen from *Table 22.3.*

11 **The effects of electrical fields on colloids**. The apparatus used to investigate the effect of an electrical field is shown in *Figure 22.2.* When a potential difference is set up between the cathode and the anode, the colloidal particles **move towards the cathode**

Table 22.3 A comparison of lyophilic and lyopholic sols

Lyophilic sols	*Lyophobic sols*
Stability due to adsorption of the dispersion medium	Stability due to electrical charges
Prepared by direct mixing and are reversible sols	Prepared by indirect methods and are irreversible sols
Commonly form gels	Do not usually form gels
Undergo electrophoresis only if the particles are charged	Undergo electrophoresis
Exhibit the Tyndall effect weakly	Exhibit the Tyndall effect clearly
Co-agulated by high concentrations of electrolyte	Co-agulated by low concentrations of electrolyte
Purified by dialysis	Purified by dialysis

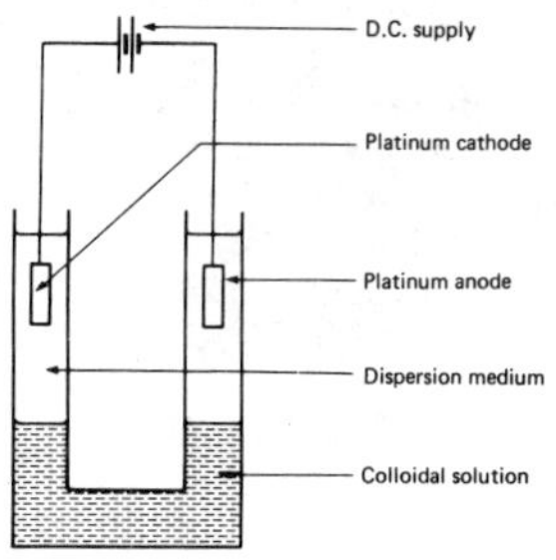

Figure 22.2

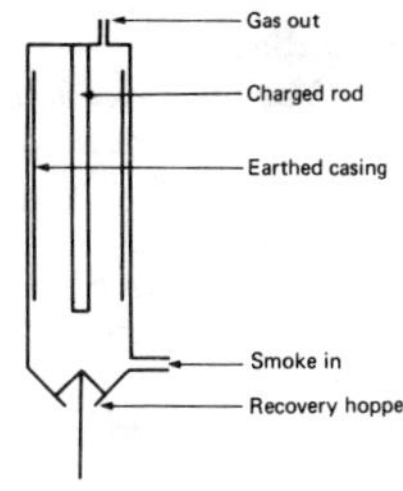

Figure 22.3

or the anode depending upon the charge carried by the disperse phase. The movement of particles can be observed by setting up a boundary between the colloid and its dispersion medium. If the colloidal solution is coloured the effect can be more easily observed. Examples of negatively charged colloids are starch, clay and metallic particles in water. Examples of positively charged colloids are iron (III) hydroxide and aluminium hydroxide solutions. This process of the **movement of charged particles** is called **electrophoresis**.

A second property of colloids in electrical fields is the **electrostatic precipitation** of solid particles from a smoke or mist. The passage of smoke or mist up a chimney to which a high tension d.c. supply has been fitted, can be used industrially to recover solids which otherwise would be lost to the atmosphere. A diagram showing an **electrostatic precipitator** is shown in *Figure 22.3*.

12 **Dialysis of colloids**. When a solvent is carefully placed on top of a true solution, the solute will diffuse up into the solvent layer until eventually the solution becomes homogeneous.

If a colloidal solution is placed in contact with its dispersion medium, the boundary between the two layers will remain definite for a much longer period of time. This is due to the much larger size of the disperse phase particles. This **difference in diffusion rates can be used to purify colloidal solutions**. For example, when iron (III) hydroxide is prepared from iron (III) chloride and water, the colloidal solution is contaminated by hydrochloric acid. If this mixture is placed in **a cellophane tube** and then immersed in running water, the hydrochloric acid which is a true solution **diffuses** more rapidly than the iron (III) hydroxide which is a colloidal solution. This results in the removal of the hydrochloric acid leaving a purified iron (III) hydroxide sol. This process is called **dialysis**.

13 **The effect of a light beam on colloids** When a beam of light is passed through a true solution it does not undergo any dispersion effects. However, if the same beam of light is passed through a lyophobic sol, the passage of the light is easily defined. Car headlights in fog or a cine projection through a smoky room are good examples of this effect, which is called the **Tyndall effect**.

The **dispersion of light** is caused by the **difference in refractive indices** between the disperse phase and the dispersion medium. Lyophilic sols tend to be more homogenous than lyophobic sols and the difference in refractive indices tends to be much smaller resulting in only slight, if any dispersion effects. Thus a beam of light can be used to identify a lyophobic sol, but not always a lyophilic sol.

14 **The effect of adding electrolytes to colloids** The addition of an electrolyte to a lyophobic sol results in the **coagulation of the sol**. The addition of a negatively charged ion causes the coagulation of a positively charged colloidal particle and vice versa. The amount of coagulation caused by a particular electrolyte is due to the **valency** of the **coagulating ion**. Thus for equal concentrations, an aluminium ion Al^{3+}, which is **trivalent** will coagulate a negatively charged colloidal solution more effectively than say a sodium ion, Na^+ which is monovalent. Similarly, phosphate ions PO_4^{3-} will be more effective in coagulating a positively charged sol than a chloride ion. This is called the **Hardy-Schultze rule**.

Examples of the commercial use of coagulating agents are, the use of aluminium sulphate in treating dirty water, the coagulation of rubber latex using methanoic acid and the curdling of milk by 2-hydroxypropanoic acid $CH_3CHOHCOOH$. An

Table 22.4 Some methods of preparing colloidal systems

Preparation of lyophobic sols	*Preparation of lyophilic sols*
Sulphur sol $Na_2S_2O_3(aq) + 2HCl(aq) =$ $2NaCl(aq) + SO_2(g) + H_2O(l) + S(s)$	**Milling Method** Crush the solid into fine particles and disperse into the continuous phase
Iron (III) hydroxide sol $FeCl_3(s) + 3H_2O(l) =$ $Fe(OH)_3(s) + 3HCl(aq)$	**Arc Method** Striking an arc between metallic electrodes immersed in a dispersion medium
Silver sol Reduce ammoniacal silver nitrate solution with a reducing sugar	

example of coagulation in nature is the formation of deltas in rivers when river water meets the higher concentration of sodium chloride in sea-water. Lyophilic sols can only be precipitated or coagulated by the addition of high concentrations of electrolytes.

15 Different types of colloids are prepared in different ways, lyophobic sols by indirect methods usually involving chemical reactions and lyophilic sols by direct mixing after obtaining a disperse phase of the appropriate size. Some examples are given in *Table 22.4*.

23 Pressure in fluids

1 The **pressure** acting on a surface is defined as the perpendicular force per unit area of surface. The unit of pressure is the **pascal**, (**Pa**), where 1 pascal is equal to 1 newton per square metre. Thus

$$\textbf{pressure, } p = \frac{F}{A} \textbf{ pascals}$$

where F is the force in newtons acting at right angles to a surface of area A square metres.

When a force of 20 N acts uniformly over, and perpendicular to an area of 4 m^2, then the pressure on the area, p, is given by:

$$p = \frac{20 \text{ N}}{4 \text{ m}^2} = 5 \text{ Pa}$$

2 A **fluid** is either a liquid or a gas and there are four basic factors governing the pressure within fluids.

(a) The pressure at a given depth in a fluid is equal in all directions, see *Figure 23.1(a)*.

(b) The pressure at a given depth in a fluid is independent of the shape of the container in which the fluid is held. In *Figure 23.1(b)*, the pressure at X is the same as the pressure at Y.

(c) Pressure acts at right angles to the surface containing the fluid. In *Figure 23.1(c)*, the pressure at points A to F all act at right angles to the container.

(d) When a pressure is applied to a fluid, this pressure is transmitted equally in all directions. In *Figure 23.1(d)*, if the mass of the fluid is neglected, the pressures at points A to D are all the same.

3 The pressure, p, at any point in a fluid depends on three factors:

(a) the density of the fluid, ρ in kg/m^3;

(b) the gravitational acceleration, g, taken as 9.8 m/s^2; and

(c) the height of fluid vertically above the point, h metres.

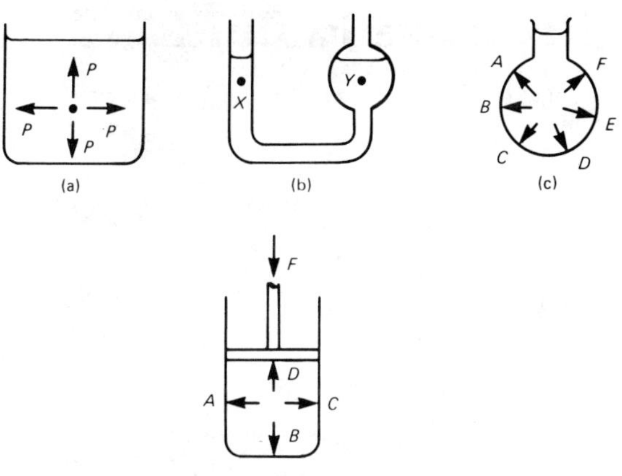

Figure 23.1

The relationship connecting these quantities is: $\boldsymbol{p = \rho g h}$ **pascals**.

When the container shown in *Figure 23.2* is filled with water of density 1000 kg/m^3, the pressure due to the water at a depth of 0.03 m below the surface is given by:

$$p = \rho g h$$
$$= (1000 \times 9.8 \times 0.03)$$
$$= \mathbf{294\ Pa}.$$

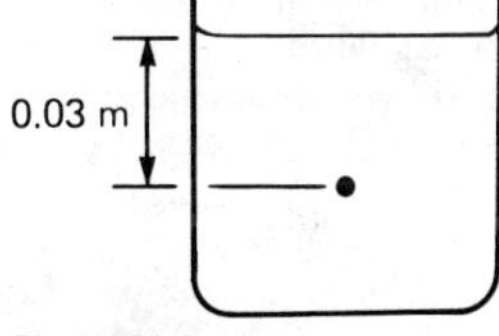

Figure 23.2

4 The air above the earth's surface is a fluid, having a density, ρ, which varies from approximately 1.225 kg/m^3 at sea level to zero in outer space. Since $p = \rho g h$, where height h is several thousands of metres, the air exerts a pressure on all points on the earth's surface. This pressure, called **atmospheric pressure**, has a value of approximately 100 kilopascals. Two terms are commonly used when measuring pressures:

(a) **absolute pressure**, meaning the pressure above that of an absolute vacuum (i.e. zero pressure), and
(b) **gauge pressure**, meaning the pressure above that normally present due to the atmosphere. Thus:

absolute pressure = atmospheric pressure + gauge pressure

Thus, a gauge pressure of 50 kPa is equivalent to an absolute pressure of (100 + 50) kPa, i.e. 150 kPa, since the atmospheric pressure is approximately 100 kPa.
Another unit of pressure, used in particular for atmospheric pressures, is the bar.

$$1 \text{ bar} = 10^5 \text{ N/m}^2 = 100 \text{ kPa}.$$

5 There are various ways of measuring pressure, and these include by:

(a) a U-tube manometer (see para. 6);
(b) a barometer (see paras 7 and 8); and
(c) a pressure gauge (see para. 9).

U-tube manometer

6 A manometer is a device used for measuring relatively small pressures, either above or below atmospheric pressure. A simple U-tube manometer is shown in *Figure 23.3*. Pressure p acting in, say, a gas main, pushes the liquid in the U-tube until equilibrium is obtained. At equilibrium: pressure in gas main, p = (atmospheric pressure, p_a) + (pressure due to the column of liquid, ρgh) i.e. $p = p_a + \rho gh$.

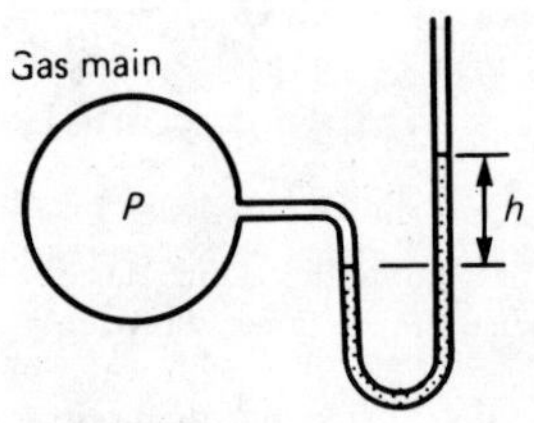

Figure 23.3

Thus, for example, if the atmospheric pressure, p_a, is 101 kPa, the liquid in the U-tube is water of density 1000 kg/m^3 and height, h is 300 mm, then

$$\text{absolute gas pressure} = (101\,000 + 1000 \times 9.8 \times 0.3) \text{ Pa}$$
$$= (101\,000 + 2940) \text{ Pa}$$
$$103\,940 \text{ Pa} = 103.94 \text{ kPa}.$$

The gauge pressure of the gas is 2.94 kPa.

By filling the U-tube with a more dense liquid, say mercury having a density of 13 600 kg/m^3, for a given height of U-tube, the pressure which can be measured is increased by a factor of 13.6.

By inclining one limb of the U-tube, as shown in *Figure 23.4* greater sensitivity is achieved, that is, there is a larger movement of the liquid for a given change in pressure when compared with a U-tube having vertical limbs. An inclined manometer normally has a reservoir of sufficient area to give virtually a constant level in the

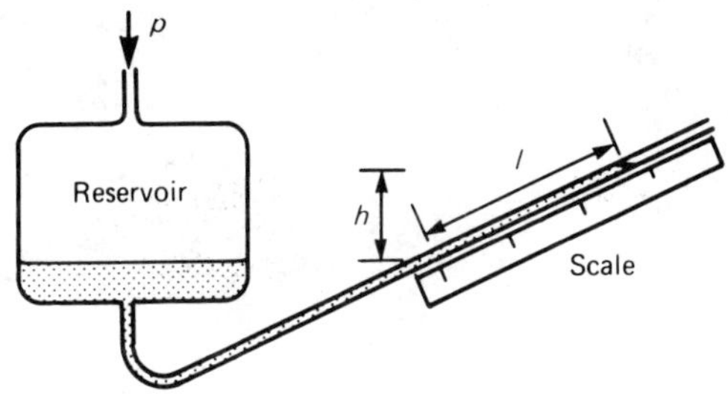

Figure 23.4

left-hand limb. From *Figure 23.4*, it can be seen that pressure p applied to the reservoir causes a scale change of 'l' in the inclined manometer compared with 'h' in a normal manometer.

Simple barometer

7 A simple barometer consists of a length of glass tubing, approximately 800 mm long and sealed at one end, which is filled with mercury and then inverted in a beaker of mercury, as shown in *Figure 23.5*. At equilibrium, the atmospheric pressure, p_a, is tending to force the mercury up the tube, whilst the force due to the column of mercury is tending to force the mercury out of the tube, i.e. $p_a = \rho gh$. As the atmospheric pressure varies, height h varies, giving an indication on the scale of the atmospheric pressure.

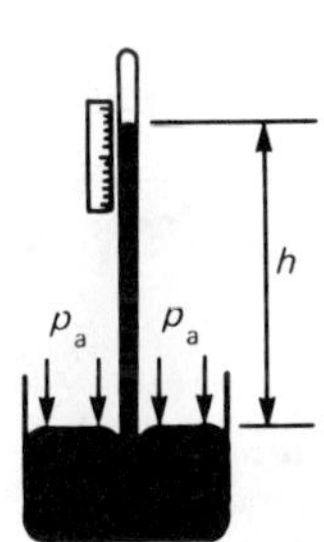

Figure 23.5

Fortin barometer

8 The Fortin barometer is as shown in *Figure 23.6*. Mercury is contained in a leather bag at the base of the mercury reservoir, and height, H, of the mercury in the reservoir can be adjusted using the screw at the base of the barometer to depress or release the leather bag. To measure the atmospheric pressure, the screw is adjusted until the pointer at H is just touching the surface of the mercury and the height of the mercury column is then read using

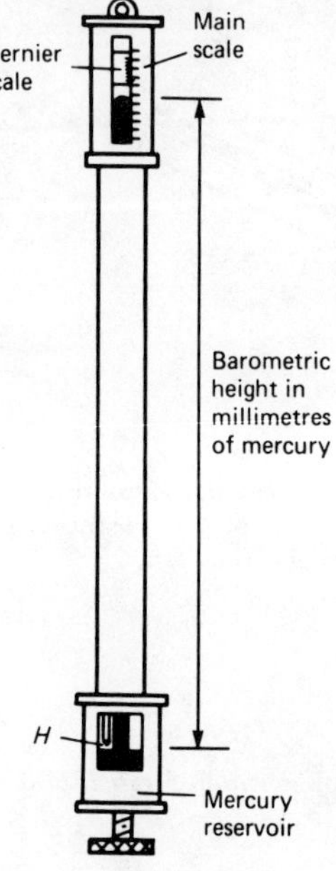

Figure 23.6

the main and vernier scales. The measurement of atmospheric pressure using a Fortin barometer is achieved much more accurately than by using a simple barometer.

Bourdon pressure gauge

9 The main components of a Bourdon pressure gauge are shown in *Figure 23.7*. When pressure, p, is applied to the curved phosphor bronze tube, which is sealed at A, it tends to straighten, moving A to the right. Conversely, a decrease in pressure below that due to the atmosphere moves point A to the left. When A moves to the right, B moves to the left, rotating the pointer across a scale. This type of pressure gauge can be used to measure large pressures and pressures both above and below atmospheric pressure. The Bourdon pressure gauge indicates gauge pressure and is very widely used in industry for pressure measurements.

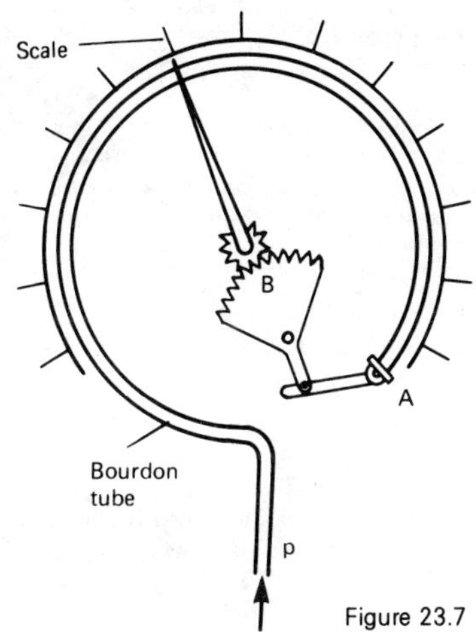

Figure 23.7

Hydrostatic pressure

10 The pressure at the base of the tank shown in *Figure 23.8(a)* is:

$$p = \rho g h = wh,$$

where w is the specific weight, i.e. the weight per unit volume, its unit being N/m^3.

The pressure increases to this value uniformly from zero at the free surface. The pressure variation is shown in *Figure 23.8(b)*.

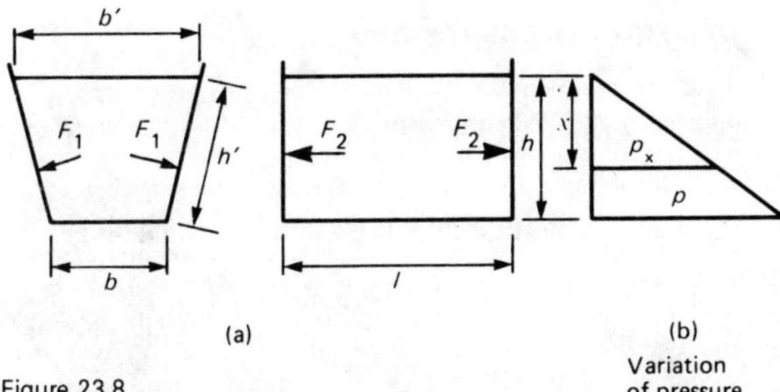

Figure 23.8

At any intermediate depth x the pressure is

$$p_x = \rho g x = w x.$$

It may be shown that the average pressure on any wetted plane surface is the pressure at the centroid, the centre of area. The sloping sides of the tank of *Figure 23.8* are rectangular and therefore the average pressure on them is the pressure at half depth: $\frac{p}{2} = \frac{\rho g h}{2} = \frac{wh}{2}$

The force on a sloping side is the product of this average pressure and the area of the sloping side:

$$F_1 = \frac{\rho g h}{2} \times l h' = \frac{whlh'}{2}$$

where h' is the slant height. The pressure and consequently the forces F_1 are at right angles to the sloping sides as shown in *Figure 23.8.*

The average pressure on the vertical trapezoidal ends of the tank is not the pressure at half depth. This is because the centroid of an end is not at half depth — it is rather higher. The depth of the centroid is given by:

$$\frac{h}{3}\left(\frac{2b+b'}{b+b'}\right),$$ (see *Figure 23.8*)

The average pressure on an end is therefore:

$$\frac{\rho g h}{3}\left(\frac{2b+b'}{b+b'}\right) = \frac{wh}{3}\left(\frac{2b+b'}{b+b'}\right)$$

The forces F_2 on the vertical trapezoidal ends of the tank are horizontal forces given by the product of this average pressure and the area of the trapezium,

$$\left(\frac{b+b'}{2}\right)h.$$

The force on the base of the tank is $(\rho g h) \times$ (area of base)

$$= \rho g h l b = w h l b$$

For a tank with vertical sides this is the weight of liquid in the tank.

11 In any vessel containing homogeneous liquid at rest and in continuous contact, the pressure must be the same at all points at the same level. In a U-tube, as shown in *Figure 23.9*, with the liquid in the lower part at rest, the pressure must be the same on

both sides for all levels up to X_1X_2. The pressure at X_1, however, is greater than the pressure at Y_1 by an amount $p_1 = w_1h = \rho_1gh$, where w_1, ρ_1 are the specific weight and density respectively of the liquid, or gas, between X_1 and Y_1.

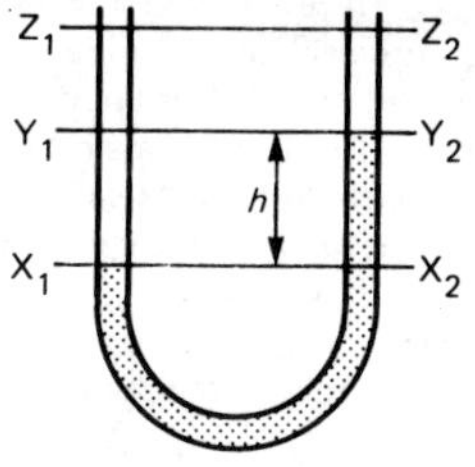

Similarly, the pressure at X_2 is greater than the pressure at Y_2 by an amount given by
$p_2 = w_2h = \rho_2gh$,
where w_2, ρ_2 are the specific weight and density respectively of the liquid in the bottom of the U-tube.

Figure 23.9

For practical reasons ρ_2 must be greater than ρ_1 and the pressure at Y_1 will exceed that at Y_2 by

$$p_2 - p_1 = (w_2 - w_1)h = (\rho_2 - \rho_1)gh.$$

If the upper limbs of the U-tube contain air or any other gas or gas mixture w_1 and ρ_1 can reasonably be ignored, giving

$$p_2 - p_1 = w_2h = \rho_2gh$$

If the upper limbs contain a lighter liquid, then the pressure difference may be expressed as

$$p_2 - p_1 = \rho_1(d-1)gh = (d-1)w_1h \text{ where } d = \frac{\rho_2}{\rho_1}$$

A common arrangement is mercury and water, in which case d is the relative density of mercury approximately 13.6. This gives:

$$p_2 - p_1 = 12.6\rho gh = 12.6wh,$$

ρ and w being respectively, the density and specific weight of water. The pressure difference at Z_1Z_2 will be the same as at Y_1Y_2 if both limbs contain the same liquid between these levels. This follows from the fact that the pressure increase from Z_1 to Y_1 is the same as the increase from Z_2 to Y_2.

Archimedes' principle

12 If a solid body is immersed in a liquid, the apparent loss of weight is equal to the weight of liquid displaced. If V is the volume of the body below the surface of the liquid, then the apparent loss of weight is $W = Vw = V\rho g$, where w, ρ are respectively the specific weight and density of the liquid.

If ρ is known and W obtained from a simple experiment, V can be calculated. Hence the density of the solid body can be calculated. If V is known and W obtained, ρ can be calculated.

If a body floats on the surface of a liquid all of its weight appears to have been lost. The weight of liquid displaced is equal to the weight of the floating body.

For example, a body weighs 2.760 N in air and 1.925 N when completely immersed in water of density 1000 kg/m^3.

Hence the apparent loss of weight is

$$2.760 \text{ N} - 1.925 \text{ N} = 0.835 \text{ N}.$$

This is the weight of water displaced, i.e. $V\rho g$, where V is the volume of the body and ρ is the density of water.

Thus $0.835 \text{ N} = (V)(1000 \text{ kg/m}^3)(9.81 \text{ m/s}^2)$

from which, volume, $V = \dfrac{0.835}{9810} = \mathbf{8.512 \times 10^{-5}\ m^3}$

$$\text{The density of the body} = \frac{\text{mass}}{\text{volume}} = \frac{\text{weight}}{gV}$$

$$= \frac{2.760 \text{ N}}{(9.81 \text{ m/s}^2)(8.512 \times 10^{-5} \text{ m}^3)}$$

$$= \mathbf{3305\ kg/m^3}$$

$$\text{Relative density} = \frac{\text{density}}{\text{density of water}} = \frac{3305}{1000} = \mathbf{3.305}$$

24 Surface tension and viscosity

Surface tension

1 The force of attraction between molecules in a liquid gives rise to what is termed surface tension.

The **surface tension** γ of a liquid is the force per unit length acting in the surface perpendicular to one side of a line in the surface.

The **free surface energy** σ is the energy required to create an additional unit area of surface against the attractive forces of the molecules. The surface tension γ and the free surface energy σ are numerically the same as shown below.

Consider a wire frame as shown in *Figure 24.1* on which there is a soap film. XY is a sliding wire. The length of soap film in

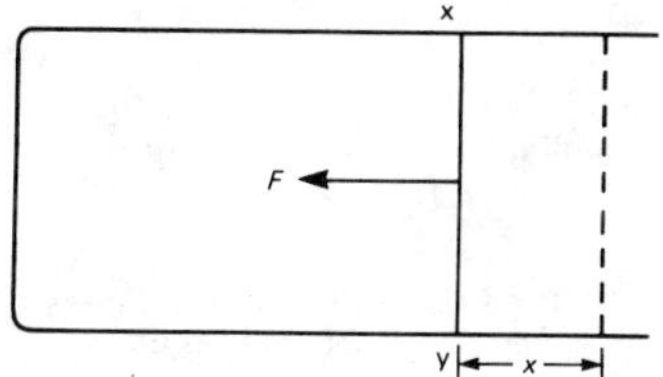

Figure 24.1

contact with the sliding wire is l. The force F due to surface tension on the wire XY is $2\gamma l$, the factor 2 occurring because there are two surfaces to the soap film. γ is the surface tension of the soap film. If the wire is moved a distance x to the right the work done against the force of surface tension is $2\gamma lx$.

Thus the increase in the surface area of the film is $2lx$.

Hence the energy required to create an additional unit

area of film is $\dfrac{2\gamma lx}{2lx}=\gamma$

But this is the definition of free surface energy σ.

Thus $\sigma = \gamma$ numerically.

2 Because of differences between the cohesive force between molecules of liquid and the adhesive force between molecules of liquid and molecules of solids, a liquid surface is usually curved where it makes contact with a solid.

For example, the surface of water in a glass tube is concave and the surface of mercury in a glass tube is convex. (See *Figure 24.2.*)

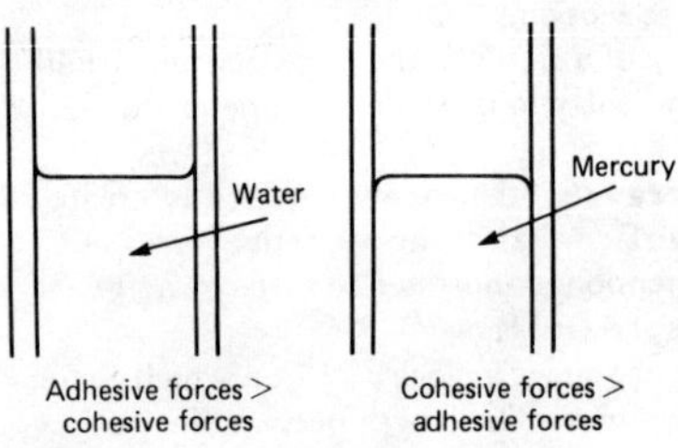

Figure 24.2

The angle of contact θ is defined as the angle between the solid surface and the tangent to the liquid surface. θ is measured through the liquid as shown in *Figure 24.3.* If $\theta < 90°$ the liquid is said to 'wet' the solid surface.

3 Liquids for which $\theta < 90°$ rise in a tube with a small internal diameter (such as a capillary tube). *Figure 24.4* shows a liquid which has risen a height h up a capillary tube of radius r.

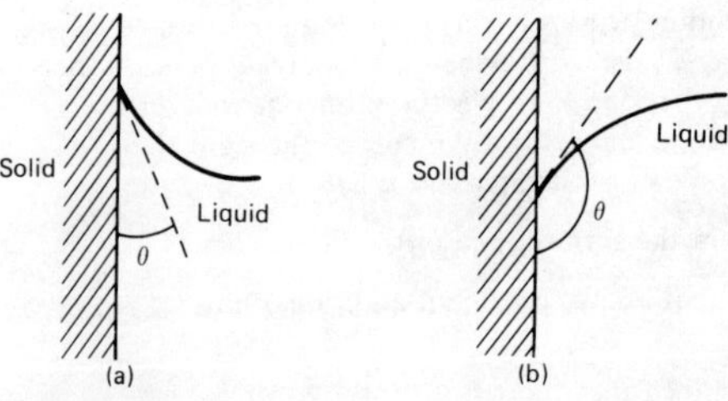

Figure 24.3

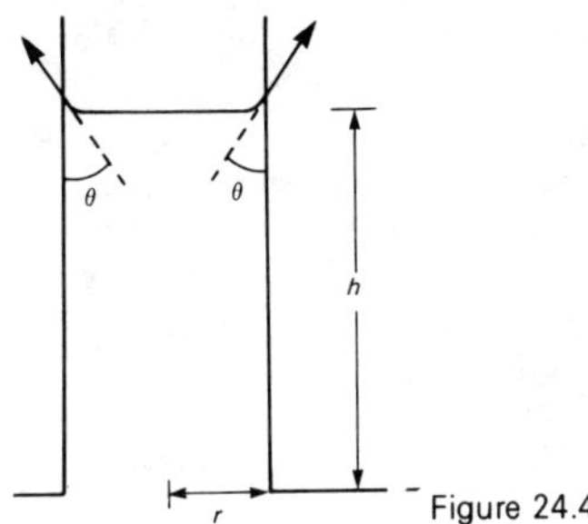

Figure 24.4

The force due to the surface tension acting on the meniscus depends upon the circumference of the meniscus and the surface tension γ.

The upward vertical component of the force due to surface tension is

$$\gamma \times \text{circumference} \times \cos\theta = \gamma(2\pi r)\cos\theta$$

The downward vertical force on the column of liquid is due to

weight of the liquid
$= \text{volume} \times \text{density} \times \text{earths gravitational field}$
$= (\pi r^2 h)\rho g$

These two forces are equal

Thus $\gamma(2\pi r)\cos\theta = \pi r^2 h\rho g$

from which, **height** $\boldsymbol{h = \dfrac{2\gamma\cos\theta}{r\rho g}}$

Thus, for example, if the surface tension of mercury at 20°C is 0.465 N m^{-1} and its angle of contact with glass is 140°, the capillary rise h of the mercury in a capillary tube of internal radius 2 mm is given by:

$$h = \frac{2\gamma\cos\theta}{r\rho g} = \frac{2(0.465)(-0.766)}{(2\times10^{-3})(13.6\times10^{3})(9.81)}$$

$= -2.67\times10^{-3}$ m $=$ **−2.67 mm.**

(The negative sign indicates that the mercury level in the capillary tube falls.)

4 It may be shown that there is a pressure inside a spherical drop of liquid which exceeds the surrounding air pressure by an

amount equal to $2\gamma/R$ where R is the radius of the drop. This is called the **excess pressure**.

(i) For a spherical drop of liquid in air the excess pressure is $2\gamma/R$.
(ii) For a bubble of gas in a liquid the excess pressure is $2\gamma/R$.
(iii) For a soap bubble in air the excess pressure is $4\gamma/R$ since a soap bubble has two surfaces.

5 From a knowledge of the free surface energy of a liquid an approximate value of the energy needed to break an intermolecular bond may be found. A molecule moving to the surface of a liquid as a new surface is created has its number of near neighbours decreased from ten to five. If n is the number of molecules per unit area of surface, then $5n/2$ bonds are broken for each unit area of surface produced and $5n\varepsilon/2$ is the energy needed if ε is the energy required to break one bond. Thus, as σ, the force surface energy, is the energy required to produce unit area of surface, we have $\sigma=5n\varepsilon/2$, from which $\varepsilon=2\sigma/5n$.

The molar latent heat of vaporisation of a substance is the energy required to evaporate 1 mole of the substance at standard pressure. If a solid which has a hexagonal close packed crystal structure is considered each atom has twelve near neighbours. In the liquid form each atom has about ten near neighbours and bonds must be broken as the solid turns into a liquid and loses two near neighbours per atom. A vapour has no near neighbours and thus the liquid loses ten near neighbours per atom in evaporating.

In one mole there are 6×10^{23} atoms and thus if each atom has ten near neighbours the number of bonds that need to be broken is

$$\frac{10\times 6\times 10^{23}}{2}.$$

The divisor of 2 is present because each bond connects two atoms. Thus the molar latent heat is $5N_A\varepsilon$, where N_A is the number of atoms in a mole, i.e.

$$\text{molar latent heat}=5N_A\left(\frac{2\sigma}{5n}\right)=\frac{2N_A\sigma}{n}$$

Therefore liquids with high values of σ should have high values of molar latent heat of vaporisation. This is reasonably confirmed by experiment.

Viscosity

6 Liquids (and gases) in contact with a solid surface stick to that surface. If a liquid flows on a solid surface we can consider the

liquid to consist of layers. The bottom layer remains in contact with the solid and at rest. The other layers slide on one another and travel with velocities which increase the further the layer is from the solid (see *Figure 24.5*). This is a description of streamline flow. If the velocity increases to beyond a critical value the flow becomes **turbulent** and the description in terms of layers no longer applies. In *Figure 24.5*, the arrows indicate the velocities of different layers. This condition will exist when the liquid is subjected to a **shear** force. The opposition to this is called the **viscosity** of the liquid.

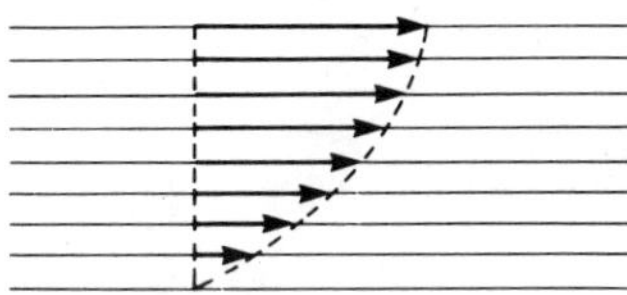

Figure 24.5

Consider two parallel layers of liquid separated by a distance Δy travelling at velocities v and $v+\Delta v$. The lower layer tends to impede the flow of the upper layer and exerts a retarding force F on it, whereas the lower layer itself experiences an accelerating force F exerted on it by the upper layer.

The tangential stress between the two layers is F/A where A is the area of contact between the layers.

The ratio $\Delta v/\Delta y$ is called the velocity gradient.

Newton realised that for some fluids:

tangential stress $\propto$ velocity gradient and thus

$$\frac{F}{A}=\eta\frac{\Delta v}{\Delta y}$$

where η is a constant called the **coefficient of viscosity**.

thus the coefficient of viscosity, $\eta=\dfrac{\textbf{tangential stress}}{\textbf{velocity gradient}}$.

η usually decreases with increasing temperature although 'viscostatic' oils are almost temperature independent.

The units of the coefficient of viscosity are N s m^{-2} or, alternatively, kg m^{-1} s^{-1} (since 1 N=1 kg m s^{-2}).

7 **Poiseuille's formula** for streamline flow through a circular

pipe gives an expression for the volume V of liquid passing per second:

$$V=\frac{\pi p r^4}{8\eta l}$$

where r is the radius of the pipe;
p is the pressure difference between the ends of the pipe;
l is the length of the pipe; and
η is the coefficient of viscosity of the liquid.

For example, in *Figure 24.6* water flows from a tank through a tube of length 1 m and internal radius 2 mm. If the viscosity of water is

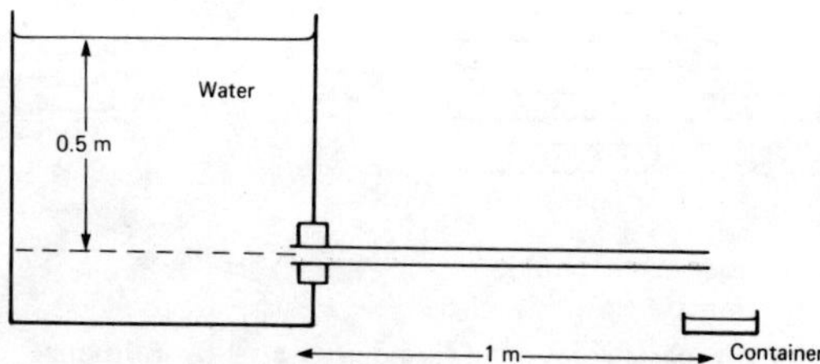

Figure 24.6

1×10^{-3} kg m^{-1} s^{-1}, the rate at which water is collected in the small container is determined as follows:

From Poiseuille's formula the volume collected per second, V, is given by

$$V=\frac{\pi p r^4}{8\eta l}$$

In this case the pressure difference between the ends of the tube,

$p=h\rho g$, i.e. $p=(0.5\times10^3)(9.81)$ Pa.

$$\text{Hence } V=\frac{\pi(0.5\times10^3)(9.81)(2\times10^{-3})^4}{8(1\times10^{-3})(1)}$$

$$=\mathbf{3.08\times10^{-5}\ m^3\ s^{-1}}$$

This rate will not be maintained because the water level in the tank will fall and pressure p will decrease.

8 **Stokes Law** gives an expression for the force F due to

viscosity acting on a sphere moving with streamline flow through a fluid.

Force, $F = 6\pi\eta rv$

where r is the radius of the sphere and v its velocity.

For example, if a steel ball-bearing of radius 1 mm falls through water its terminal velocity v is determined as follows (assuming the density of steel is 7.8×10^3 kg m^{-3}, the density of water is 1×10^3 kg m^{-3}, the viscosity of water is 1×10^{-3} kg m^{-1} s^{-1} and that streamline flow is maintained).

The volume of the sphere is $\frac{4}{3}\pi r^3$, thus the weight of the sphere is

$$\tfrac{4}{3}\pi r^3 \rho g$$

where ρ is the density of the sphere and g is the earth's gravitational field.

The volume of the liquid displaced is $\frac{4}{3}\pi r^3$. Thus the upthrust (i.e. weight of liquid displaced) is

$$\tfrac{4}{3}\pi r^3 \rho_0 g$$

where ρ_0 is the density of the liquid.

When the terminal velocity is reached there is no resultant force on the sphere.

Thus the weight of the sphere = viscous drag + upthrust

i.e. $$\tfrac{4}{3}\pi r^3 \rho g = 6\pi\eta rv + \tfrac{4}{3}\pi r^3 \rho_0 g$$

Therefore $$6\pi\eta rv = \tfrac{4}{3}\pi r^3 (\rho - \rho_0) g$$

and terminal velocity

$$v = \frac{\frac{4}{3}\pi r^3 (\rho - \rho_0) g}{6\pi\eta r} = \frac{2r^2 g(\rho - \rho_0)}{9\eta}$$

$$= \frac{2(1\times 10^{-3})^2 (9.81)(7.8\times 10^3 - 1\times 10^3)}{9(1\times 10^{-3})} = \mathbf{14.8\ m\ s^{-1}}.$$

25 Ideal gas laws

1 The relationship which exists between pressure, volume and temperature of a gas are given in a set of laws called the **gas laws**.

2 (i) **Boyle's law states:**

'the volume V of a fixed mass of gas is inversely proportional to its absolute pressure p at constant temperature'

i.e. $p \alpha \frac{1}{V}$ or $p = \frac{k}{V}$ or $pV = k$, at constant temperature,

where

p = absolute pressure in pascals (Pa),
V = volume in m^3, and
k = a constant.

(ii) Changes which occur at constant temperature are called **isothermal** changes.
(iii) When a fixed mass of gas at constant temperature changes from pressure p_1 and volume V_1 to pressure p_2 and volume V_2 then:

$$\boxed{p_1 V_1 = p_2 V_2}$$

3 (i) **Charles' law states:**

'for a given mass of gas at constant pressure, the volume V is directly proportional to its thermodynamic temperature T'

i.e. $V \alpha T$ or $V = kT$ or $\frac{V}{T} = k$, at constant pressure,

where T = thermodynamic temperature in kelvin (K).

(ii) A process which takes place at constant pressure is called an **isobaric process**.
(iii) The relationship between the Celsius scale of temperature and the thermodynamic or absolute scale is given by:

kelvin = degrees Celsius + 273

i.e. **K = °C + 273** or **°C = K − 273**

(iv) If a given mass of gas at constant pressure occupies a volume V_1 at a temperature T_1 and a volume V_2 at temperature T_2, then

$$\frac{V_1}{T_1}=\frac{V_2}{T_2}$$

For example, a gas occupies a volume of 1.2 litres at 20°C. If the pressure is kept constant, the volume it occupies at 130°C is determined from

$$\boxed{\frac{V_1}{T_1}=\frac{V_2}{T_2},}$$

$$\text{i.e. } V_2=V_1\left(\frac{T_2}{T_1}\right)=(1.2)\,\frac{(130+273)}{(20+273)}$$

$$=\frac{(1.2)(403)}{(293)}=\mathbf{1.65\ litres}.$$

4 (i) **The Pressure law states:**

'the pressure p of a fixed mass of gas is directly proportional to its thermodynamic temperature T at constant volume'

i.e. $p \propto T$ or $p=kT$ or $\frac{P}{T}=k$

(ii) When a fixed mass of gas at constant volume changes from pressure p_1 and temperature T_1, to pressure p_2 and temperature T_2 then:

$$\boxed{\frac{p_1}{T_1}=\frac{p_2}{T_2}}$$

5 (i) **Dalton's law of partial pressure states:**

'the total pressure of a mixture of gases occupying a given volume is equal to the sum of the pressures of each gas, considered separately, at constant temperature'.

(ii) The pressure of each constituent gas when occupying a fixed volume alone is known as the **partial pressure** of that gas.

6 An **ideal gas** is one which completely obeys the gas laws given in paras 2 to 5. In practice no gas is an ideal gas, although air is very close to being one. For calculation purposes the difference between an ideal and an actual gas is very small.

7 (i) Frequently, when a gas is undergoing some change, the

pressure, temperature and volume all vary simultaneously. Provided there is no change in the mass of a gas, the above gas laws can be combined giving:

$$\boxed{\frac{p_1V_1}{T_1}=\frac{p_2V_2}{T_2}=k,}$$ where k is a constant.

(ii) For an ideal gas, constant $k=mR$, where m is the mass of the gas in kg, and R is the **characteristic gas constant**,

i.e. $\frac{pV}{T}=mR$ or $\boldsymbol{pV=mRT}$

This is called the characteristic gas equation. In this equation p = absolute pressure in pascals, V = volume in m^3, m = mass in kg, R = characteristic gas constant in J/(kg K) and T = thermodynamic temperature in kelvin.
(iii) Some typical values of the characteristic gas constant R include: air, 287 J/(kg K), hydrogen 4160 J/(kg K), oxygen 260 J/(kg K) and carbon dioxide 184 J(kg K). For example, some air at a temperature of 40°C and pressure 4 bar occupies a volume of 0.05 m^3. The mass of air is determined from

$$pV=mRT$$

$$\text{Hence, mass } m=\frac{pV}{RT}=\frac{(4\times 10^5\text{ Pa})(0.05\text{ m}^3)}{(287\text{ J/(kg K)}(40+273)\text{ K}}$$

$$=\mathbf{0.223\ kg\ or\ 223\ g}.$$

8 **Standard temperature and pressure (i.e. STP)** refers to a temperature of 0°C, i.e. 273 K, and normal atmospheric pressure of 101.325 kPa.

Kinetic theory of gases

9. It is shown in Chapter 46 that a gas occupying a volume V at pressure p and containing n molecules each of mass m moving at an average velocity of c,

$$\boxed{pV=\tfrac{1}{3}mnc^2}$$

Also, the kinetic energy of the molecules of a gas is proportional to its thermodynamic temperature.

10 When a liquid evaporates molecules with sufficient kinetic energy escape from the liquid's surface. The higher the

temperature of the liquid the greater the average kinetic energy of the molecules and the greater the number of molecules which are able to escape. Since it is the molecules with the highest kinetic energy which escape the average kinetic energy of the remaining molecules decreases and thus the liquid cools.

11 If a liquid evaporates a **vapour** is formed. When a vapour exists in the presence of its own liquid a **saturated vapour** is formed. If all the liquid evaporates an **unsaturated vapour** is produced. The higher the temperature the greater the number of molecules which escape to form the vapour. These molecules bombard the walls of the container and thus exert a pressure.

The **saturated vapour pressure** depends only on the temperature of the vapour. The saturated vapour pressure of water at various temperatures is shown in *Table 25.1*.

Table 25.1

Temperature (°C)	*Saturated vapour pressure of water* (10^3 Pa)
0	0.61
10	1.23
20	2.33
30	4.23
40	7.35
50	12.3
60	19.9
70	31.2
80	47.4
90	70.2
100	101
150	476
200	1550

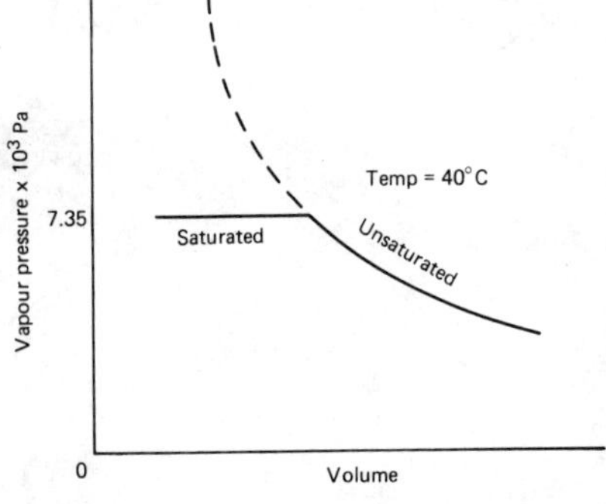

Figure 25.1

A liquid boils at a temperature when its saturated vapour pressure is equal to the atmospheric pressure. Thus water will boil at a temperature greater than 100°C if the atmospheric pressure is increased. This is the principle of the pressure cooker.

12 **A saturated vapour does not obey the gas laws since its** pressure depends only on temperature. An unsaturated vapour will obey the gas laws fairly closely as long as it remains unsaturated. If an unsaturated vapour at a particular temperature is decreased in volume its pressure will rise in accordance with Boyle's law until it reaches the saturated vapour pressure at that particular temperature (see *Figure 25.1*). When the vapour pressure at 40°C reaches 7.35×10^3 Pa the vapour becomes saturated as it starts to liquify.

26 Simple harmonic motion and resonance

1 A body is said to move with **simple harmonic motion (SHM)** if its acceleration, a, is proportional to its displacement, s, and opposite in direction.

This is normally expressed by the equation

$$a = -\omega^2 s$$

where ω is a constant which depends on the characteristic of the system involved. It is often referred to as the SHM constant. A body moving in this way will oscillate such that the period T (the time for one oscillation) is independent of the amplitude A (the maximum displacement).

The graph of displacement against time is shown in *Figure 26.1* and is similar to that of $\sin\theta$ plotted against θ. In this graph time is measured from the instant when $s=0$. A cosine graph represents the motion if $t=0$ when $s=A$.

2 The equation of the graph shown in *Figure 26.1* is:

$$s = A \sin \frac{2\pi t}{T} \quad (1)$$

The velocity v at any time is given by the gradient of the

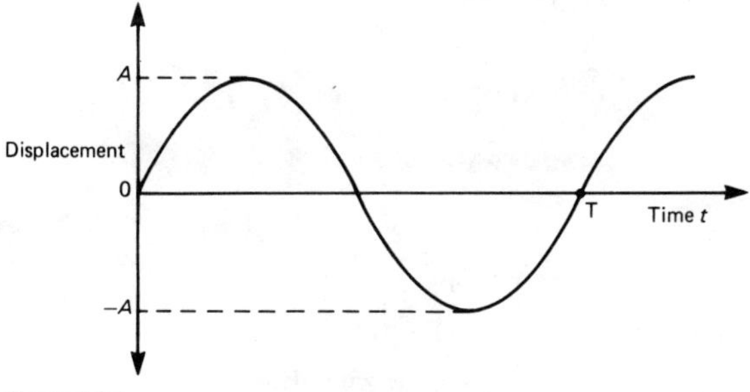

Figure 26.1

displacement time graph, i.e. ds/dt.

$$\frac{ds}{dt}=A.\frac{2\pi}{T}\cos\frac{2\pi t}{T}$$

Therefore, $v=A\frac{2\pi}{T}\cos\frac{2\pi t}{T}$ (2)

The acceleration, a, at any time is given by the gradient of the velocity time graph, i.e. dv/dt.

$$\frac{dv}{dt}=-A\left(\frac{2\pi}{T}\right)^2\sin\frac{2\pi t}{T}$$

Therefore $a=-A\left(\frac{2\pi}{T}\right)^2\sin\frac{2\pi t}{T}$

but $s=A\sin\frac{2\pi t}{T}$

Therefore $a=-\left(\frac{2\pi}{T}\right)^2 s$

Thus the constant ω (see para 1) $=\frac{2\pi}{T}$

showing that the period is a constant since ω is a constant.

From equation (1) $s=A\sin\frac{2\pi t}{T}$ and $\frac{2\pi}{T}=\omega$

Therefore $\sin\omega t=\frac{s}{A}$ and $\sin^2\omega t=\frac{s^2}{A^2}$

From equation (2) $v=A.\frac{2\pi}{T}\cos\frac{2\pi t}{T}$ and $\frac{2\pi}{T}=\omega$

Therefore $\cos\omega t=\frac{v}{A\omega}$

and $\cos^2\omega t=\frac{v^2}{A^2\omega^2}$

Hence $\frac{v^2}{A^2\omega^2}+\frac{s^2}{A^2}=\cos^2\theta+\sin^2\theta=1$

Therefore $v^2+s^2\omega^2=A^2\omega^2$

and $v^2=\omega^2(A^2-s^2)$

from which **velocity** $v=\omega\sqrt{(A^2-s^2)}$

For example, if a body moves with SHM with a period of 0.5 s and an amplitude of 40 mm, the maximum acceleration and velocity may be determined as follows:

SHM constant $\omega = \frac{2\pi}{T}$, (where T=period)

$$= \frac{2\pi}{0.5} = 4\pi \text{ s}^{-1}$$

Acceleration $a = -\omega^2 s$, where s=displacement

Hence maximum acceleration $a = -(4\pi)^2(40\times10^{-3}) = \mathbf{6.32\ m\ s^{-2}}$ towards the centre.

Maximum speed, v_{MAX}, occurs when $s=0$ in the equation.

$$v_{MAX} = \omega\sqrt{(A^2 - s^2)}$$

i.e. $v_{MAX} = \omega A = (4\pi)(40\times10^{-3})$

$$= \mathbf{0.50\ m\ s^{-1}}.$$

3 Simple harmonic motion may be represented by a **phasor**. A phasor is a rotating line of constant length whose projection on to a fixed axis represents the displacement of the body performing simple harmonic motion. The anticlockwise angular velocity of the rotating line OP in *Figure 26.2* is ω and its length is A.
OP has rotated through an angle ωt in time t.
The projection of OP on to the line XX′ is OQ

$OQ = A \sin \omega t$
but $s = A \sin \omega t$

gives the displacement after time t of a body moving with SHM. This means that the point Q moves with SHM and OQ is the displacement, s.

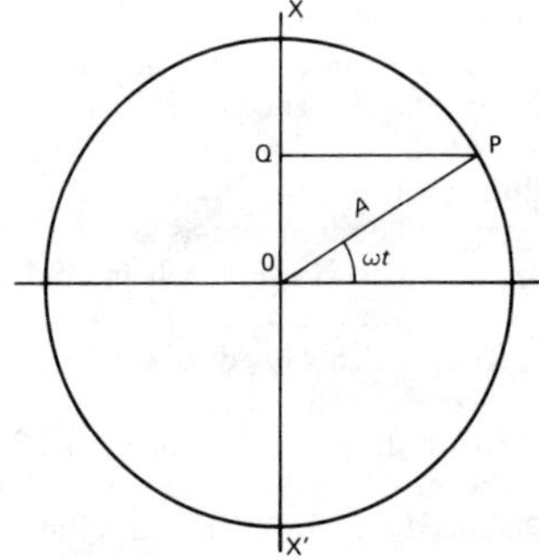

Figure 26.2

4 The natural frequency, f, of a mass, m, at the end of a tethered spring of force constant, k, is given by:

$$f = \frac{1}{2\pi}\sqrt{\left(\frac{k}{m}\right)}$$

Many mechanical systems can be likened to the mass on a spring system mentioned above in that they have a natural frequency of oscillation which is given by:

$$f = \frac{1}{2\pi}\sqrt{\left(\frac{\textbf{stiffness factor}}{\textbf{inertia factor}}\right)}$$

A mechanical system may be forced into vibration. The amplitude of these oscillations will be determined by the frequency of the forcing oscillations. The closer the forcing frequency is to the natural frequency the greater is the amplitude of oscillation of the mechanical system. When the forcing frequency equals the natural frequency the amplitude becomes a maximum. This phenomenon is called **resonance** and the frequency is called the **resonant frequency**.

Mechanical systems have a natural frequency of oscillation which depends on such factors as their mass and their stiffness. Systems may be forced to oscillate at some other frequency by the action of some outside periodic force. If this 'driving' frequency matches the natural frequency of oscillation of the system energy is absorbed and the amplitude of the oscillations will increase. An example of this is the child on a swing. If his parent pushes the swing with the natural frequency of the swing and child the oscillations will build up. Another common example is the vibration of a car's rear view mirror at some critical engine speed.

There are a number of examples where the outcome may be dangerous. It is well known that a marching column of soldiers is supposed to break step when crossing a suspension bridge in case the bridge starts to resonate. In the case of the Tacoma Narrows Bridge in Seattle in 1940 the damage was indeed serious. The cause in this case was the wind although it was not the wind's speed that directly caused the damage. The speed was as low as 19 m s^{-1} and the force exerted was not high. However when wind is incident upon a horizontal barrier vortices are produced alternately above and below the barrier which caused upward and downward forces on it. For the Tacoma Narrows Bridge the frequency of these forces matched one of the natural frequencies of oscillation of the bridge, resonance occurred and the bridge collapsed. The Tay Bridge in 1878 and the Chain Pier at Brighton in 1834 collapsed for the same reason.

Fractures of materials which resonate continually and become fatigued are not unknown. The BOAC Comet crashes of 1954 were probably caused by this effect.

Other accidents have probably been caused by resonance. Collisions of helicopters with overhead power cables have been attributed to the fact that the frequency of vibrations produced by the rotors matched the natural frequency of oscillation of the pilots eyeballs. The resultant impaired vision proved fatal.

5 Diatomic molecules may be thought of as a mass connected to a tethered spring. When infra-red radiation of a suitable frequency is incident upon such molecules they may be caused to resonante. Measurement of the resonant frequency yields information about the stiffness of the interatomic bond.

If the resonant frequency of the diatomic molecule is f the mass of the lighter atom is m, and k is the force constant of the interatomic bond they are related by the equation

$$f \approx \frac{1}{2\pi}\sqrt{\left(\frac{k}{m}\right)}$$

The greater the difference between the masses of the atoms the closer the approximation becomes to the truth.

27 Basic d.c. circuit theory

Standard symbols for electrical components

1 Symbols are used for components in electrical circuit diagrams and some of the more common ones are shown in *Figure 27.1.*

2 (i) All substances are made from **elements** and the smallest particle to which an element can be reduced is called an **atom**.

(ii) An atom consists of **electrons** which can be considered to be orbiting around a central **nucleus** containing **protons** and **neutrons**.

(iii) An electron possesses a **negative charge**, a proton a **positive charge** and a neutron has no charge.

(iv) There is a force of **attraction** between oppositely charged bodies and a force of **repulsion** between similarly charged bodies.

(v) The **force** between two charged bodies depends on the amount of charge on the bodies and their distance apart.

(vi) **Conductors** are materials having electrons that are loosely connected to the nucleus and can easily move through the material from one atom to another. **Insulators** are materials whose electrons are held firmly to their nucleus.

(vii) A drift of electrons in the same direction constitutes an **electric current**.

(viii) The unit of charge is the **coulomb**, **C**, and when 1 coulomb of charge is transferred in 1 second a current of 1 ampere flows in the conductor. Thus electric current I is the rate of flow of charge in a circuit. The unit of current is the **ampere**, **A**.

(x) For a continuous current to flow between two points in a circuit a **potential difference (p.d.)** or **voltage**, **V**, is required between them; a complete conducting path is necessary to and from the source of electrical energy. The unit of p.d. is the **volt**, **V**.

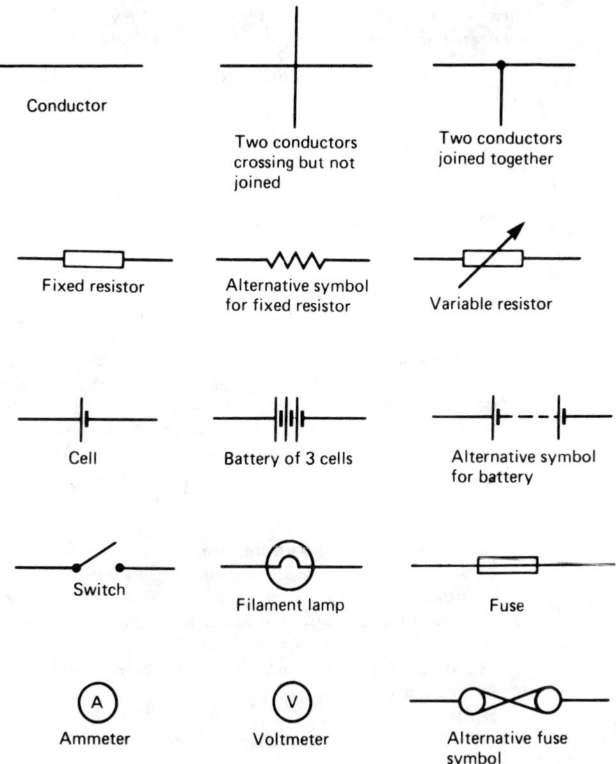

Figure 27.1

(xi) *Figure 27.2* shows a cell connected across a filament lamp. Current flow, by convention, is considered as flowing from the positive terminal of the cell, around the circuit to the negative terminal.

3 The flow of electric current is subject to friction. This friction, or opposition, is called **resistance** R and is the property of a conductor that limits current. The unit of resistance is the ohm, Ω. 1 ohm is defined as the resistance which will have a

current of 1 ampere flowing through it when 1 volt is connected across it,

$$\text{i.e. } \textbf{resistance } R = \frac{\textbf{potential difference}}{\textbf{current}}, \text{ i.e. } R = \frac{V}{I}$$

4 The reciprocal of resistance is called **conductance** and is measured in siemens (S). Thus

conductance in siemens, $\boldsymbol{G} = \dfrac{\mathbf{1}}{\boldsymbol{R}}$ where R is the resistance in ohms.

Electrical measuring instruments

5 (i) An **ammeter** is an instrument used to measure current and must be connected **in series** with the circuit. *Figure 27.2* shows an ammeter connected in series with the lamp to measure the current flowing through it. Since all the current in the circuit passes through the ammeter it must have a very **low resistance**.

(ii) A **voltmeter** is an instrument used to measure p.d. and must be connected **in parallel** with the part of the circuit whose p.d. is required. In *Figure 27.2*, a voltmeter is connected in parallel with the lamp to measure the p.d. across it. To avoid a significant current flowing through it a voltmeter must have a very **high resistance**.

(iii) An **ohmmeter** is an instrument for measuring resistance.

(iv) A **multimeter**, or universal instrument, may be used to measure voltage, current and resistance. An 'Avometer' is a typical example.

(v) The **cathode ray oscilloscope (CRO)** may be used to observe waveforms and to measure voltages and currents. The display of a CRO involves a spot of light moving across a screen. The amount by which the spot is deflected from its initial position depends on the p.d. applied to the terminals of the CRO and the range selected. The displacement is calibrated in 'volts per cm'. For example, if the spot is deflected 3 cm and the volts/cm switch is on 10 V/cm then the magnitude of the p.d. is 3 cm × 10 V/cm, i.e., 30 V. (See page 256).

Linear and non-linear devices

6 *Figure 27.3* shows a circuit in which current I can be varied by the variable resistor R_2. For various settings of R_2, the current

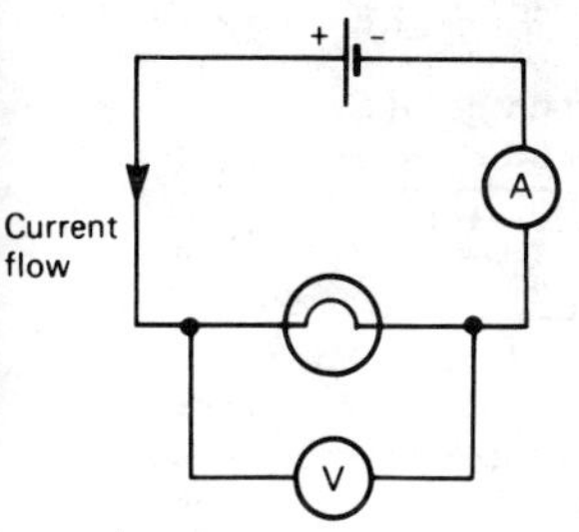

Figure 27.2

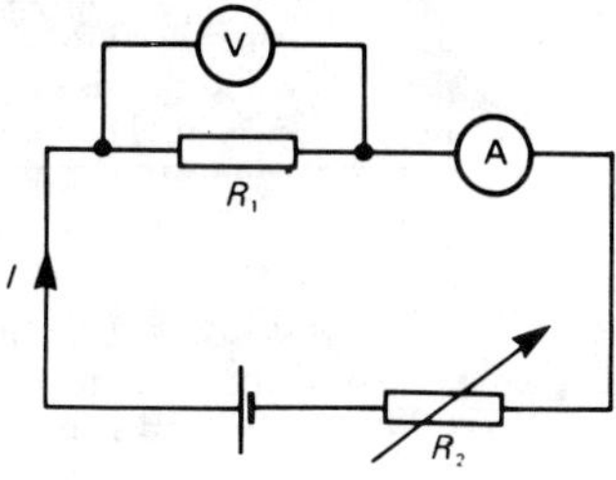

Figure 27.3

flowing in resistor R_1, displayed on the ammeter, and the p.d. across R_1, displayed on the voltmeter, are noted and a graph is ploted of p.d. against current. The result is shown in *Figure 27.4(a)* where the straight line graph passing through the origin indicates that current is directly proportional to the p.d. Since the gradient

$$\text{i.e. } \frac{\text{p.d.}}{\text{current}}$$

is constant, resistance R_1 is constant. A resistor is thus an example of a **linear device.**

If the resistor R_1 in *Figure 27.3* is replaced by a component such as a lamp then the graph shown in *Figure 27.4(b)* results when values of p.d. are noted for various current readings. Since the gradient is changing the lamp is an example of a **non-linear device**.

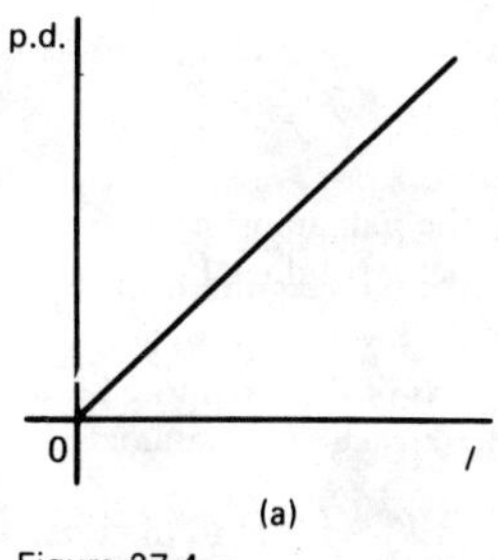

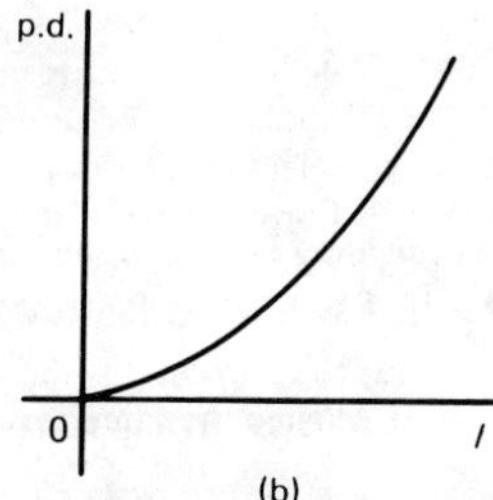

Figure 27.4

7 **Ohm's law** states that the current *I* flowing in a circuit is directly proportional to the applied voltage *V* and inversely proportional to the resistant *R*, provided the resistance remains constant. Thus:

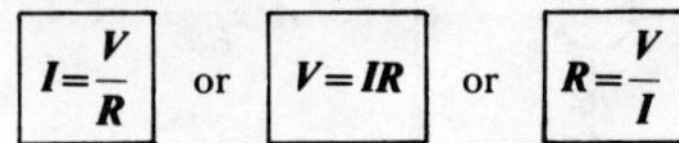

$$I=\frac{V}{R} \quad \text{or} \quad V=IR \quad \text{or} \quad R=\frac{V}{I}$$

8 (i) A **conductor** is a material having a low resistance which allows electric current to flow in it. All metals are conductors and some examples include copper, aluminium, brass, platinum, silver, gold and also carbon.
(ii) An **insulator** is a material having a high resistance which does not allow electric current to flow in it. Some examples of insulators include plastic, rubber, glass, porcelain, air, paper, cork, mica, ceramics and certain oils.

Series circuit

9 *Figure 27.5* shows three resistors R_1, R_2 and R_3 connected end to end, i.e., in series with a battery source of *V* volts. Since the

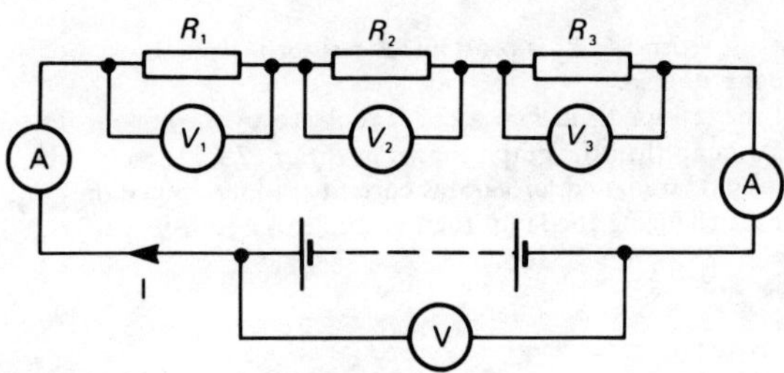

Figure 27.5

circuit is closed a current *I* will flow and the p.d. across each resistor may be determined from the voltmeter readings V_1, V_2 and V_3. In a series circuit:

(a) the current *I* is the same in all parts of the circuit and hence the same reading is found on each of the ammeters shown, and
(b) the sum of the voltages V_1, V_2 and V_3 is equal to the total

applied voltage, V, i.e.

$$V = V_1 + V_2 + V_3.$$

From Ohm's law:

$$V_1 + IR_1,\ V_2 = IR_2,\ V_3 = IR_3 \text{ and } V = IR$$

where R is the total circuit resistance.

Since $V = V_1 + V_2 + V_3$

then $IR = IR_1 + IR_2 + IR_3$.

Dividing throughout by I gives

$R = R_1 + R_2 + R_3$.

Thus for a series circuit, the total resistance is obtained by adding together the values of the separate resistances.

10 The voltage distribution for the circuit shown in *Figure 27.6(a)* is given by:

$$V_1 = \left(\frac{R_1}{R_1 + R_2}\right)V$$

$$V_2 = \left(\frac{R_2}{R_1 + R_2}\right)V$$

The circuit shown in *Figure 27.6(b)* is often referred to as a **potential divider** circuit. Such a circuit can consist of a number of similar elements in series connected across a voltage source, voltages being taken from connections in-between the elements. Frequently the potential divider consists of two resistors as shown

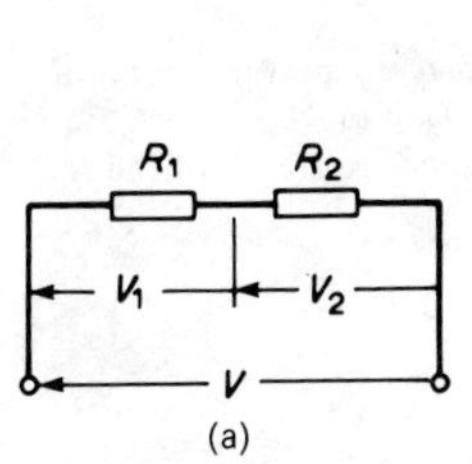

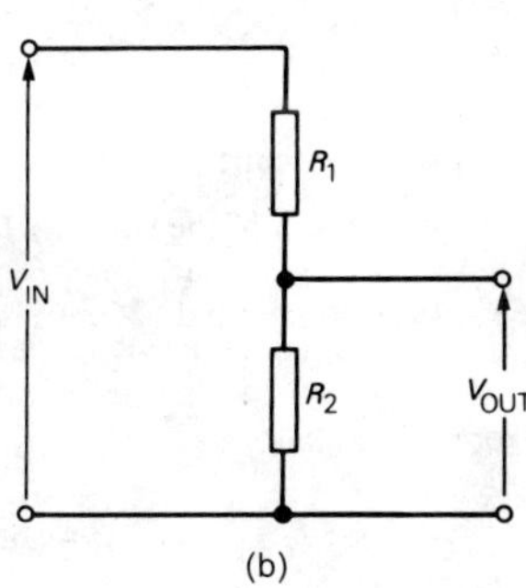

Figure 27.6

in *Figure 27.6(b)* where

$$V_{OUT} = \left(\frac{R_2}{R_1 + R_2}\right) V_{IN}$$

Where a continuously variable voltage is required from a fixed supply a single resistor with a sliding contact is used. Such a device is known as a **potentiometer**.

Parallel circuit

11 *Figure 27.7* shows three resistors, R_1, R_2 and R_3 connected across each other, i.e. in parallel, across a battery source of V volts.

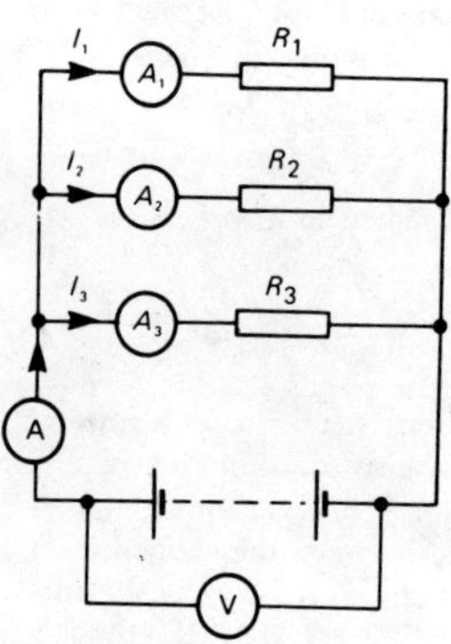

Figure 27.7

In a parallel circuit:

(a) the sum of the currents I_1, I_2 and I_3 is equal to the total circuit current, I, i.e. $I = I_1 + I_2 + I_3$, and
(b) the source p.d., V volts, is the same across each of the resistors.

From Ohm's law:

$$I_1 = \frac{V}{R_1},\ I_2 = \frac{V}{R_2},\ I_3 = \frac{V}{R_3} \text{ and } I = \frac{V}{R}$$

where R is the total circuit resistance.

Since $I = I_1 + I_2 + I_3$

Then $\frac{V}{R} = \frac{V}{R_1} + \frac{V}{R_2} + \frac{V}{R_3}$

Dividing throughout by V gives:

$$\frac{1}{R} = \frac{1}{R_1} + \frac{1}{R_2} + \frac{1}{R_3}$$

This equation must be used when finding the total resistance R of a parallel circuit.

12 For the special case of two resistors in parallel:

$$\frac{1}{R} = \frac{1}{R_1} + \frac{1}{R_2} = \frac{R_2 + R_1}{R_1 R_2}$$

Hence

$$R = \frac{R_1 R_2}{R_1 + R_2} \left(\text{i.e. } \frac{\text{product}}{\text{sum}}\right)$$

13 The current division for the circuit shown in *Figure 27.8* is given by:

$$I_1 = \left(\frac{R_2}{R_1 + R_2}\right) I$$

$$I_2 = \left(\frac{R_1}{R_1 + R_2}\right) I$$

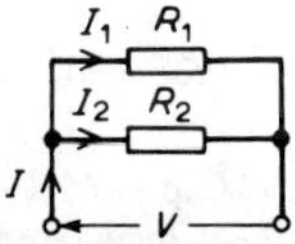

Figure 27.8

Wiring lamps in series and in parallel

Series connection

14 *Figure 27.9* shows three lamps, each rated at 240 V, connected in series across a 240 V supply.

(i) Each lamp has only $\frac{240}{3}$ V, i.e. 80 V across it and thus each lamp glows dimly.

(ii) If another lamp of similar rating is added in series with the other three lamps then each lamp now has $\frac{240}{4}$ V, i.e. 60 V across it and each now glows even more dimly.

(iii) If a lamp is removed from the circuit or if a lamp develops a fault (i.e. an open circuit) or if the switch is opened then the circuit is broken, no current flows, and the remaining lamps will not light up.

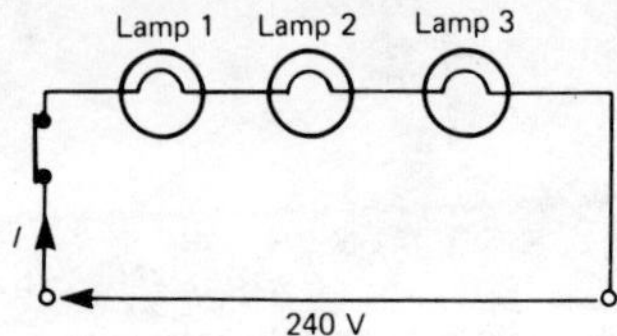

Figure 27.9

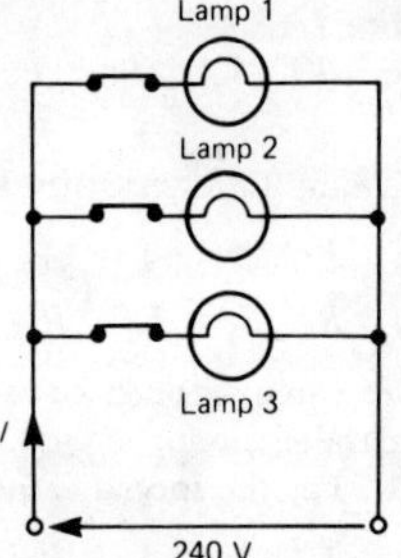

Figure 27.10

(iv) Less cable is required for a series connection than for a parallel one.

The series connection of lamps is usually limited to decorative lighting such as for Christmas tree lights.

Parallel connection

Figure 27.10 shows three similar lamps, each rated at 240 V, connected in parallel across a 240 V supply.

(i) Each lamp has 240 V across it and thus each will glow brilliantly at their rated voltage.
(ii) If any lamp is removed from the circuit or develops a fault (open circuit) or a switch is opened, the remaining lamps are unaffected.
(iii) The addition of further similar lamps in parallel does not affect the brightness of the other lamps.
(iv) More cable is required for parallel connection than for a series one.

The parallel connection of lamps is the most widely used in electrical installations.

15 **Power** P in an electrical circuit is given by the product of potential difference V and current I. The unit of power is the **watt, W**.

Hence $\boldsymbol{P = V \times I}$ **watts.** (1)

From Ohm's law, $V = IR$
Substituting for V in (1) gives:

$P = (IR) \times I$
i.e. $\boldsymbol{P = I^2 R}$ **watts.**

Also, from Ohms law, $I=\dfrac{V}{R}$

Substituting for I in (1) gives:

$$P=V\times\left(\frac{V}{R}\right) \text{ i.e. } P=\frac{V^2}{R} \text{ watts.}$$

There are thus three possible formulae which may be used for calculating power.

16 **Electrical energy = power × time.**
If the power is measured in watts and the time in seconds then the unit of energy is watt-seconds or **joules**. If the power is measured in kilowatts and the time in hours then the unit of energy is kilowatt-hours, often called the 'unit of electricity'. The 'electricity meter' in the home records the number of kilowatt-hours used and is thus an energy meter.

$$\begin{aligned}1 \text{ kWh} &= 1000 \text{ watt hours}\\ &= 1000\times 3600 \text{ watt seconds or joules}\\ &= 3\,600\,000 \text{ J.}\end{aligned}$$

17 (i) The **three main effects of an electric current** are:
(a) magnetic effect;
(b) chemical effect;
(c) heating effect.

(ii) Some practical applications of the effects of an electric current include:
Magnetic effect: bells, relays, motors, generators, transformers, telephones, car-ignition and lifting magnets.
Chemical effect: primary and secondary cells and electroplating.
Heating effect: cookers, water heaters, electric fires, irons, furnaces, kettles and soldering irons.

18 A **fuse** is used to prevent overloading of electrical circuits.
The fuse, which is made of material having a low melting point, utilizes the heating effect of an electric current. A fuse is placed in an electrical circuit and if the current becomes too large the fuse wire melts and so breaks the circuit. A circuit diagram symbol for a fuse is shown in *Figure 27.1*, page 151.

Resistance variation

19 The resistance of an electrical conductor depends on 4 factors, these being:

(a) the length of the conductor;
(b) the cross-sectional area of the conductor;

(c) the type of material; and
(d) the temperature of the material.

20 (i) Resistance, R, is directly proportional to length, l, of a conductor, i.e. $R\alpha l$. Thus, for example, if the length of a piece of wire is doubled, then the resistance is doubled.
(ii) Resistance (R) is inversely proportional to cross-sectional area, a, of conductor, i.e. $R\alpha(1/a)$. Thus, for example, if the cross-sectional area of a piece of wire is doubled then the resistance is halved.
(iii) Since $R\alpha l$ and $R\alpha(1/a)$ then $R\alpha(1/a)$. By inserting a constant of proportionality into this relationship the type of material used may be taken into account. The constant of proportionality is known as the **resistivity** of the material and is given the symbol ρ (rho).

Thus, resistance, $$\boldsymbol{R=\frac{\rho l}{a}}\ \textbf{ohms}$$

ρ is measured in ohm metres (Ωm). The value of the resistivity is that resistance of a unit cube of the material measured between opposite faces of the cube.
(iv) Resistivity varies with temperature and some typical values of resistivities measured at about room temperature are given below:

Copper, 1.7×10^{-8} Ωm (or 0.017 $\mu\Omega$m)
Aluminium 2.6×10^{-8} Ωm (or 0.026 $\mu\Omega$m)
Carbon (graphite) 10×10^{-8} Ωm (0.10 $\mu\Omega$m)
Glass 1×10^{10} Ωm (or 10^4 $\mu\Omega$m)
Mica 1×10^{13} Ωm (or 10^7 $\mu\Omega$m).

Note that good conductors of electricity have a low value of resistivity and good insulators have a high value of resistivity.

21 (i) In general, as the temperature of a material increases, most conductors increase in resistance, insulators decrease in resistance whilst the resistance of some special alloys remain almost constant.
(ii) The **temperature coefficient of resistance** of a material is the increase in the resistance of a 1 Ω resistor of that material when it is subjected to a rise of temperature of 1°C. The symbol used for the temperature coefficient of resistance is α (alpha). Thus, if some copper wire of resistance 1 Ω is heated through 1°C and its resistance is then measured as 1.0043 Ω then

$\alpha=0.0043$ Ω/Ω°C for copper.

The units are usually expressed only as 'per °C', i.e., $\alpha = 0.0043/°C$ for copper. If the 1 Ω resistor of copper is heated through 100°C then the resistance at 100°C would be

$$1 + 100 \times 0.0043 = 1.43\ \Omega.$$

(iii) If the resistance of a material at 0°C is known, the resistance at any other temperature can be determined from:

$$\boxed{R_\theta = R_0(1 + \alpha_0 \theta)}$$

where R_0 = resistance at 0°C
R_θ = resistance at temperature θ°C;
α_0 = temperature coefficient of resistance at 0°C.

(iv) If the resistance at 0°C is not known, but is known a some other temperature θ_1, then the resistance at any temperature can be found as follows:

$$R_1 = R_0(1 + \alpha_0 \theta_1) \text{ and } R_2 = R_0(1 + \alpha_0 \theta_2)$$

Dividing one equation by the other gives:

$$\boxed{\frac{R_1}{R_2} = \frac{1 + \alpha_0 \theta_1}{1 + \alpha_0 \theta_2}}$$

where R_2 = resistance at temperature θ_2.

(v) If the resistance of a material at room temperature (approximately 20°C), R_{20}, and the temperature coefficient of resistance at 20°C, α_{20} are known then the resistance R_θ at temperature θ°C is given by:

$$\boxed{R_\theta = R_{20}[1 + \alpha_{20}(\theta - 20)]}$$

(vi) Some typical values of temperature coefficient of resistance measured at 0°C are given below:

Copper	0.0043/°C
Aluminium	0.0038/°C
Nickel	0.0062/°C
Carbon	−0.00048/°C
Constantan	0
Eureka	0.00001/°C

(Note that the negative sign for carbon indicates that its resistance falls with increase of temperature.)

28 D.C. circuit analysis

1 The laws which determine the currents and voltage drops in d.c. networks are

(a) Ohm's law;
(b) the laws for resistors in series and in parallel; and
(c) Kirchhoff's laws.

Kirchhoff's laws

2 Kirchhoff's laws state:

(a) Current law

At any junction in an electric circuit the total current flowing towards that junction is equal to the total current flowing away from the junction, i.e. $\Sigma I = 0$

Thus, referring to *Figure 28.1*:

$$I_1 + I_2 = I_3 + I_4 + I_5$$
$$\text{or} \quad I_1 + I_2 - I_3 - I_4 - I_5 = 0$$

(b) Voltage law

In any closed loop in a network, the algebraic sum of the voltage drops (i.e. products of current and resistance) taken around the loop is equal to the resultant emf acting in that loop.

Thus referring to *Figure 28.2*:

$$E_1 - E_2 = IR_1 + IR_2 + IR_3$$

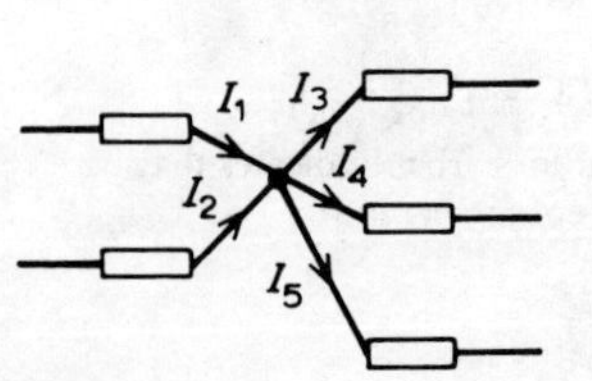

Figure 28.1

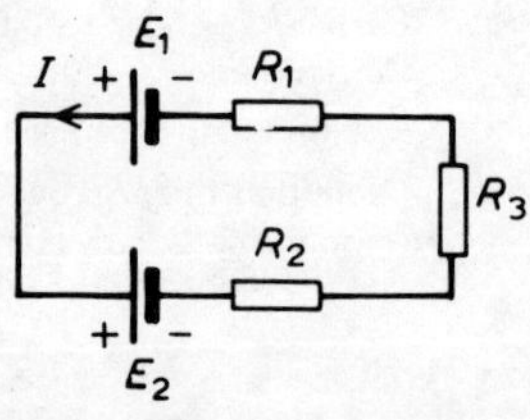

Figure 28.2

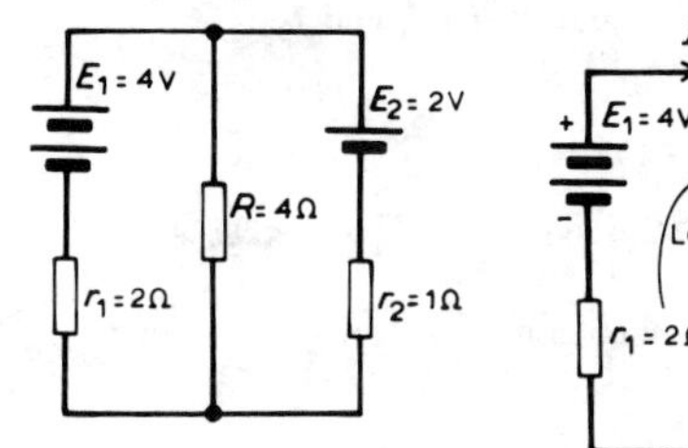

Figure 28.3

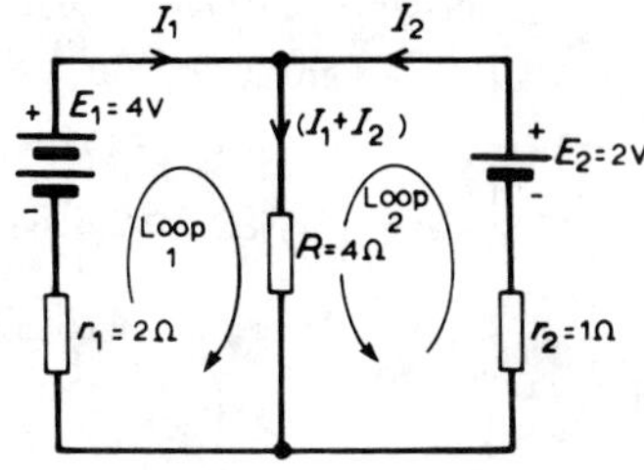

Figure 28.4

(Note that if current flows away from the positive terminal of a source, that source is considered by convention to be positive. Thus moving anticlockwise around the loop of *Figure 28.2*, E_1 is positive and E_2 is negative.)

For example, using Kirchhoff's laws to determine the current flowing in each branch of the network shown in *Figure 28.3*, the procedure is as follows:

(i) Use Kirchhoff's current law and label current directions on the original circuit diagram. The directions chosen are arbitrary, but it is usual, as a starting point, to assure that current flows from the positive terminals of the batteries. This is shown in *Figure 28.4* where the three branch currents are expressed in terms of I_1 and I_2 only, since the current through R is $I_1 + I_2$.
(ii) Divide the circuit into two loops and apply Kirchoff's voltage law to each. From loop 1 of *Figure 28.4*, and moving in a clockwise direction as indicated (the direction chosen does not matter) gives

$$E_1 = I_1 r_1 + (I_1 + I_2)R,$$
i.e. $4 = 2I_1 + 4(I_1 + I_2)$,
i.e. $6I_1 + 4I_2 = 4$ (1)

From loop 2 of *Figure 28.4*, and moving in an anticlockwise direction as indicated (once again, the choice of direction does not matter; it does not have to be in the same direction as that chosen from the first loop), gives:

$$E_2 = I_2 r_2 + (I_1 + I_2)R,$$
i.e. $2 = I_2 + 4(I_1 + I_2)$,
i.e. $4I_1 + 5I_2 = 2$ (2)

(iii) Solve equations (1) and (2) for I_1 and I_2.

$2\times(1)$ gives: $12I_1+8I_2=8$ (3)

$3\times(2)$ gives: $12I_1+15I_2=6$ (4)

$(3)-(4)$ gives: $-7I_2=2$ and $I_2=-\dfrac{2}{7}=$ $\mathbf{-0.286\ A}$

(i.e. I_2 is flowing in the opposite direction to that shown in *Figure 28.4*).

From (1) $\quad 6I_1+4(-0.0286)=4$
$\qquad 6I_2=4+1.144$

Hence $\quad I_1=\dfrac{5.144}{6}=0.857\text{ A}$

Current flowing through $R=I_1+I_2=0.857+(-0.286)$
$=0.571$ A.

General hints on simple d.c. circuit analysis

3 (i) The open-circuit voltage, E, across terminals AB in *Figure 28.5* is equal to 10 V, since no current flows through the 2 ohm resistor and hence no voltage drop occurs.

(ii) The open-circuit voltage, E, across terminals AB in *Figure 28.6(a)* is the same as the voltage across the 6 ohm resistor. The circuit may be redrawn as shown in *Figure 28.6(b)*.

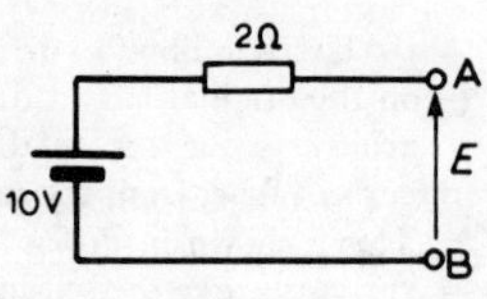

Figure 28.5

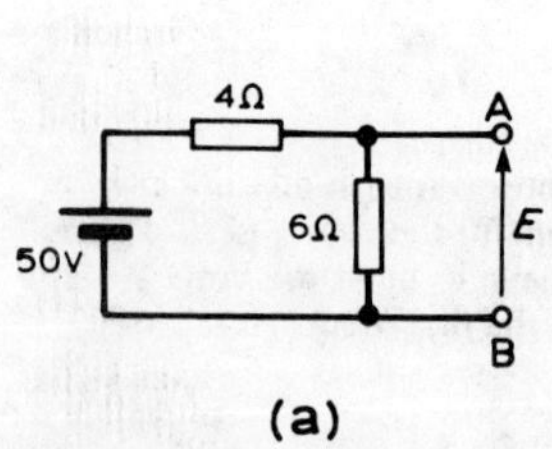

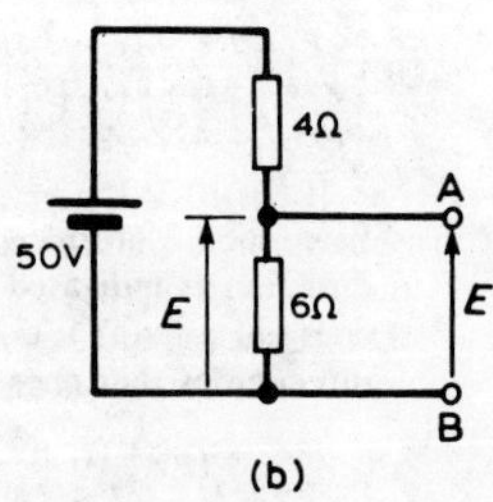

Figure 28.6

$$E=\left(\frac{6}{6+4}\right)(50)$$

by voltage division in a series circuit,

i.e. $E=30$ V.

(iii) For the circuit shown in *Figure 28.7(a)* representing a practical source supplying energy, $V=E-Ir$ where E is

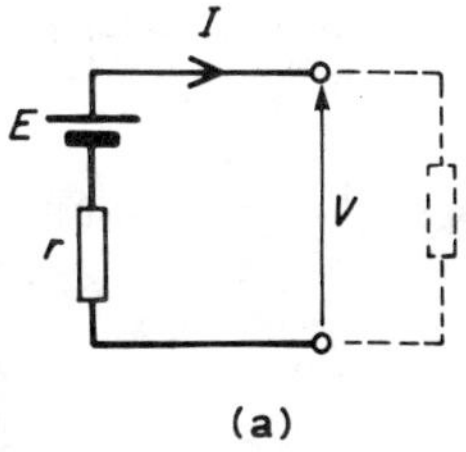

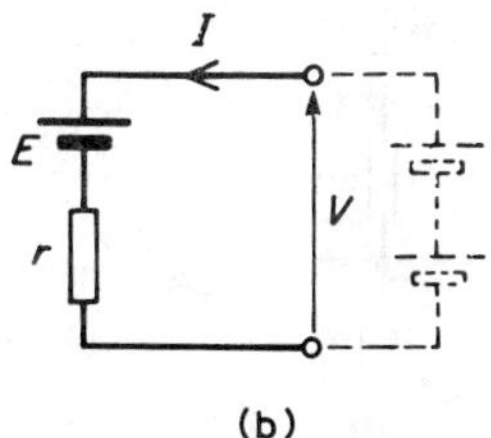

Figure 28.7

the battery emf, V is the battery terminal voltage and r is the internal resistance of the battery. For the circuit shown in *Figure 28.7(b)*,

$$V=E-(-I)r,\ \text{i.e.}\ V=E+Ir$$

(iv) The resistance 'looking-in' at terminals AB in *Figure 28.8(a)* is obtained by reducing the circuit in stages as shown in *Figures 28.8(b)* to *(d)*. Hence the equivalent resistance across AB is 7Ω.

(v) For the circuit shown in *Figure 28.29(a)*, the 3Ω resistor carries no current and the p.d. across the 20Ω resistor is 10 V. Redrawing the circuit gives *Figure 28.29(b)*, from which

$$E=\left(\frac{4}{4+6}\right)\times 10=4\ \text{V}.$$

(vi) If the 10 V battery in *Figure 28.9(a)* is removed and replaced by a short-circuit, as shown in *Figure 28.9(c)*, then the 20Ω resistor may be removed. The reason for this is that a short-circuit has zero resistance, and 20Ω in parallel with zero ohms gives an equivalent resistance of

$$\frac{20\times 0}{20+0},\ \text{i.e.}\ 0\Omega.$$

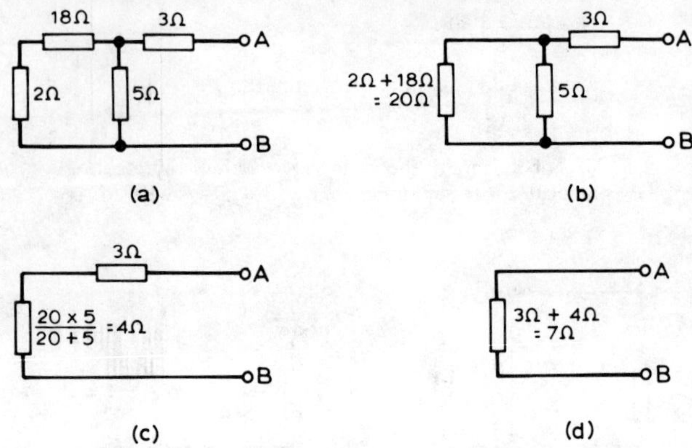

Figure 28.8

The circuit is then as shown in *Figure 28.9(d)*, which is redrawn in *Figure 28.9(c)*. From *Figure 28.9(e)*, the equivalent resistance across AB,

$$r=\frac{6\times 4}{6+4}+3=2.4+3=5.4\Omega$$

(vii) To find the voltage across AB in *Figure 28.10*.

Since the 20 V supply is across the 5Ω and 15Ω resistor in series then, by voltage division, the voltage drop across AC,

$$V_{AC}=\left(\frac{5}{5+15}\right)(20)=5\text{ V}.$$

Similarly,

$$V_{CB}=\left(\frac{12}{12+3}\right)(20)=16\text{ V}$$

V_C is at a potential of $+20$ V.

$$V_A=V_C-V_{AC}=+20-5=15\text{ V}$$

and $V_B=V_C-V_{BC}=+20-16=4$ V.

Hence the voltage between AB is

$$V_A-V_B=15-4=11\text{ V}$$

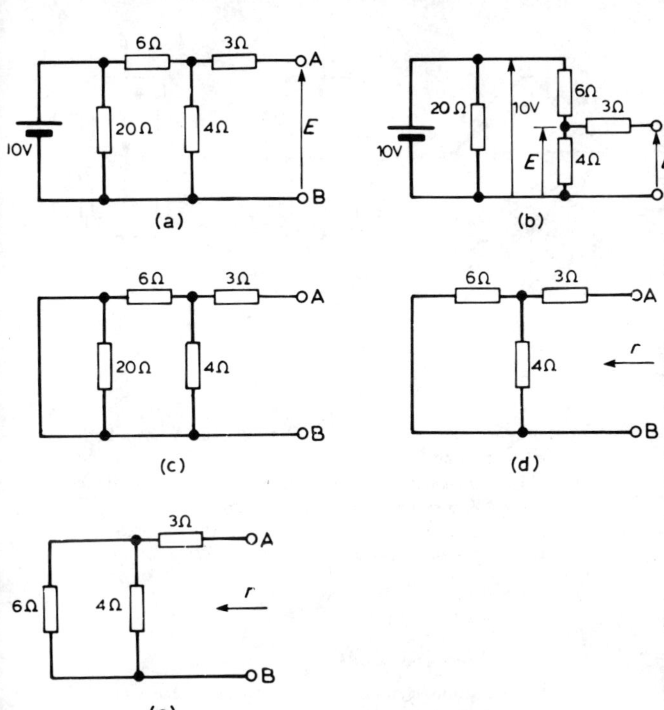

Figure 28.9

A
5Ω
15Ω
C
D
12Ω
3Ω
B
20V

Figure 28.10

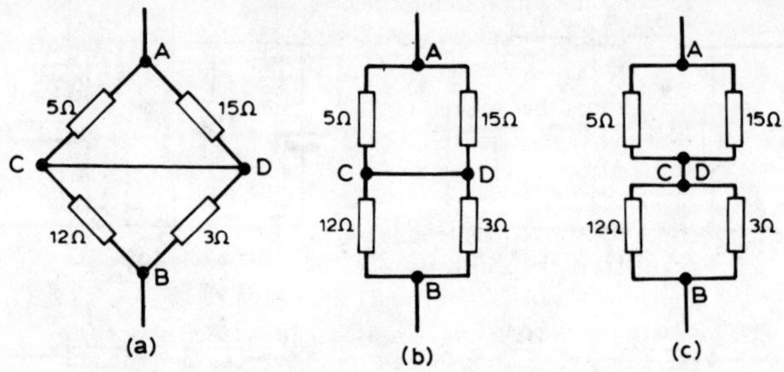

Figure 28.11

and current would flow from A to B since A has a higher potential than B.

(viii) In *Figure 28.11(a)*, to find the equivalent resistance across AB, the circuit may be redrawn as in *Figures 28.11(b)* and *(c)*. From *Figure 28.11(c)*, the equivalent resistance across AB

$$=\frac{5\times 15}{5+15}+\frac{12\times 3}{12+3}$$

$$=3.75+2.4=6.15\Omega$$

4 There are a number of circuit theorems which have been developed for solving problems in d.c. electrical networks. These include:

(i) the superposition theorem;
(ii) Thévenins theorem;
(iii) Norton's theorem, and
(iv) the maximum power transfer theorem.

5 **The superposition theorem** states:

'In any network made up of linear resistances and containing more than one source of emf, the resultant current flowing in any branch is the algebraic sum of the currents that would flow in that branch if each source was considered separately, all other sources being replaced at that time by their respective internal resistances.'

For example, to determine the current in each branch of the network shown in

Figure 28.12 using the superposition theorem, the procedure is as follows.

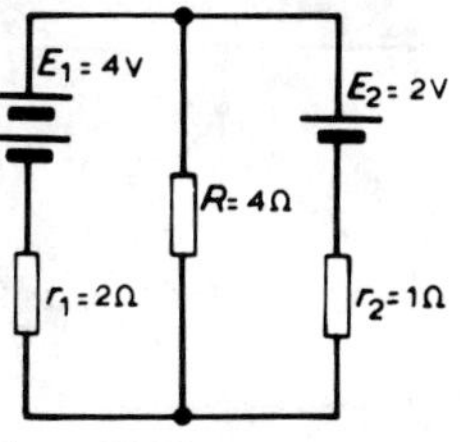

Figure 28.12

(i) Redraw the original circuit with source E_2 removed, being replaced by r_2 only, as shown in *Figure 28.13(a)*.

(ii) Label the currents in each branch and their directions as shown in *Figure 28.13(a)* and determine their values. (Note that the choice of current directions depends on the battery polarity, which, by convention is taken as flowing from the positive battery terminal as shown.) R in parallel with r_2 gives an equivalent resistance of

$$\frac{4 \times 1}{4+1} = 0.8\Omega.$$

From the equivalent circuit of *Figure 28.13(b)*.

$$I_1 = \frac{E_1}{r_1 + 0.8} = \frac{4}{2 + 0.8} = 1.429\text{A}$$

From *Figure 28.13(a)*,

$$I_2 = \left(\frac{1}{4+1}\right) I_1 = \frac{1}{5}(1.429) = 0.286 \text{ A and}$$

$$I_3 = \left(\frac{4}{4+1}\right) I_1 = \frac{4}{5}(1.429) = 1.143 \text{ A}$$

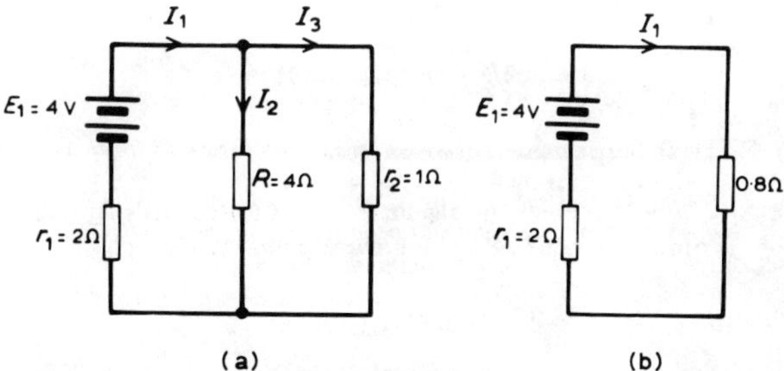

Figure 28.13

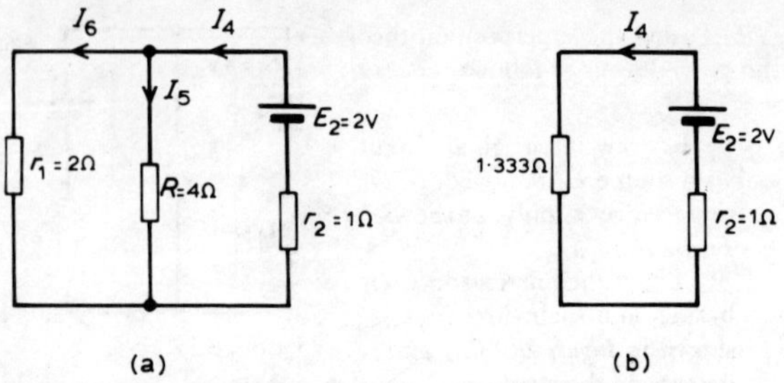

Figure 28.14

(iii) Redraw the original circuit with source E_1 removed, being replaced by r_1 only, as shown in *Figure 28.14(a)*.
(iv) Label the currents in each branch and their directions as shown in *Figure 28.14(a)* and determine their values. r_1 in parallel with R gives an equivalent resistance of

$$\frac{2\times 4}{2+4}=\frac{8}{6}=1.333\Omega$$

From the equivalent circuit of *Figure 28.14(b)*:

$$I_4=\frac{E_2}{1.333+r_2}=\frac{2}{1.333+1}$$

$$=0.857\text{ A}$$

From *Figure 28.14(a)*

$$I_5=\left(\frac{2}{2+4}\right)I_4=\frac{2}{6}(0.857)=0.286\text{ A}$$

$$I_6=\left(\frac{4}{2+4}\right)I_4=\frac{4}{6}(0.857)=0.571\text{ A}$$

(v) Superimpose *Figure 28.14(a)* onto *Figure 28.13(a)* as shown in *Figure 28.15*.
(vi) Determine the algebraic sum of the currents flowing in each branch. Resultant current flowing through source 1, i.e.

$$I_1-I_6=1.429-0.571$$

$$=\mathbf{0.858\text{ A (discharging)}}$$

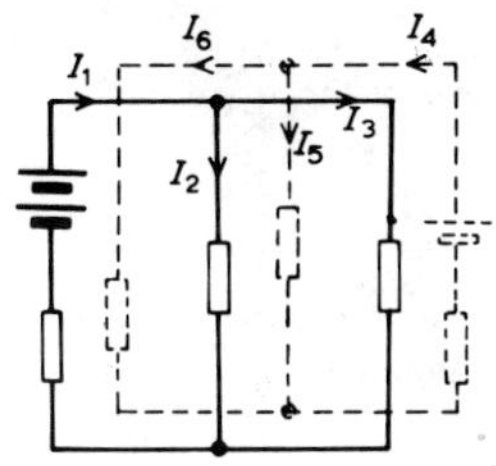

Figure 28.15

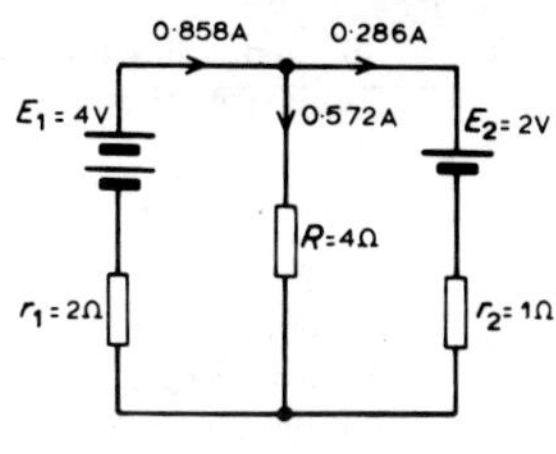

Figure 28.16

Resultant current flowing through source 2, i.e.

$$I_4 - I_3 = 0.857 - 1.143$$

$$= -\mathbf{0.286\ A\ (changing)}$$

Resultant current flowing through resistor R, i.e.

$$I_2 + I_5 = 0.286 + 0.286$$

$$= \mathbf{0.572\ A.}$$

The resultant currents with their directions are shown in *Figure 28.16.*

6 (a) **Thévenin's theorem** states:

'The current in any branch of a network is that which would result if an emf, equal to the p.d. across a break made in the branch, were introduced into the branch, all other emf's being removed and represented by the internal resistances of the sources.'

(b) The procedure adopted when using Thévenin's theorem is summarised below. To determine the current in any branch of an active network (i.e. one containing a source of emf):
(i) remove the resistance R from that branch,
(ii) determine the open-circuit voltage, E, across the break,
(iii) remove each source of emf and replace them by their internal resistances and then determine the resistance, r, 'looking-in' at the break.

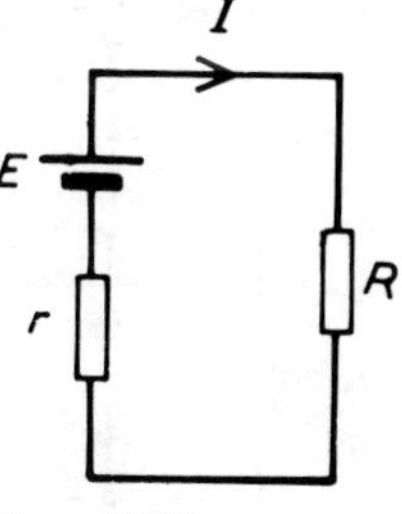

Figure 28.17

(iv) determine the value of the current from the equivalent circuit shown in *Figure 28.17*,

i.e. $I = \dfrac{E}{R + r}$

For example, using Thévenin's theorem to determine the current flowing in the 4Ω resistor shown in *Figure 28.18(a)*, using the above procedure:

(i) The 4Ω resistor is removed from the circuit as shown in *Figure 28.18(b)*.

(ii) Current $I_1 = \dfrac{E_1 - E_2}{r_1 + r_2} = \dfrac{4-2}{2+1} = \dfrac{2}{3}$A.

P.d. across AB, $E = E_1 - I_1 r_1 = 4 - \dfrac{2}{3}(2) = 2\dfrac{2}{3}$ V

$\left(\text{Alternatively, p.d. across AB, } E = E_2 - I_1 r_2 = 2 - \left(-\dfrac{2}{3}\right) = 2\dfrac{2}{3}\text{V}\right).$ (1)

(iii) Removing the sources of emf gives the circuit shown in *Figure 28.18(c)*, from which resistance $r = \dfrac{2 \times 1}{2+1} = \dfrac{2}{3}$Ω.

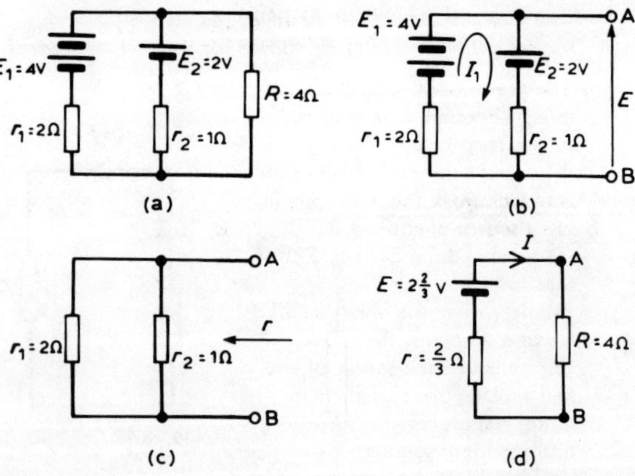

Figure 28.18

(iv) The equivalent Thévenin's circuit is shown in *Figure 28.18(d)* from which current

$$I=\frac{E}{r+R}=\frac{2\frac{2}{3}}{\frac{2}{3}+4}=\frac{8/3}{14/3}=\frac{8}{14}=0.571 \text{ A}.$$

(c) Thévenin's theorem can be used to analyse part of a circuit, and in complicated networks the principle of replacing the supply by a constant voltage source in series with a resistance is very useful.

7 (a) **Norton's theorem** states:

'The current that flows in any branch of a network is the same as that which would flow in the branch if it was connected across a source of electricity, the short-circuit current of which is equal to the current that would flow in a short-circuit across the branch, and the internal resistance of which is equal to the resistance which appears across the open-circuited branch terminals.'

(b) The procedure adopted when using Norton's theorem is summarised below. To determine the current in any branch AB of an active network:

(i) short-circuit that branch,
(ii) determine the short-circuit current, I_{SC},
(iii) remove each source of emf and replace them by their internal resistances (or, if a current source exists replace with an open circuit), then determine the resistance, R, "looking-in" at a break made between A and B,
(iv) determine the value of the current from the equivalent circuit shown in *Figure 28.19*,

$$\text{i.e. } I=\left(\frac{R}{R+R_L}\right)I_{SC}.$$

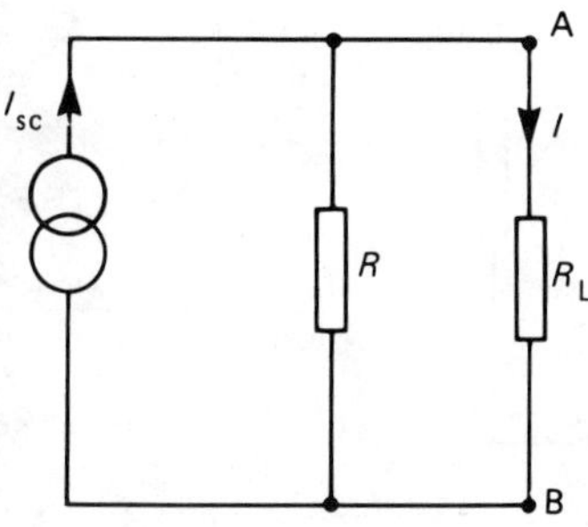

Figure 28.19

For example, to determine the current flowing in the 4Ω resistor of *Figure 28.18(a)* using Norton's theorem by the above procedure:

(i) the branch containing the 4Ω resistor is short-circuited as shown in *Figure 28.20*.

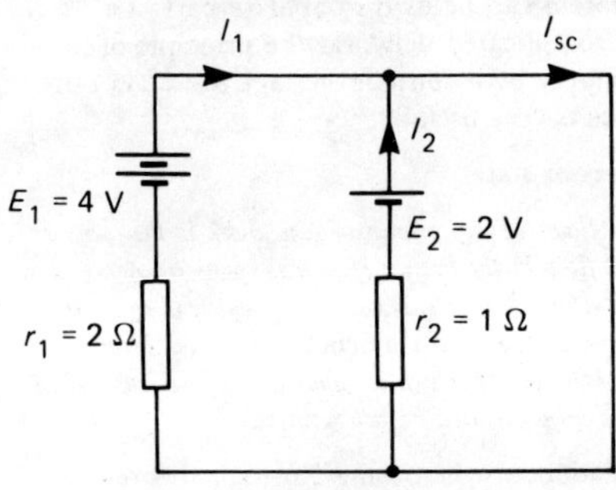

Figure 28.20

(ii) the short-circuit current I_{sc} is given by:

$$I_{SC} = I_1 + I_2 = \frac{4}{2} + \frac{2}{1} = 2 + 2 = 4 \text{ A}.$$

(iii) resistance $R = \frac{2}{3}\Omega$ (same as procedure (iii) of para. 6)

(iv) from the equivalent Norton circuit shown in *Figure 28.21*.

$$\text{current } I = \left(\frac{R}{R+R_L}\right) I_{sc} = \left(\frac{\frac{2}{3}}{\frac{2}{3}+4}\right)(4) = 0.571 \text{ A}.$$

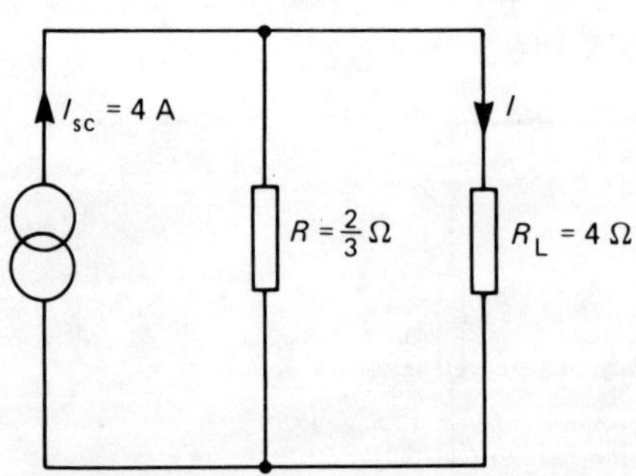

Figure 28.21

8 A Thévenin equivalent circuit having emf E and internal resistance r can be replaced by a Norton equivalent circuit containing a current generator I_{sc} and internal resistance R, where:

$$R = r,\ E = I_{sc}R \text{ and } I_{sc} = \frac{E}{r}.$$

Thus,

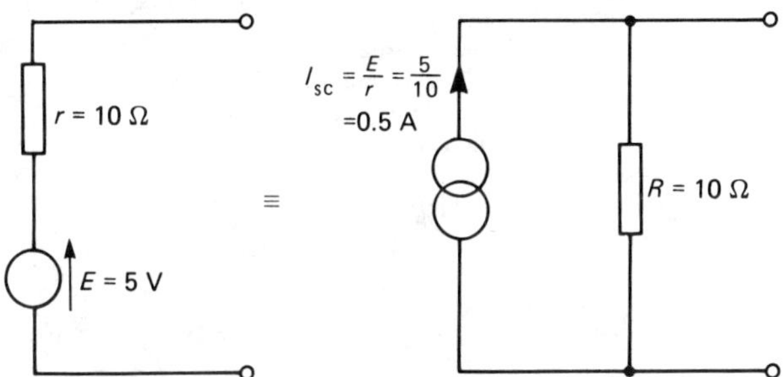

Figure 28.22

and

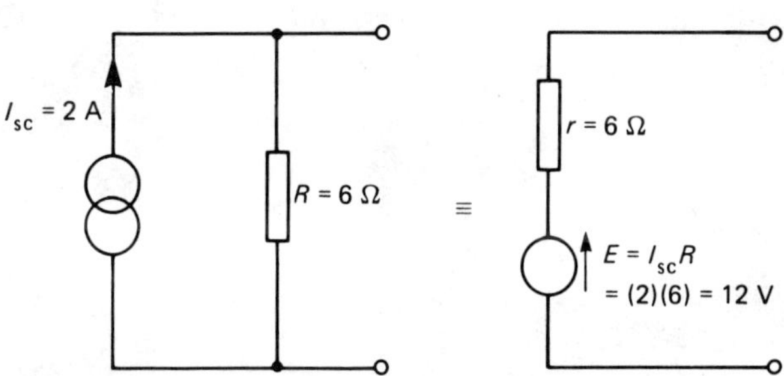

Figure 28.23

9 (a) **The maximum power transfer theorem** states:

'The power transferred from a supply source to a load is at its maximum when the resistance of the load is equal to the internal resistance of the source.'

Hence, in *Figure 28.24*, when $R = r$ the power transferred from the source to the load is a maximum.

(b) Varying a load resistance to be equal, or almost equal; to the source internal resistance is called **resistance matching**. Examples where resistance matching is important include coupling an aerial to a transmitter or receiver, or in coupling a loudspeaker to an amplifier where coupling transformers may be used to give maximum power transfer.

With d.c. generators or secondary cells, the internal resistance is usually very small. In such cases, if an attempt is made to make the load resistance as small as the source internal resistance, overloading of the source results.

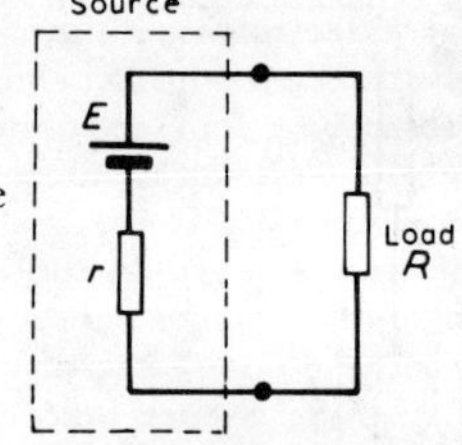

Figure 28.24

29 Electrolysis

1 **Electrolysis** is the name given to the **chemical changes which occur** when an **electric current is passed through an electrolyte**. An **electrolyte** is defined as a substance which in **its molten state** or **in solution** with water **can conduct a current of electricity**. The resistivity of electrolytes vary depending upon their composition. **Strong electrolytes** which **ionise completely** in aqueous solution have a value of approximately 10^{-1} ohms-metre and **poor electrolytes** 10^3 ohms-metre.

2 In the **molten state** the ions which make up the electrolyte are **relatively free moving ions** which can act as **electron carriers**. For example,

$$KCl(s) \xrightarrow[\text{molten}]{\text{heat until}} K^+(l) + Cl^-(l)$$

In the **solution state** the ions are separated from each other and become **hydrated**, the **freely moving aqueous ions** can then act as **electron carriers**. For example,

$$NaCl(s) \xrightarrow[\text{in water}]{\text{dissolve}} Na^+(aq) + Cl^-(aq)$$

In solution, water is always present, and exists in the **equilibrium state** as

$$2H_2O(l) \rightleftharpoons H_3^+O(aq) + OH^-(aq)$$

Although in pure water only **one molecule in ten million** is split up into ions, by disturbing the equilibrium, both **$H_3^+O(aq)$** and **$OH^-(aq)$** are very **important entities** in electrolysis reactions.

3 Examples of electrolytes include the salts formed by reaction between an acid and a base, and between acids and alkaline solutions, all of which contain hydrated ions in solution with water.

4 A typical arrangement of an apparatus for performing electrolysis in the laboratory is shown in *Figure 29.1*. This arrangement of apparatus is called an **electrolysis circuit**.

5 **The electrolysis circuit** The circuit shown in *Figure 29.1* is composed of the following components.

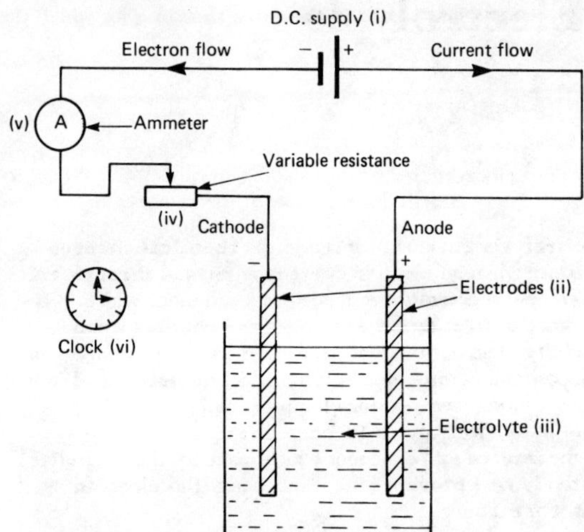

Figure 29.1 An electrolysis circuit

(i) **A supply of direct current**. This is important because the flow of electrons through the cell must take place in a **constant direction** for the chemical reactions to take place.
(ii) **The electrodes**. The electrode which **supplies electrons** to the electrolyte is called the **cathode** because it attracts the **cations** (positively charged ions) of the electrolyte. The electrode which **removes electrons** from the electrolyte is called the **anode** because it attracts the **anions** (negatively charged ions) of the electrolyte. The electrodes which **take no part** in the electrolysis reactions are usually **carbon** or **platinum**. For other purposes **dissolving anodes** can be used which are the **same metal** as that which is contained in the electrolyte; for example, copper anodes in copper (II) sulphate solution.
(iii) **The electrolyte**. The products of electrolysis depend to a certain extent on the **concentration of the electrolyte** in solution.
(iv) **The variable resistance**. The resistance of the circuit is **constantly changing** and the **variable resistance** is needed to keep the d.c. current a **constant value**.
(v) **The ammeter**. This instrument is used to monitor the current which is passed through the electrolyte.

vi) **The clock**. This is used to **measure the time** for which the current passes through the electrolyte.

Electrode reactions

6 (i) **Cathode reactions** The cathode in electrolysis is **negatively charged** and attracts cations towards itself. When these **positively charged ions** arrive at the cathode, electrons are supplied to the cation which can **most readily accept the electrons**. If only one cation is present in the electrolyte the electrode reaction is straightforward and can be summarised as

$$\text{Cation}^{+n} + ne^- = \text{Element liberated.}$$

If more than one cation is present then **one of them will be liberated with preference to the other**. This is called **selective discharge of an ion**. The ion most likely to be liberated is the cation which **occurs lowest in the electrode potential series** given in part in *Table 29.1*.

Table 29.1 Standard electrode potentials (V)

Cations			*Anions*		
Ion	Element	$E^\ominus$	Ion	Element	$E^\ominus$
K^+	K	−2.92	OH^-	O_2	+0.40
Ca^{2+}	Ca	−2.87	I^-	I_2	+0.54
Na^+	Na	−2.81	Br^-	Br_2	+1.07
Al^{3+}	Al	−1.66	Cl^-	Cl_2	+1.36
Zn^{2+}	Zn	−0.76	SO_4^{-2}	None	+2.01
Fe^{2+}	Fe	−0.44			
Pb^{2+}	Pb	−0.13	F^-	F_2	+2.87
H_3^+O	H_2	0			
Cu^{2+}	Cu	+0.34			
Ag^+	Ag	+0.80			

For example, if a solution contains hydrated sodium ions and hydrated hydrogen ions, because **H_3^+O is below Na^+** in the series, the **hydrogen ion** undergoes **selective discharge** at the cathode.
(ii) **Anode reactions** The anode in electrolysis is **positively charged** and attracts anions towards itself. When these **negatively charged ions** arrive at the anode, electrons are released by the anion which can **most easily lose electrons** to the anode. If

only one anion is present the electrode reaction can be summarised as

$$\text{Anion}^{-n} = \text{Element} + ne^-$$

If the electrolyte contains more than one anion the ion which is **selectively discharged** is the ion which gives up electrons most easily and is the anion which is **highest in the electrode potential series** (*Table 29.1*). For example, in a solution which contains hydrated hydroxide ions and hydrated sulphate ions, the hydroxide ions are selectively discharged to liberate oxygen gas.
(iii) The simple rules given above can be applied in most cases, but other factors may have to be taken into consideration in certain cases which are outside the scope of this book.

The electrolysis of dilute solutions of acids, alkalis and salts

7 An apparatus which can be used to investigate the products of the electrolysis of dilute solutions is the **Hofman voltameter** shown in *Figure 29.2*.

The **gas burettes** can be used to measure the **relative volumes of gases** liberated during electrolysis. Some examples of these electrolysis reactions are:

(i) Electrolysis of dilute sulphuric acid using platinum electrodes

The **products of electrolysis** are best considered as the reactions occurring on the cathode and the anode which can be set out in the following way:

Cathode reactions

Cations $H_3^+O(aq)$

$H_3^+O(aq) + 1e^- = H(g) + H_2O(l)$
atomic hydrogen

$H(g) + H(g) = H_2(g)$
molecular hydrogen evolved

Anode reactions

Anions $OH^-(aq)$ $SO_4^{-2}(aq)$

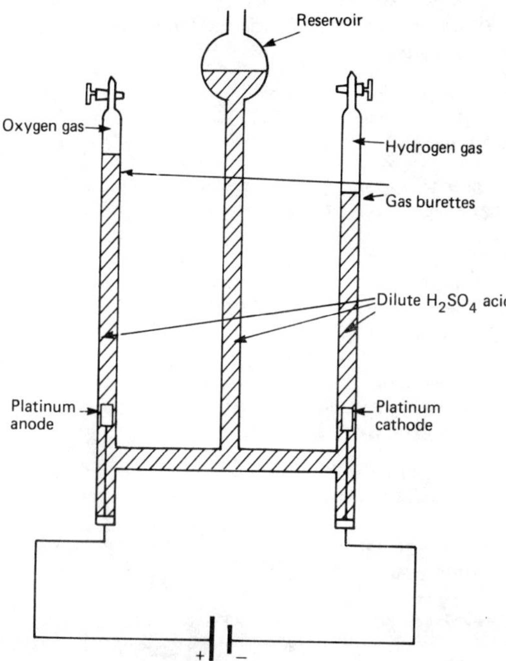

Figure 29.2 A voltameter

Selective discharge

$$OH^-(aq) = OH(aq) + 1e^-$$

hydroxide radicals

$$4OH(aq) = O_2(g) + 2H_2O(l)$$

molecular oxygen evolved

The reaction at the anode produces 4 electrons for every molecule of oxygen, and at the cathode the 4 electrons must produce 2 molecules of hydrogen. Hence **the volume of hydrogen evolved is double the volume of oxygen**.

(ii) Dilute sodium hydroxide using platinum electrodes

Cathode reaction

Cations $H_3^+O(aq)$ $Na^+(aq)$

Selective discharge

$H_3^+O(aq) + 1e^- = H(g) + H_2O(l)$

atomic hydrogen

$H(g) + H(g) = H_2(g)$

molecular hydrogen evolved

Anode reaction

Anions $OH^-(aq)$

$OH^-(aq) = OH(aq) + 1e^-$

hydroxide radicals

$4OH(aq) = O_2(g) + 2H_2O(l)$

molecular oxygen evolved

(iii) Dilute sodium chloride using platinum electrodes

Cathode reaction

Cations $H_3^+O(aq)$ $Na^+(aq)$

Selective discharge

$H_3^+O(aq) + 1e^- = H(g) + H_2O(l)$

atomic hydrogen

$H(g) + H(g) = H_2(g)$

molecular hydrogen evolved

Anode reaction

Anions $OH^-(aq)$ $Cl^-(aq)$

Selective discharge

$OH^-(aq) = OH(aq) + le^-$

hydroxide radicals

$4PH(aq) = O_2(g) + 2H_2O(l)$

molecular
oxygen
evolved

(iv) Dilute copper sulphate using carbon electrodes

Cathode reaction

Cations $H_3^+O(aq)$ $Cu^{+2}(aq)$

Selective discharge

$Cu^{+2}(aq) + 2e^- = Cu(s)$

metallic copper

Anode reaction

Anions $OH^-(aq)$ $SO_4^{-2}(aq)$

Selective discharge

$OH^-(aq) = OH(aq) + 1e^-$

hydroxide radical

$4OH^-(aq) = O_2(g) + 2H_2O(1)$

molecular
oxygen
evolved

The electrolysis of molten salts and concentrated solutions

3 (i) Molten sodium chloride using carbon electrodes

Cathode reaction

Cation $Na^+(l)$

$Na^+(l) + le^- = Na(l)$

molten sodium metal

Anode reaction

Anion $Cl^-(l)$

$Cl^-(l) = Cl(g) + 1e^-$

atomic chlorine

$Cl(g) + Cl(g) = Cl_2(g)$

molecular chlorine evolved

(ii) **Concentrated hydrochloric acid using carbon electrodes**

Cathode reaction

Cation $H_3^+O(aq)$

$H_3^+O(aq) + 1e^- = H(g) + H_2O(l)$

atomic hydrogen

$H(g) + H(g) = H_2(g)$

molecular hydrogen evolved

Anode reaction

Anions $OH^-(aq)$ $Cl^-(aq)$

Selective discharge overcome by concentration effects

$Cl^-(aq) = Cl(g) + 1e^-$

atomic chlorine

$Cl(g) + Cl(g) = Cl_2(g)$

molecular chlorine evolved

Electrolysis using a dissolving anode (electroplating) Copper (II) sulphate solution using copper electrodes

9 **Cathode reaction**

Cations $Cu^{2+}(aq)$ $H_3^+O(aq)$

$Cu^{2+}(aq) + 2e^- = Cu(s)$

copper metal deposited on the cathode.

Anode reaction
$Cu(s) = Cu^{2+}(aq) + 2e^-$
copper anode dissolves maintaining the concentration of the $CuSO_4$ solution.

Industrial applications of electrolysis

10 (i) **Extraction of metals** The **group I**, **group II** and certain other metals are **too reactive** with the elements of the environment to be extracted from their naturally occurring ores by **reduction** methods. In these cases electrolysis is used to extract and refine the metal; for example, the extraction of aluminium from its naturally occurring ores of bauxite, $Al_2O_3 . 2H_2O$ and cryolite Na_3AlF_6. A diagram of the electrolysis cell used for the extraction is shown in *Figure 29.3.*

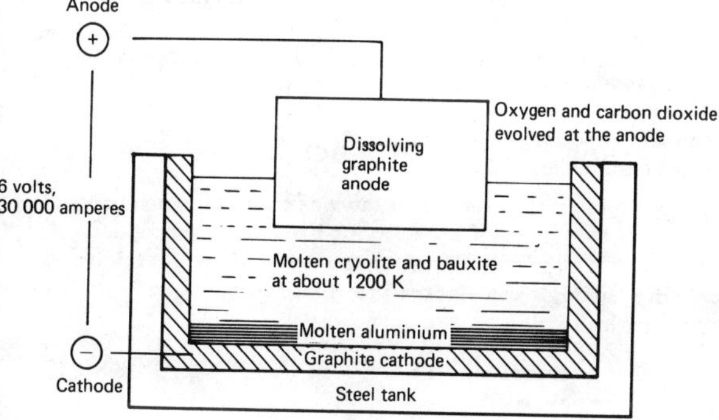

Figure 29.3 The extraction of aluminium by electrolysis

(ii) **Refining of metals** Copper metal can be obtained in an impure state fairly easily. By making this **impure copper the anode** of an electrolysis cell and making **pure copper metal the cathode**, in a copper (II) sulphate solution as electrolyte, the copper is obtained at about a 99.95% purity. A typical arrangement is shown in *Figure 29.4*.
(iii) **Electroplating** Electroplating can be used for decorative effects, for example, **silver plating**, or for protective effects, for example, **nickel** or **chromium plating**. The development of

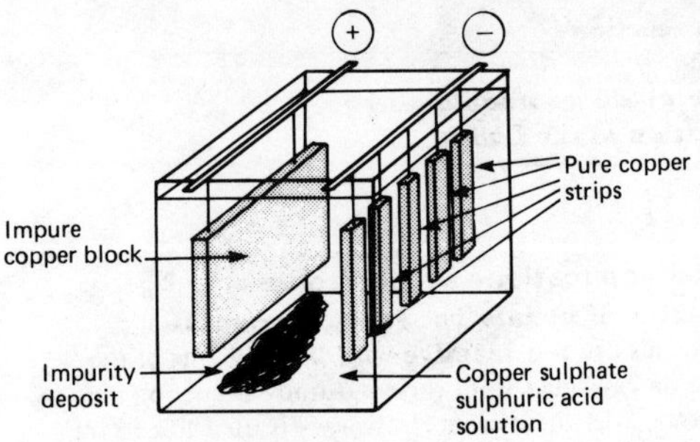

Figure 29.4 The referring of copper

electrolysis which allow thin stable films of one metal to adhere to another metal has been given a great deal of attention. Mixtures of electrolytes are usually used to give the best results; for example, for the silver plating electrolysis cell similar to that shown in *Figure 29.4*, the solution contains AgCN, KCN, K_2CO_3, Ag and CS_2.

Faraday's laws of electrolysis

11 **Faraday** put forward **two laws of electrolysis**. The first law states that **the mass of an element liberated during electrolysis is proportional to the amount of electricity passing through the electrolyte**. This can be stated mathematically as:

Mass deposited, (M) $\propto$ current, (I) $\times$ time, (t)
or **$M \propto It$.**

The second law states that **the masses of various elements liberated by the passage of a constant quantity of electricity through different electrolytes are proportional to the chemical equivalents of the ions undergoing reaction**. Expressing this as an equation:

Mass deposited, $M \propto E$ where

$$E = \frac{\text{the mass of 1 mole of the ion}}{\text{the charge on that ion}}$$

For example, when a quantity of electricity is passed through the series of electrolysis cells as shown in *Figure 29.5*.

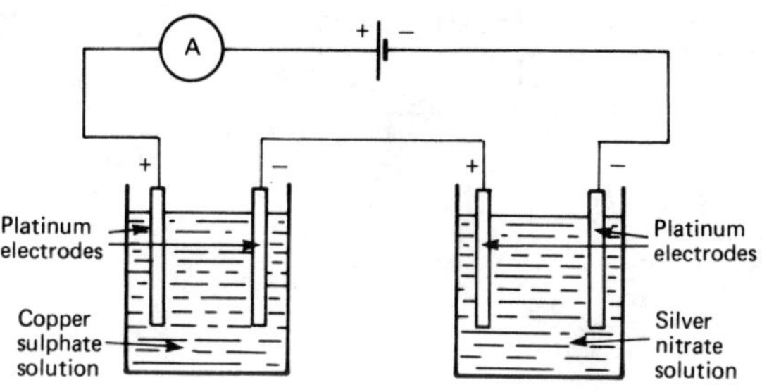

Figure 29.5

The mass of silver deposited is $\propto \frac{108}{1}$ and

The mass of copper deposited is $\propto \frac{63.5}{2}$

and hence silver and copper will be deposited in the ratio

$$\frac{108}{1} : \frac{63.5}{2} \text{ or } 3.4:1.$$

12 **The Faraday constant** The combination of the two laws

M$\propto$It and **M$\propto$E**

gives the relationship $M \propto ItE$ or **M=(constant) ItE**
To deposit **1 mole of an ion with a charge of +1**, the constant was found to be **1 Faraday**, for **1 mole of an ion of charge +2**, **2 Faradays** and **1 mole of an ion charge +3**, **3 Faradays**. This allowed the constant and **E** to be combined into a single constant called the **electrochemical equivalent**, **z** and the combined law stated as $M = zIt$, where **z is the mass liberated by 1 ampere in 1 second**. Some values for **z** are given in *Table 29.2*.

Faraday's laws and the mole concept

13 The **Faraday** is the **quantity of electricity** which can supply **1 mole of electrons** to a conductor. The quantity of electricity which corresponds to **1 ampere** being passed for **1 second** is **1 coulomb**, and the **number of coulombs** in **1 Faraday is 96500**. This means that **1 mole of electrons is equivalent to 96500 coulombs**. Thus, if the number of coulombs

Table 29.2 Some electrochemical equivalents

Element	z kg A^{-1} s^{-1}
Silver	1118.3×10^{-9}
Chlorine	376.6×10^{-9}
Hydrogen	10.45×10^{-9}
Potassium	40.53×10^{-9}
Sodium	238.4×10^{-9}
Copper	329.5×10^{-9}
Zinc	338.8×10^{-9}
Oxygen	82.9×10^{-9}
Aluminium	93.6×10^{-9}

of electricity is known, the amount of an element liberated can be found. For example, if 10 amperes is passed for 1 minute through a solution of silver nitrate the quantity of electricity is $10 \times 1 \times 60 = 600$ coulombs. Silver ions have a charge of $+1$, and 1 mole of silver, that is 108 g would be liberated by 96500 coulombs.

$$\text{Mass of silver will be } 108 \times \frac{600}{96500} = 0.67 \text{ g.}$$

The use of this concept is often more useful than applying Faraday's laws to calculate the masses of elements liberated during electrolysis.

30 **Electrode potentials, cells and corrosion**

1 When a metal is immersed in a solution of its most stable ions, there are two possible reactions which may occur. These are: (a) the **metal ions** may **leave the surface** of the metal lattice and **go into solution**, resulting in the development of a **negative charge on the metal surface** (see *Figure 30.1(a)*), or (b) the **metal ions** from the solution may **be deposited on the metal** surface, resulting in the development of a **positive charge on the metal surface** (see *Figure 30.1(b)*). Both *(a)* and *(b)* result in a **potential difference** existing between the metal and the solution.

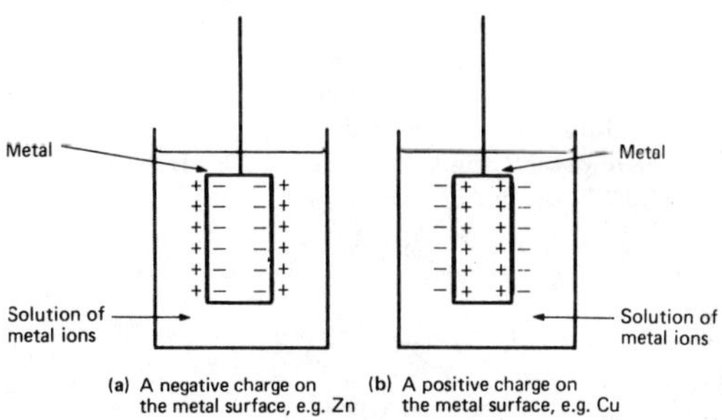

Figure 30.1

2 The reactions can be expressed by the general equation

$$M \rightleftharpoons M^{n+} + ne^-$$

where M is any metal and n is the oxidation state of the metal ion When equilibrium is displaced on the right, the reaction corresponds to reaction (a) described in para 1. Displacement of the reaction to the left corresponds to reaction (b) described in para 1.

3 Any metal in contact with a solution of its ions is called a **half-cell**, and the potential difference of the half-cell is called the

electrode potential of the half-cell. The symbol denoting the electrode potential is E, and the quantity is measured in volts.

4 The electrode potential of a half-cell cannot be determined independently, but is measured by **placing it in series** with a different half-cell. These are joined together by a **salt bridge**, and a voltmeter is connected across the complete circuit to measure the voltage produced.

5 The electrode potential of a half-cell is dependent on **three factors**. These are:

(a) the **concentration** of the solution of the ions;
(b) the **absolute temperature** of the solution; and
(c) the **oxidation number** of the metal ions in solution.

Because of this dependence, a set of **standard conditions** have been selected for use in the measurement of electrode potentials. These are:

(a) the solutions of ions are exactly **one molar in concentration**; and
(b) the measurements are made at **298 K**.

When measured under these conditions the values become **standard electrode potentials** and are given the symbol $E^{\ominus}$, the units of which are also in volts.

6 Since the electrode potentials of half-cells can only be determined when placed in series with a known potential half-cell, **a standard electrode** has been selected for this purpose. The **standard reference electrode** is the **standard hydrogen electrode** which is shown in *Figure 30.2*.

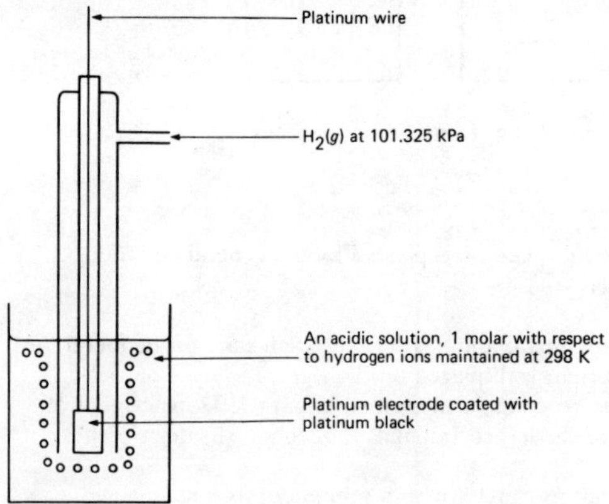

Figure 30.2

The electrode consists of a piece of platinum foil coated with platinum black embedded in the walls of a glass tube. This foil is connected by means of a platinum wire to the external lead. The **platinum black** material is **very finely divided platinum** and it **catalyses** the reaction which occurs between hydrogen gas and hydrogen ions as shown in the equation

$$\tfrac{1}{2}\mathbf{H_2(g)} \rightleftharpoons \mathbf{H^+(aq)} + \mathbf{e^-}$$

The hydrogen gas is bubbled through the electrode at **a pressure of 101.325 kPa**, and the concentration of the acid solution must be exactly **1.00 molar strength** with respect to hydrogen ions.

During experimental work to determine standard electrode potentials, the acid solution must be thermostatically controlled at 298 K. By mixing together the hydrogen gas and the hydrogen ions in the presence of the platinum electrode, the electrode potential is developed between the element (hydrogen), and a solution of its ions (hydrogen ions). When the electrode is set up in this way, the **standard electrode potential which is developed is defined as zero**.

7 The standard hydrogen electrode requires considerable time and care to set up, and it is difficult to move from one experiment to another. Consequently, **alternative reference electrodes** have been devised, an example being the **calomel reference electrode** which is shown in *Figure 30.3*.

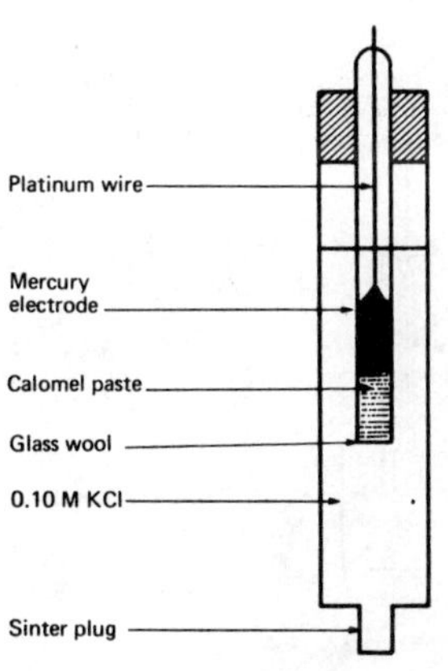

Figure 30.3

In this standard electrode, the platinum electrode provides the contact between the mercury and the calomel paste, which contains mercury (I) chloride, Hg_2Cl_2 as the source of Hg^+ ions. The concentration of the mercury (I) chloride is maintained by placing it in contact with 0.1 molar potassium chloride solution. The electrode potential of this electrode, with respect to the standard hydrogen electrode, is **+0.336 V**. Thus when the standard calomel electrode is used, +0.336 V must be added to any standard electrode potential which is measured.

8 When the standard hydrogen electrode is connected in series to another standard half-cell as shown in *Figure 30.4* the **direction of the current flow** is dependent on the particular metal/metal

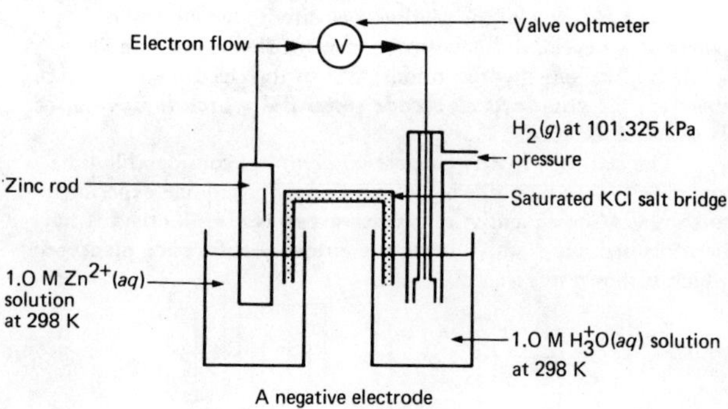

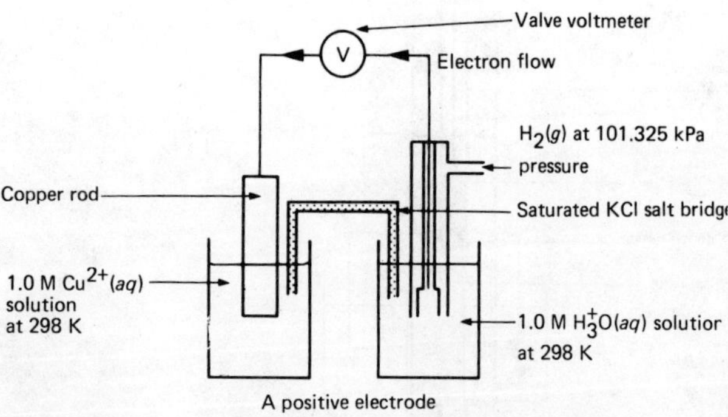

Figure 30.4

ion electrode system which is used. The convention which has been adopted is that, **metals which release hydrogen gas from a solution of hydrogen ions are designated as negative (−ve) with respect to the standard electrode. Metals which do not release hydrogen gas are designated as positive (+ve) with respect to the standard electrode.** Examples of some $E^{\ominus}$ values are shown in *Table 29.10*.

9 When two different standard half-cells are connected in series in a circuit, they form a **galvanic cell** as shown in *Figure 30.5*. The resultant potential difference of the two standard half-cells is called the **standard electromotive force of** the galvanic cell, and is symbolised by $E^{\ominus}_{cell}$.

10 A galvanic cell can be represented systematically in the form of a **cell equation**. Any two half-cells can be represented by $A^{x+} | A$ **and** $B^{y+} | B$ which when forming a galvanic cell can be represented by the cell equation

$$A|A^{x+} \vdots B^{y+}|B$$

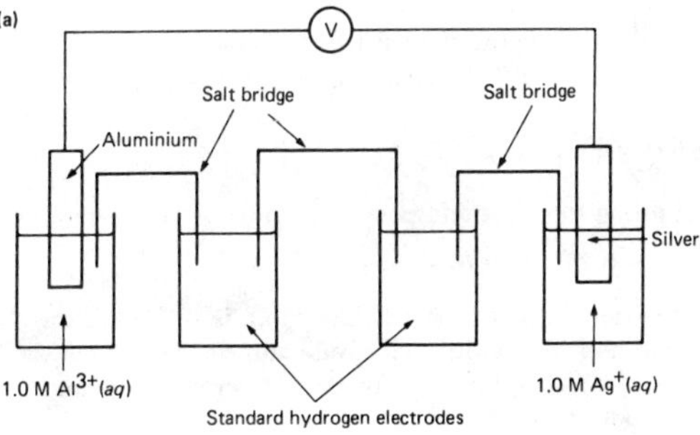

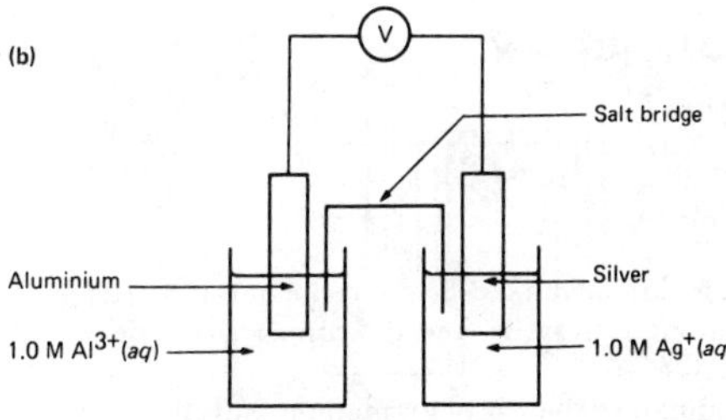

Figure 30.5

In this equation the **solid vertical line** represents the **solid electrodes** and the **dotted line** represents the **salt bridge** joining the half-cells together. The standard e.m.f. of the cell is defined as:

The standard e.m.f. of the cell $E^{\ominus}_{cell} = E^{\ominus}$ (Right Hand Electrode) $- E^{\ominus}$ (Left Hand Electrode), which is simplified to:

$$E^{\ominus}_{cell} = E^{\ominus}_{R.H.E.} - E^{\ominus}_{L.H.E.}$$

For the general cell equation given above, $A|\,A^{x+}$ is the **left hand electrode**, and $B^{y+}\,|\,B$ is the **right hand electrode**. The **polarity of the right hand electrode is defined as being the same sign as the e.m.f. of the galvanic cell**, i.e. if the $E^{\ominus}_{cell}$ value is positive, then the right hand electrode has positive polarity. For example, when the zinc half-cell and the hydrogen half-cell are connected together to form a galvanic cell as shown in *Figure 30.4* the cell equation is written as

$$\mathrm{Zn}|\,\mathrm{Zn^{2+}}\,(\mathrm{aq}) \;\vdots\; \mathrm{H^{+}}\,(\mathrm{aq})|\mathrm{H_2},\ \mathrm{Pt}$$

LHE – 0.76 V RHE 0 V

where LHE and RHE signify the left hand electrode and the right hand electrode respectively. The e.m.f. of this cell is calculated using the equation

$$E^{\ominus}_{cell} = E^{\ominus}_{RHE} - E^{\ominus}_{LHE} = E^{\ominus}_{H_2} - E^{\ominus}_{Zn}$$

and by substituting the electrode potential values gives

$$E^{\ominus}_{cell} = 0 - (-0.76) = +0.76\ \mathrm{V}$$

This means that for this galvanic cell, the e.m.f. is +0.76 V, the hydrogen electrode is the positive electrode and the zinc electrode is the negative electrode. Similarly, for the copper electrode and hydrogen electrode the cell equation is

$$\mathrm{Cu}|\,\mathrm{Cu^{2+}}\,(\mathrm{aq}) \;|\; \mathrm{H^{+}}\,(\mathrm{aq})|\,\mathrm{H_2Pt}$$

LHE + 0.34 V RHE 0 V

and the *e.m.f.* is given by

$$E^{\ominus}_{cell} = E^{\ominus}_{RHE} - E^{\ominus}_{LHE} = E^{\ominus}_{H_s} - E^{\ominus}_{Cu} =$$

$$0 - 0.34 = -0.34\ \mathrm{V}$$

This means that for this galvanic cell, the cell e.m.f. is –0.34 V, the hydrogen electrode is negative and the copper electrode is positive.

11 When the standard half-cells of aluminium and silver are combined into a galvanic cell using hydrogen electrodes as shown

in *Figure 30.5(a)*. The cell equation for this arrangement is

$$\mathbf{Al|Al^{3+}(aq)\vdots H^{+}(aq)|H_2, Pt \quad Pt, H_2|H^{+}(aq)\vdots Ag^{+}(aq)|Ag}$$

From *Table 29.1* the $E^{\ominus}$ values for the aluminium and silver arc -1.67 V and $+0.80$ V respectively. The circuit in *Figure 30.5(a)* is a combination of the two galvanic cells in series which can be represented by the cell equations

(i) $\mathbf{Al|Al^{3+}(aq)\vdots H^{+}(aq)|H_2Pt,}$
$E^{\ominus}_{cell} = E^{\ominus}_{H_2} - E^{\ominus}_{Al} = \mathbf{0-(-1.67)=+1.67\ V}$

and

(ii) $\mathbf{Pt,H_2|H^{+}(aq)\vdots Ag^{+}(aq)|Ag.}$
$E^{\ominus}_{cell} = E^{\ominus}_{Ag} - E^{\ominus}_{H_2} = \mathbf{+0.80-0=+0.80\ V}$

The total *e.m.f.* of the two galvanic cells in series is given by

$E^{\ominus}_{Total} = E^{\ominus}_{cell\ (i)} + E^{\ominus}_{cell\ (ii)}$ or

$E^{\ominus}_{Total} = +1.67\ V + 0.80\ V + 2.47\ V.$

The *e.m.f.* produced is $+2.47$ V.

Similarly, the cell equation for the circuit shown in (b) is

$$Al|Al^{3+}(aq)\vdots Ag^{+}(aq)|Ag$$

Applying the equation

$$E^{\ominus}_{cell} = E^{\ominus}_{RHE} - E^{\ominus}_{LHE}$$

gives

$$E^{\ominus}_{cell} = E^{\ominus}_{Ag} - E^{\ominus}_{Al} = 0.80 - (1.67)\ V = 2.47\ V.$$

The e.m.f. of the cell is $+2.47$ V.

From the results it can be seen that when a galvanic cell is formed by two half-cells, it is unnecessary to consider the standard hydrogen electrodes when calculating values of e.m.f. It should be noted that when drawing a diagram of the circuit or writing a cell equation for a galvanic cell, it is usual by convention to select the more positive electrode as the right hand electrode.

12 The list of values of standard electrode potentials shown in *Table 29.1* with the most negative at the top, is also called the **electrochemical series**. In the table, any element above another can theoretically be used to **displace the lower element from a solution of its ions**. For example, **magnesium is above zinc** in the table and the theoretical reaction is

$$Mg(s) + Zn^{2+}(aq) = Mg^{2+}(aq) + Zn(s)$$

Similarly, **zinc is above copper** and the equation of reaction is

$$Zn(s) + Cu^{2+}(aq) = Zn^{2+}(aq) + Cu(s)$$

Such reactions, where a metal is deposited from its solution by another metal are called **displacement reactions**.

13 A **displacement reaction** can be considered as **two reactions taking place simultaneously**. For the magnesium and zinc reaction shown above the reactions taking place are

(a) $Mg(s) \rightarrow Mg^{2+}(aq) + 2e^-$ and
(b) $Zn^{2+}(aq) + 2e^- \rightarrow Zn(s)$

Reaction (a) involves a **loss of electrons** by magnesium and this process is called **oxidation**. Reaction (b) involves a **gain in electrons** by zinc ions and this process is called **reduction**. The reactions can be called **oxidation/reduction reactions** or more commonly **redox reactions**, where **red-** is taken from **reduction** and **-ox** from **oxidation**. It is important to remember that a reduction can only take place accompanied by an oxidation, and vice versa.

14 When a galvanic cell is formed the **positive electrode undergoes reduction (gaining electrons)**, whilst the **negative electrode undergoes oxidation (losing electrons)**. Since the reaction at the positive electrode involves the attraction of positively charged ions, or cations, the electrode is called the cathode. The other electrode undergoing electron loss is called the anode.

15 **Applications of galvanic cells** The production of an e.m.f. by chemical change offers a form of portable electricity. The production of electricity in amounts suitable for domestic and commercial purposes is achieved by **arranging a series of cells into a battery**. The simple cell shown in *Figure 30.6* has two faults (a) **polarisation effects** and (b) **local action effects**.

(a) If the simple cell shown in *Figure 30.6* is left connected for some time, the current I decreases fairly rapidly. This is because of the formation of **a film of hydrogen bubbles** on the copper anode. This effect is known as the **polarization of the cell**. The hydrogen prevents full contact between the copper electrode and the electrolyte and this **increases the internal resistance** of the cell. The effect can be overcome by using a **chemical depolarizing agent** or **depolarizer**, such as potassium dichromate which removes the hydrogen bubbles as they form. This allows the cell to deliver a steady current.

(b) When commercial zinc is plated in dilute sulphuric acid, hydrogen gas is liberated from it and the zinc dissolves.

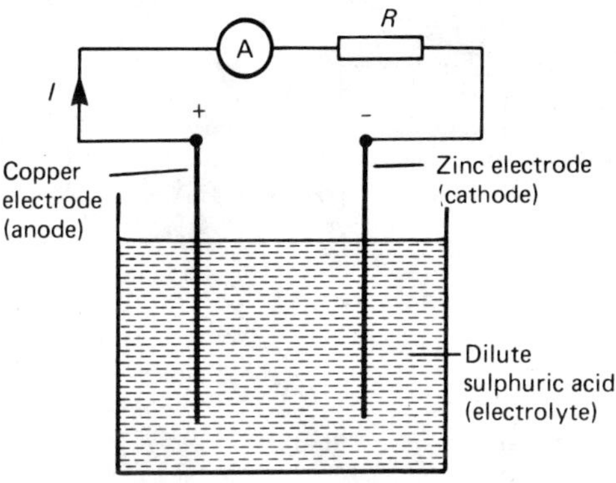

Figure 30.6

The reason for this is that impurities, such as traces of iron, are present in the zinc which set up **small primary cells** with the zinc. These small cells are short-circuited by the electrolyte, with the result that **localized currents flow causing corrosion**. This action is known as **local action** of the cell. This may be prevented by rubbing a small amount of mercury on the zinc surface, which forms a protective layer on the surface of the electrode.

16 When a galvanic cell is set up, the p.d. between its terminals when not connected to a load is the **electromotive force** of the cell. The e.m.f. of a cell is measured by using a **high resistance voltmeter** connected in parallel with the cell. The voltmeter must have a high resistance otherwise it will pass current and the cell will not be on no-load. For example, if the resistance of a cell is 1 Ω and that of a voltmeter is 1 MΩ then the equivalent resistance of the circuit is 1 MΩ+ 1 Ω, i.e. approximately 1 MΩ, hence no current flows and the cell is not loaded.

17 The voltage available at the terminals of a cell falls when a load is connected. This is caused by the **internal resistance of the cell** which is the opposition of the material of the cell to the flow of current. The internal resistance acts in series with other resistances in the circuit. *Figure 30.7* shows a cell of e.m.f. E volts and internal resistance r, XY represents the terminals of the cell.

When a load (shown as resistance R) is not connected, no current flows and the terminal p.d., $V=E$. When R is connected a

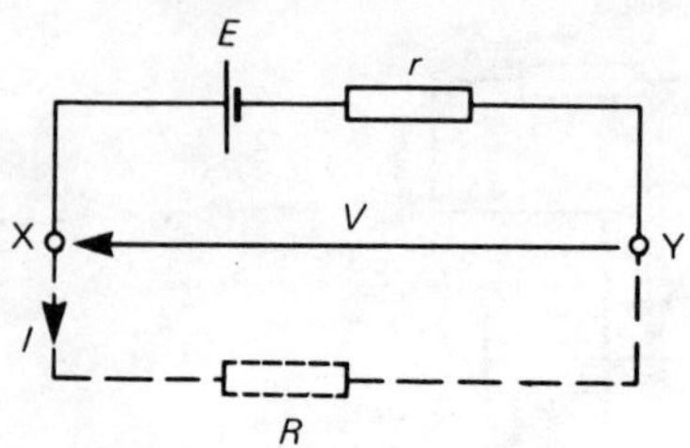

Figure 30.7

current I flows which causes a voltage drop in the cell, given by Ir. The p.d. available at the cell terminals is less than the e.m.f. of the cell and is given by $V=E-Ir$. Thus if a battery of *e.m.f.* 12 V and internal resiatance 0.01 Ω delivers a current of 100 A, the terminal p.d.,

$$V=12-(100)(0.01)=12-1=11 \text{ V}.$$

When a current is flowing in the direction shown in *Figure 30.7* the cell is said to be discharging ($E>V$), and when a current flows in the opposite direction to that shown in *Figure 30.7* the cell is said tc be charging ($V<E$).

18 A **battery** is a combination of more than one cell. The cells in a battery may be connected in series or in parallel.

(i) **For cells connected in series:**
Total e.m.f. = sum of cell's e.m.f.'s
Total internal resistance
= sum of cell's internal resistances.

(ii) **For cells connected in parallel:**
If each cell has the same e.m.f. and internal resistance:
Total e.m.f. = e.m.f. of one cell.
Total internal resistance of n cells

$$=\frac{1}{n}\times \text{internal resistance of one cell.}$$

19 There are two main types of cell — primary cells and secondary cells.

(i) **Primary cells** cannot be recharged, that is, the conversion of chemical energy is irrefersible and the cell cannot be used once the chemicals are exhausted. Examples of primary cells include the **Leclanché** cell and the **mercury cell**. A typical **dry Leclanché** cell is shown in *Figure 30.8*. Such a cell has an e.m.f. of about 1.5 V when new, but this falls rapidly if in continuous use due to

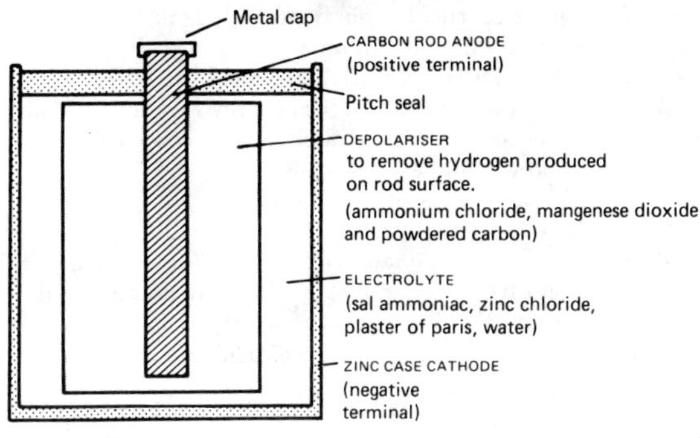

Figure 30.8

polarisation. The hydrogen film on the carbon electrode forms faster than can be dissipated by the depolariser.

The Leclanché cell is suitable only for intermittent use, applications including torches, transistor radios, bells, indicator circuits, gas lighters, controlled switchgear and so on. The cell is the most commonly used of primary cells, is cheap, requires little maintenance and has a shelf life of about two years. A typical mercury cell is shown in *Figure 30.9*. Such a cell has an e.m.f. of about 1.3 V which remains constant for a relatively long time. Its main advantages over the Leclanché cell is its smaller size and its

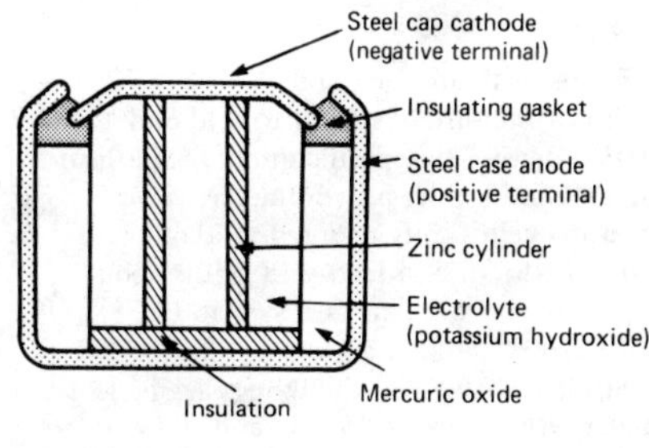

Figure 30.9

long shelf life. Typical practical applications include hearing aids, medical electronics and for guided-missiles.
(ii) **Secondary cells** can be recharged after use, that is, the conversion of chemical energy to electrical energy is reversible and the cell may be used many times. Examples of secondary cells include the lead-acid cell and alkaline cells.

A typical lead-acid cell is constructed of:
(i) A container made of glass, ebonite or plastic;
(ii) Lead plates. The negative plate (cathode) consists of spongy lead and (b) the positive plate (anode) is formed by pressing lead peroxide into the lead grid. The plates are interleaved as shown in the plan view of *Figure 30.10* to increase their effective cross-sectional area and to minimise internal resistance;

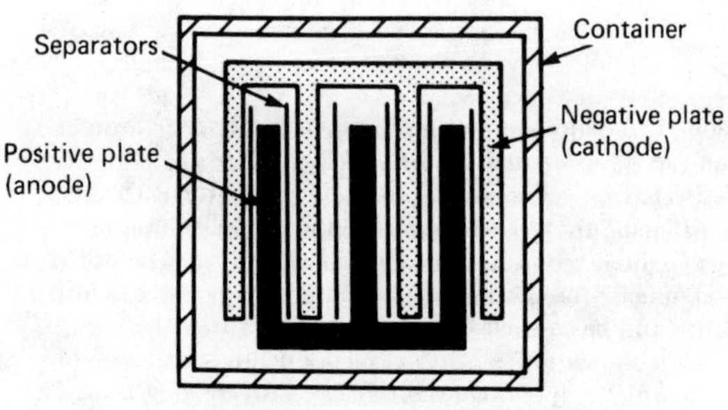

Figure 30.10

(iii) Separators made of glass, celluloid or wood;
(iv) An electrolyte which is a mixture of sulphuric acid and distilled water. The relative density (or specific gravity) of a lead-acid cell, which may be measured using a hydrometer, varies between about 1.26 when the cell is fully charged to about 1.19 when discharged. The terminal p.d. of a lead-acid cell is about 2 V. When a cell supplies current to a load it is said to be **discharging**. During discharge:
(i) The lead peroxide (positive plate) and the spongy lead (negative plate) are converted into lead sulphate, and
(ii) The oxygen in the lead peroxide combines with hydrogen in the electrolyte to form water. The electrolyte is therefore weakened

and the relative density falls. The terminal p.d. of a lead-acid cell when fully discharged is about 1.8 V.

A cell is **charged** by connecting a d.c. supply to its terminals, the positive terminal of the cell being connected to the positive terminal of the supply. The charging current flows in the reverse direction to the discharge current and the chemical action is reversed. During charging:

(i) The lead sulphate on the positive and negative plates is converted back to lead peroxide and lead respectively, and
(ii) The water content of the electrolyte decreases as the oxygen released from the electrolyte combines with the lead of the positive plate. The relative density of the electrolyte thus increases.

The colour of the positive plate when fully charged is dark brown and when discharged is light brown. The colour of the negative plate when fully charged is grey and when discharged is light grey. Practical applications of such cells include car batteries, telephone circuits and for traction purposes — such as milk delivery vans and fork lift trucks.

There are two main types of alkaline cell the nickel-iron cell and the nickel-cadmium cell. In both types the positive plate is made of nickel hydroxide enclosed in finely perforated steel tubes, the resistance being reduced by the addition of pure nickel or graphite. The tubes are assembled into nickel-steel plates.

In the **nickel-iron cell**, (sometimes called the **Edison cell** or **nife cell**), the negative plate is made of iron oxide, with the resistance being reduced by a little mercuric oxide, the whole being enclosed in perforated steel tubes and assembled in steel plates. In the nickel-cadmium cell the negative plate is made of cadmium. The electrolyte in each type of cell is a solution of potassium hydroxide which does not undergo any chemical change and thus the quantity can be reduced to a minimum. The plates are separated by insulating rods and assembled in steel containers which are then enclosed in a non-metallic crate to insulate the cells from one another. The average discharge p.d. of an alkaline cell is about 1.2 V.

Advantages of an alkaline cell (for example, a nickel-cadmium cell or a nickel-iron cell) over a lead-acid cell include:

(i) more robust construction;
(ii) capable of withstanding heavy charging and discharging currents without damage;
(iii) has a longer life;
(iv) for a given capacity is lighter in weight;

(v) can be left indefinitely in any state of charge or discharge without damage;
(vi) is not self-discharging.

Disadvantages of an alkaline cell over a lead-acid cell include:

(i) is relatively more expensive;
(ii) requires more cells for a given e.m.f.;
(iii) has a higher internal resistance;
(iv) must be kept sealed;
(v) has a lower efficiency.

Alkaline cells may be used in extremes of temperature and also in conditions where vibration is experienced or where duties require long idle periods or heavy discharge currents. Practical examples include traction and marine work, lighting in railway carriages, military portable radios and for starting diesel and petrol engines. However, the lead-acid cell is the most common one in practical use.

20 The **capacity** of a cell is measured in ampere-hours (Ah). A fully charged 50 Ah battery rated for 10 h discharge at a steady current of 5 A for 10 h, but if the load current is increased to 10 A, then the battery is discharged in 3–4 h, since the higher the discharge current, the lower is the effective capacity of the battery. Typical discharge characteristics for a lead-acid cell are shown in *Figure 30.11*.

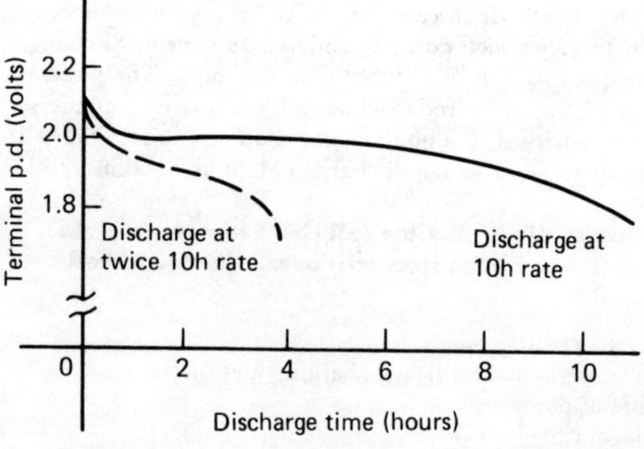

Figure 30.11

Corrosion of metals

21 When two different metals are joined together in a conducting solution, a simple cell is set up. In this simple cell, the metal highest in the electrochemical series is the anode and undergoes the decomposition reaction

$$\text{Metal(s)} \rightarrow \text{Metal ions(aq)} + xe^-$$

This change of solid metal into dissolved metal is called **corrosion**.

22 The metals used in everyday life come into contact with rain water on land, sea water at sea and oxygen from the atmosphere. Rain water contains dissolved carbon dioxide according to the equation

$$H_2O(l) + CO_2(g) \rightleftharpoons H_2CO_3(aq) \rightleftharpoons 2H^+(aq) + CO_3^{2-}(aq)$$

and in industrial areas dissolve oxides of sulphur

$$H_2O(l) + SO_3(g) \rightarrow H_2SO_4(aq) \rightarrow 2H^+(aq) + SO_4^{2-}(aq)$$

which means that **rain water can act as an electrolyte** even though it is a poor conductor. Sea water contains sodium chloride and is a good conductor of electricity. Metals in contact with these solutions in the presence of oxygen form **corrosion products**. For example, copper forms a protective layer of basic copper carbonate which is green in colour preventing further corrosion, similarly aluminium forms a surface layer of aluminium oxide and lead also forms a protective layer. These **protective layers** are effective because they are insoluble in water. However when iron or steel is corroded, the iron (III) oxide formed is **water soluble** to some extent and is granular and porous offering a continuous supply of metal to be reacted. This process is called the **rusting of iron** and is a major problem in domestic and commercial uses.

23 **The rusting of iron** Iron and steel are used extensively for purposes needing strong incompressible structures. It is a readily available material extracted from its ores by reduction with coke, a simplified equation being:

$$Fe_2O_3(s) + 3C(s) \xrightarrow{\text{heat}} 2Fe(s) + 3CO(g)$$

The production of iron and steel goods by forging, casting and other forms of machining, results in **regions of different stress** in the product. Additionally **small amounts of impurities** are present in the surface. When these areas of stress or impurity are exposed to moisture and oxygen, **rusting occurs**. The nail shown in *Figure 30.12* has **anodic regions** near the head and tip, both

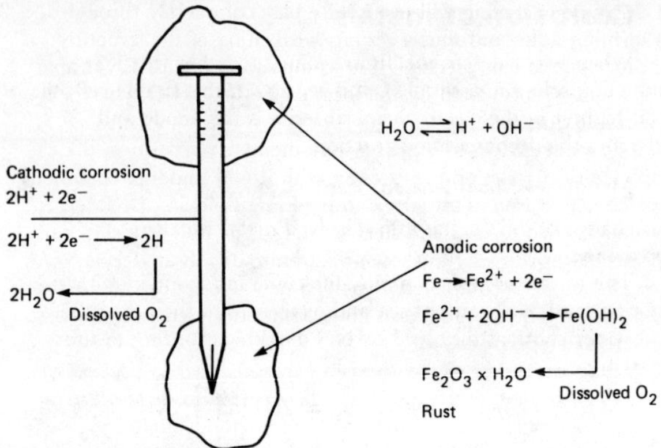

Figure 30.12 Corrosion of an iron nail

being regions of different stress originated during production.

The simplified equations of corrosion in **anodic** and **cathodic** regions are also shown in *Figure 30.12*. Eventually the head and tip will become highly corroded. A further point of interest in the rusting process is that once it begins in a particular region, it continues at that point. The problem industry faces is

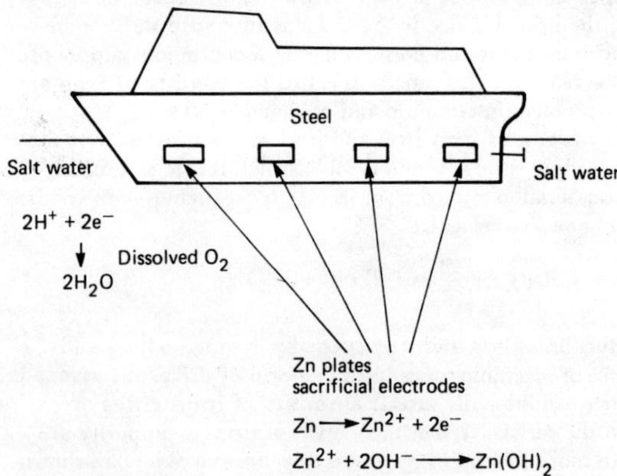

Figure 30.13 Sacrificial electrodes to reduce corrosion

that this type of rusting will eventually pass completely through the iron forming holes and thus a serious weakening of the structure. The process is called '**pitting**'. The composition of rust is variable and is usually represented by the formula $\mathbf{Fe_2O_3 . \mathit{x}H_2O}$ where x is the variable quantity.

24 **The prevention of rusting** Any metal above iron in the electrochemical series and in contact with it will undergo corrosion in preference to iron. This fact is utilised in employing **protective anodes** attached to iron and steel surfaces. The hull of a ship can be protected in this way as shown diagrammatically in *Figure 30.13.* When the anode is completely dissolved new anodes can be used to replace them. The graphic term given to anodes used in this way is '**sacrificial electrodes**'.

25 **Other methods of protection** Methods used to prevent or reduce the corrosion of metals utilise the **exclusion of moisture and oxygen** from the metal surface. Such methods include painting, immersion in oil, covering in grease, plastic coating, plating with non-corrosive metals like chrome, nickel or silver, dipping in other metals for example, galvanising and sheradising.

31 Capacitors and capacitance

1 **Electrostatics** is the branch of electricity which is concerned with the study of electrical charges at rest. An electrostatic field accompanies a static charge and this is utilised in the capacitor.

2 Charged bodies attract or repel each other depending on the nature of the charge. The rule is: **like charges repel, unlike charges attract**.

3 A **capacitor** is a device capable of storing electrical energy. *Figure 31.1* shows a capacitor consisting of a pair of parallel metal plates X and Y separated by an insulator, which could be air. Since the plates are electrical conductors each will contain a large number of mobile electrons. Since the plates are connected to a d.c. supply the electrons on plate X, which have a small negative charge, will be attracted to the positive pole of the supply and will be repelled from the negative pole of the supply on to plate Y. X will become positively charged due to its shortage of electrons whereas Y will have a negative charge due to its surplus of electrons.

The difference in charge between the plates results in a p.d. existing between them, the flow of electrons dying away and ceasing when the p.d. between the plates equals the supply voltage. The plates are then said to be **charged** and there exists an **electric field** between them.

Figure 31.2 shows a side view of the plates with the field represented by 'lines of electrical flux'. If the plates are discon-

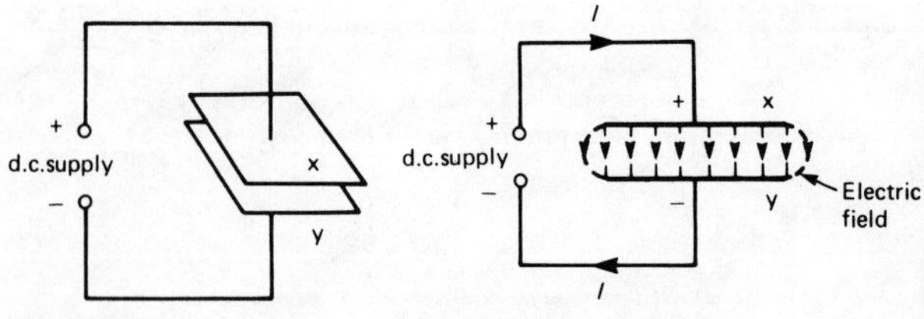

Figure 31.1 Figure 31.2

nected from the supply and connected together through a resistor the surplus of electrons on the negative plate will flow through the resistor to the positive plate. This is called **discharging**. The current flow decreases to zero as the charges on the plates reduce. The current flowing in the resistor causes it to liberate heat showing that **energy is stored in the electric field**.

4 From Chapter 27, para. 2(viii), the charge Q stored is given by:

$$\boldsymbol{Q = I \times t} \text{ coulombs}$$

where I is the current in amperes and t the time in seconds.

5 A **dielectric** is an insulating medium separating charged surfaces.

6 Electric field strength, Electric force, or voltage gradient,

$$E = \frac{\text{p.d. across dielectric}}{\text{thickness of dielectric}}, \text{ i.e. } \boldsymbol{E = \frac{V}{d}} \textbf{ volts/m.}$$

7 Charge density, $\sigma = \dfrac{\text{charge}}{\text{area of one plate}}$, i.e $\boldsymbol{\sigma = \dfrac{Q}{A}}$ $\mathbf{C/m^2}$

8 Charge Q on a capacitor is proportional to the applied voltage V,

i.e. $Q \alpha V$.

9 $Q = CV$ for $C\frac{Q}{V}$, where the constant of proportionality, C, is the **capacitance**.

10 The unit of capacitance is the **farad F** (or more usually μF $= 10^{-6}$ F or pF $= 10^{-12}$ F), which is defined as the capacitance of a capacitor when a p.d. of one volt appears across the plates when charged with one coulomb.

11 Every system of electrical conductors possess capacitance. For example, there is capacitance between the conductors of overhead transmission lines and also between the wires of a telephone cable. In these examples the capacitance is undesirable but has to be accepted, minimised or compensated for. There are other situations, such as in capacitors, where capacitance is a desirable property.

12 The ratio of charge density, σ, to electric field strength, E, is called absolute permittivity, ε, of a dielectric.

Thus $\boldsymbol{\dfrac{\sigma}{E} = \varepsilon}$.

13 Permittivity of free space is a constant, given by

$\varepsilon_0 = 8.85 \times 10^{-12}$ F/m.

14 **Relative permittivity,**

$$\varepsilon_r = \frac{\text{flux density of the field in the dielectric}}{\text{flux density of the field in vacuum}}.$$

(ε_r has no units.) Examples of the values of ε_r include: air = 1, polythene = 2.3, mica = 3–7, glass = 5–10, ceramics = 6–1000.

15 Absolute permittivity, $\varepsilon = \varepsilon_0 \varepsilon_r$.

Thus $\frac{\sigma}{E} = \varepsilon_0 \varepsilon_r$

16 For a **parallel plate capacitor**, capacitance is proportional to area A, inversely proportional to the plate spacing (or dielectric thickness) d, and depends on the nature of the dielectric and the number of plates, n.

Capacitance, $C = \frac{\varepsilon_0 \varepsilon_r A(n-1)}{d}$ F.

17 For n capacitors connected in parallel, the equivalent capacitance C_T is given by:

$C_T = C_1 + C_2 + C_3 + \ldots + C_n$ (similar to resistors connected in series)

Also total charge,

$Q_T = Q_1 + Q_2 + Q_3 + \ldots + Q_n$

18 For n capacitors connected in series, the equivalent capacitance C_T is given by:

$\frac{1}{C_T} = \frac{1}{C_1} + \frac{1}{C_2} + \frac{1}{C_3} + \ldots + \frac{1}{C_n}$ (similar to resistors connected in parallel)

The charge on each capacitor is the same when connected in series.

19 The maximum amount of field strength that a dielectric can withstand is called the **dielectric strength** of the material.

Dielectric strength, $E_{MAX} = \frac{V_{MAX}}{d}$ and $V_{MAX} = d \times E_{MAX}$.

20 The energy, W, stored by a capacitor is given by $W = \frac{1}{2} C V^2$ joules.

21 **Practical types of capacitor** are characterised by the material used for their dielectric. The main types include: variable air, mica, paper, ceramics, plastic and electrolytic.

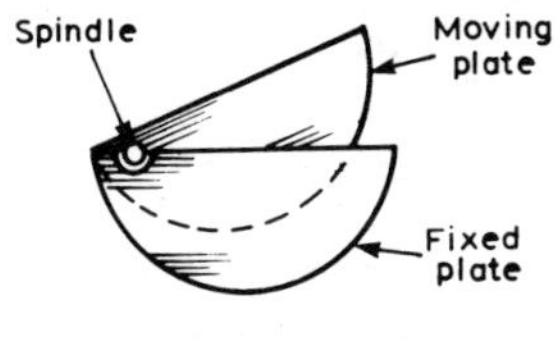

Figure 31.3

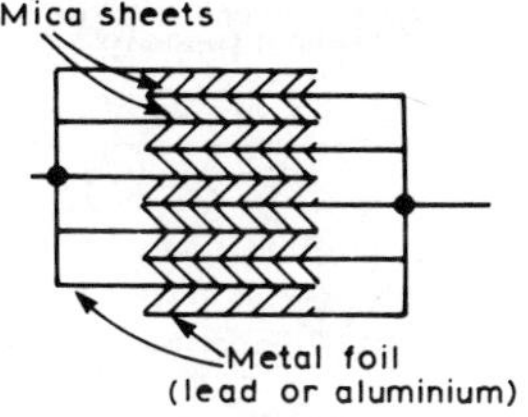

Figure 31.4

(a) **Variable air capacitors**. These usually consist of two sets of metal plates (i.e. aluminium) one fixed, the other variable. The set of moving plates rotate on a spindle as shown by the end view in *Figure 31.3*. As the moving plates are rotated through half a revolution, the meshing, and therefoe the capacitance, varies from a minimum to a maximum value. Variable air capacitors are used in radio and electronic circuits where very low losses are required, or where a variable capacitance is needed. The maximum value of such capacitors is between 500 pF and 1000 pF.

(b) **Mica capacitors** A typical older type construction is shown in *Figure 31.4*. Usually the whole capacitor is impregnated with wax and placed in a bakelite case. Mica is easily obtained in thin sheets and is a good insulator. However, mica is expensive and is not used in capacitors above about 0.1 μF. A modified form of mica capacitor is the silvered mica type. The mica is coated on both sides with a thin layer of silver which forms the plates. Capacitance is stable and less likely to change with age. Such capacitors have a constant capacitance with change of temperature, a high working voltage rating and a long service life and are used in high frequency circuits with fixed values of capacitance up to about 1000 pF.

(c) **Paper capacitors** A typical paper capacitor is shown in *Figure 31.5* where the length of the roll corresponds to the capacitance required. The whole is usually impregnated with oil or wax to exclude moisture, and then placed in a plastic or aluminium container for protection. Paper capacitors, up to about 1 μF, are made in various working voltages. Disadvantages of paper capacitors include variation in capacitance with temperature change and a shorter service life than most other types of capacitor.

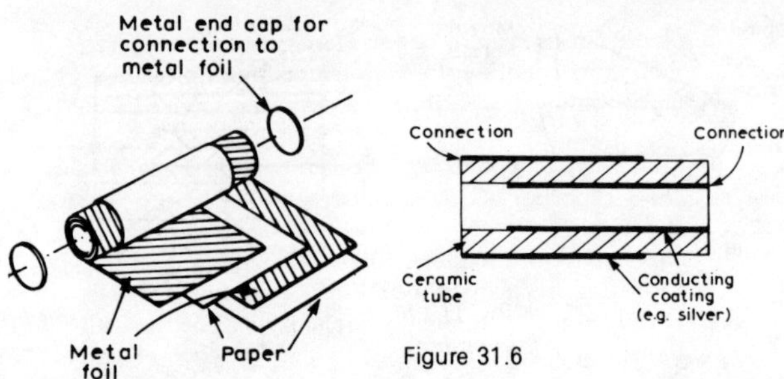

Figure 31.5

Figure 31.6

(d) **Ceramic capacitors** These are made in various forms, each type of construction depending on the value of capacitance required. For high values, a tube of ceramic material is used as shown in the cross section of *Figure 31.6*. For smaller values the cup construction is used as shown in *Figure 31.7*, and for still smaller values the disc construction shown in *Figure 31.8* is used.

Certain ceramic materials have a very high permittivity and this enables capacitors of high capacitance to be made which are of small physical size with a high working voltage rating. Ceramic capacitors are available in the range 1 pF to 0.1 μF and may be used in high frequency electronic circuits subject to a wide range of temperature.

(e) **Plastic capacitors** Some plastic materials such as polystyrene and Teflon can be used as dielectrics. Construction is similar to the paper capacitor but using a

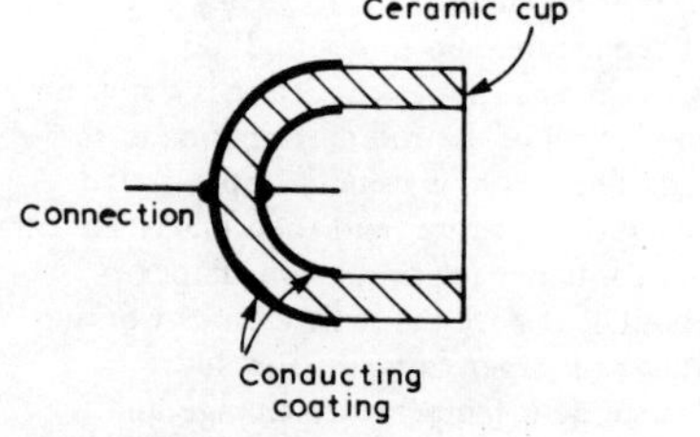

Figure 31.7

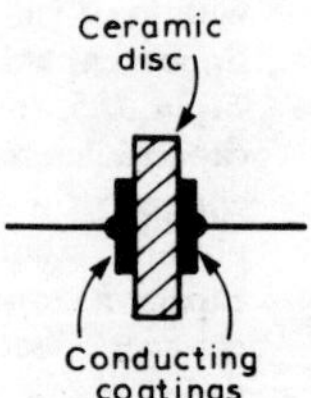

Figure 31.8

plastic film instead of paper. Plastic capacitors operate well under conditions of high temperature, provide a precise value of capacitance, a very long service life and high reliability.

(f) **Electrolytic capacitors** Construction is similar to the paper capacitor with aluminium foil used for the plates and with a thick absorbent material, such as paper, impregnated with an electrolyte (ammonium borate), separating the plates. The finished capacitor is usually assembled in an aluminium container and hermetically sealed. Its operation depends on the formation of a thin aluminium oxide layer on the positive plate by electrolytic action when a suitable direct potential is maintained between the plates. This oxide layer is very thin and forms the dielectric. (The absorbent paper between the plates is a conductor and does not act as a dielectric.) Such capacitors **must only be used on d.c.** and must be connected with the correct polarity; if this is not done the capacitor will be destroyed since the oxide layer will be destroyed. Electrolytic capacitors are manufactured with working voltages from 6 V to 500 V, although accuracy is generally not very high. These capacitors possess a much larger capacitance than other types of capacitors of similar dimensions due to the oxide film being only a few microns thick. The fact that they can be used only on d.c. supplies limits their usefulness.

22 When a capacitor has been disconnected from the supply it may still be charged and it may retain this charge for some considerable time. Thus precautions must be taken to ensure that the capacitor is automatically discharged after the supply is switched off. This is done by connecting a high value resistor across the capacitor terminals.

32 Electromagnetism and magnetic circuits

Magnetism

1 A **permanent magnet** is a piece of ferromagnetic material (such as iron, nickel or cobalt) which has properties of attracting other pieces of these materials.
2 The area around a magnet is called the **magnetic field** and it is in this area that the effects of the **magnetic force** produced by the magnet can be detected.
3 The magnetic field of a bar magnet can be represented pictorially by the 'lines of force' (or lines of 'magnetic flux' as they are called) as shown in *Figure 32.1*. Such a field pattern can be produced by placing iron filings in the vicinity of the magnet. The

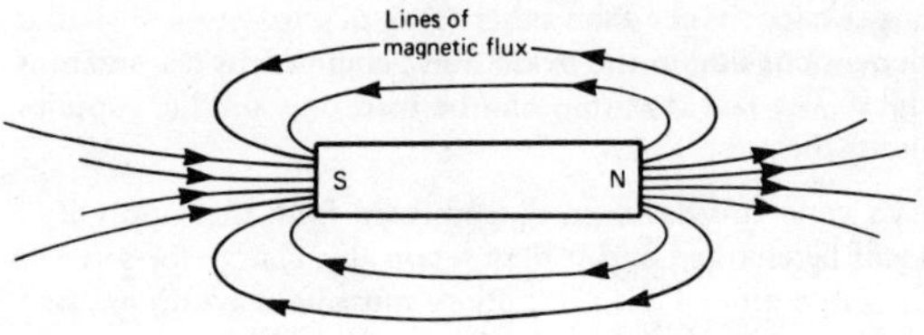

Figure 32.1

field direction at any point is taken as that in which the north-seeking pole of a compass needle points when suspended in the field. External to the magnet the direction of the field is north to south.
4 The laws of magnetic attraction and repulsion can be demonstrated by using two bar magnets. In *Figure 32.2(a)*, **with unlike poles adjacent, attraction occurs**. In *Figure 32.2(b)*, **with like poles adjacent, repulsion occurs**.

Electromagnetism

5 Magnetic fields are produced by electric currents as well as by permanent magnets. The field forms a circular pattern with the

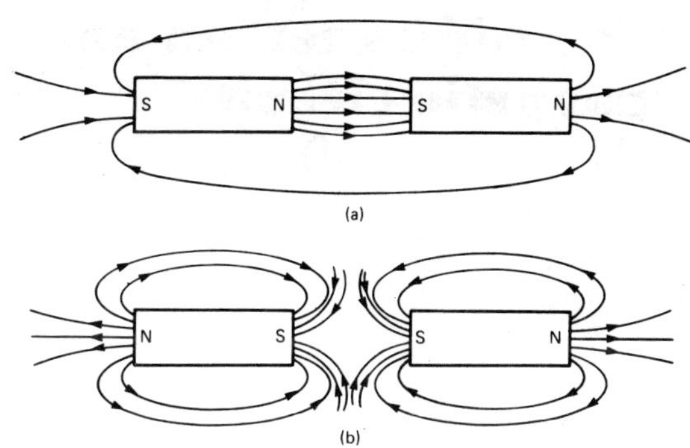

Figure 32.2

current carrying conductor at the centre. The effect is portrayed in *Figure 32.3* where the convention adopted is:

(a) current flowing away from the viewer is shown by ⊕ – can be thought of as the feather end of the shaft of an arrow
(b) current flowing towards the view is shown by ⊙ – can be thought of as the tip of an arrow.

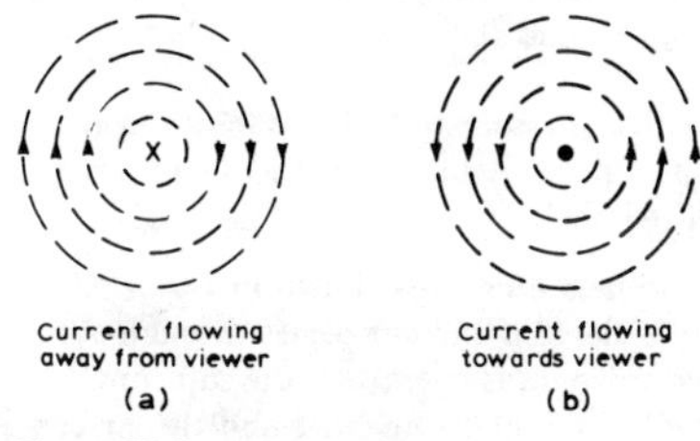

Figure 32.3

6 The **direction** of the fields in *Figure 32.3* is remembered by the **screw rule** which states: "*If a normal right-hand thread screw is screwed along the conductor in the direction of the current, the direction of rotation of the screw is in the direction of the magnetic field*".

7 A magnetic field produced by a long coil, or **solenoid**, is

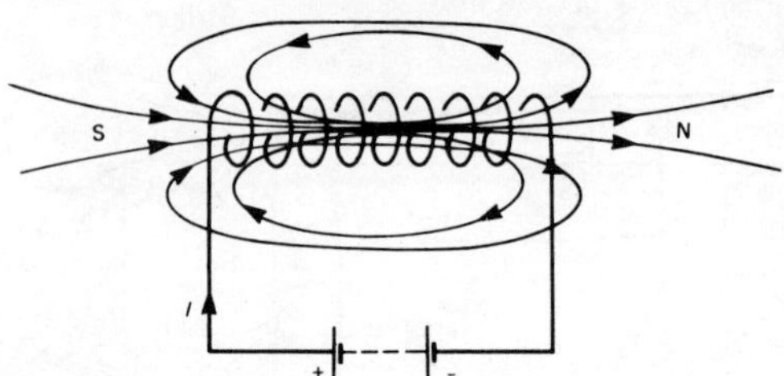

Figure 32.4

shown in *Figure 32.4* and is seen to be similar to that of a bar magnet shown in *Figure 32.1*. If the solenoid is wound on an iron bar an even stronger field is produced. The **direction** of the field produced by current *I* is determined by a compass and is remembered by either:

(a) the **screw rule**, which states that if a normal right hand thread screw is placed along the axis of the solenoid and is screwed in the direction of the current it moves in the direction of the magnetic field inside of the solenoid (i.e. points in the direction of the north pole), or
(b) the **grip rule**, which states that if the coil is gripped with the right hand with the fingers pointing in the direction of the current, then the thumb, outstretched parallel to the axis of the solenoid, points in the direction of the magnetic field inside the solenoid (i.e. points in the direction of the north pole).

8 An **electromagnet**, which is a solenoid wound on an iron core, provides the basis of many items of electrical equipment, examples including electric bells, relays and lifting magnets.

(i) A simple **electric bell circuit** is shown in *Figure 32.5*. When the switch S is closed a current passes through the coil. The iron-cored solenoid is energised, the soft iron armature is attracted to the electromagnet and the striker hits the gong. When the switch S is opened the coil becomes demagnetised and the spring steel strip pulls the armature back to its original position. This is the principle of operation of an electric bell.
(ii) A typical **relay circuit** connected to an alarm device is shown in *Figure 32.6*. When the switch S is closed a current passes through the coil and the iron-cored solenoid is energised. The hinged soft iron armature is attracted to

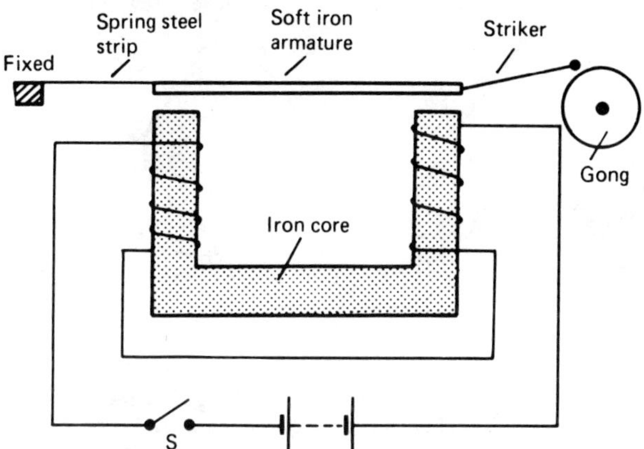

Figure 32.5

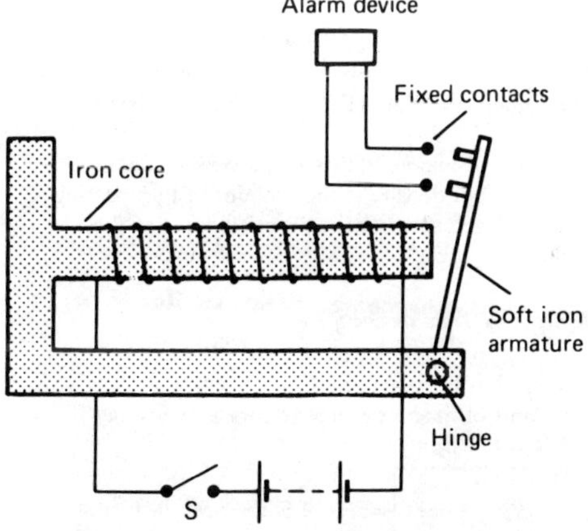

Figure 32.6

the electromagnet and pushes against the two fixed contacts so that they are connected together, thus closing the electric circuit to be controlled – in this case, an alarm circuit. The alarm sounds for as long as the current flows in the coil.

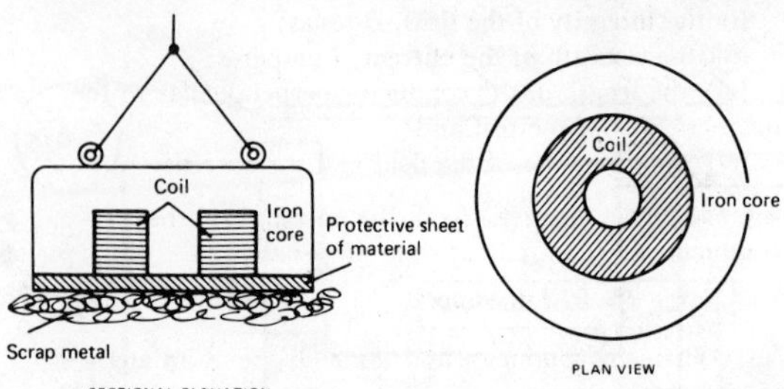

Figure 32.7

(iii) A typical scrap-metal yard **lifting magnet** showing the plan and elevation is shown in *Figure 32.7*. When current is passed through the coil, the iron core becomes magnetised (i.e., an electromagnet) and this will attract to it other pieces of magnetic material. When the circuit is broken the iron core becomes demagnetised which releases the materials being lifted.

9 (i) **Magnetic flux** is the amount of magnetic field (or the number of lines of force) produced by a magnetic source.
(ii) The symbol for magnetic flux is Φ (greek letter 'phi').
(iii) The unit of magnetic flux is the **weber, Wb**.

10 (i) Magnetic flux density is the amount of flux passing through a defined area that is perpendicular to the direction of the flux.

$$\textbf{Magnetic flux density} = \frac{\textbf{magnetic flux}}{\textbf{area}}$$

(ii) The symbol for magnetic flux density is B.
(iii) The unit of magnetic flux density is the tesla, T, where 1 T = 1 Wb/m^2.

Hence $\boxed{B = \frac{\Phi}{A} \text{ tesla,}}$ where A is the area in m^2.

11 (i) If a current carrying conductor is placed in a magnetic field produced by permanent magnets then the fields due to the current carrying conductor and the permanent magnets intersect and cause a force to be exerted on the conductor. The force on the current carrying conductor in a magnetic field depends upon:

(i) the intensity of the field, B teslas;
(ii) the strength of the current, I amperes;
(iii) the length of the conductor perpendicular to the magnetic field, l metres; and
(iv) the directions of the field and the current.

(ii) When the magnetic field, the current and the conductor are mutually at right angles then:

Force $F = BIl$ newtons

(iii) When the conductor and the field are at an angle $\theta°$ to each other then:

Force $F = BIl \sin\theta$ newtons

(iv) Since when the magnetic field, current and conductor are mutually at right angles, $F = BIl$, the magnetic flux density B may be defined by $B = \dfrac{F}{Il}$, i.e. a field intensity of 1 T is exerted on 1 m of a conductor when the conductor carries a current of 1 A.

12 If the current-carrying conductor shown in *Figure 32.3(a)* is placed in the magnetic field shown in *Figure 32.8(a)*, then the two fields interact and cause a force to be exerted on the conductor as shown in *Figure 32.8(b)*. The field is strengthened above the

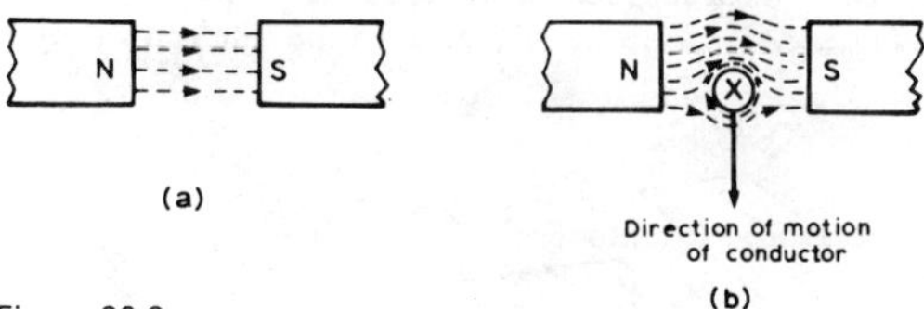

Figure 32.8

conductor and weakened below, thus tending to move the conductor downwards. This is the basic principle of operation of the electric motor (see para 14) and the moving coil instrument (see page 250).

13 The direction of the force exerted on a conductor can be predetermined by using **Fleming's left-hand rule** (often called the motor rule), which states:

'Let the thumb, first finger and second finger of the left-hand be extended such that they are all at right angles to each other, as shown in Figure 32.9. If the first finger points in the direction of the

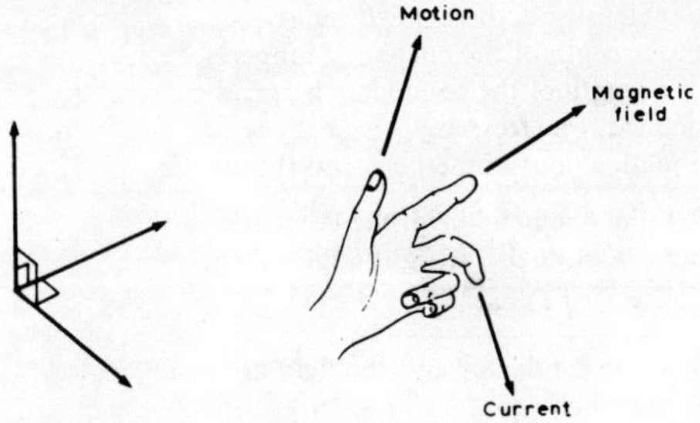

Figure 32.9

magnetic field, the second finger points in the direction of the current, then the thumb will point in the direction of the motion of the conductor.'

Summarising:

First finger	–	Field
SeCond finger	–	Current
ThuMb	–	Motion

Principle of operation of a d.c. motor

14 A rectangular coil which is free to rotate about a fixed axis is shown placed inside a magnetic field produced by permanent magnets in *Figure 32.10*. A direct current is fed into the coil via

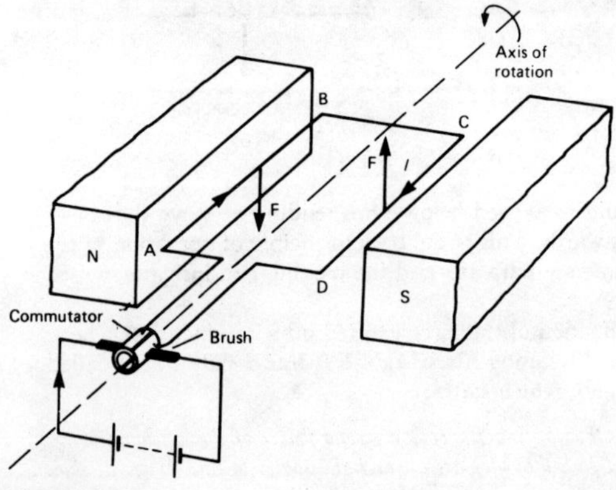

Figure 32.10

carbon brushes bearing on a commutator, which consists of a metal ring split into two halves separated by insulation. When current flows in the coil, a magnetic field is set up around the coil which interacts with the magnetic field produced by the magnets. This causes a force F to be exerted on the current carrying conductor, which, by Fleming's left-hand rule (see para 13) is downwards between points A and B and upwards between C and D for the current direction shown. This causes a torque and the coil rotates anticlockwise.

When the coil has turned through 90° from the position shown in *Figure 32.10* the brushes connected to the positive and negative terminals of the supply make contact with different halves of the commutator ring, thus reversing the direction of the current flow in the conductor. If the current is not reversed and the coil rotates past this position the forces acting on it change direction and it rotates in the opposite direction thus never making more than half a revolution. The current direction is reversed every time the coil swings through the vertical position and thus the coil rotates anticlockwise for as long as the current flows. This is the principle of operation of a d.c. motor which is thus a device that takes in electrical energy and converts it into mechanical energy.

Magnetic circuits

15 **Magnetomotive force (mmf), $F_m = NI$ ampere-turns (At)**, where N = number of conductors (or turns) and I = current in amperes.

Since 'turns' has no units, the SI unit of mmf is the ampere, but to avoid any possible confusion 'ampere-turns', (A t) are used in this chapter.

16 **Magnetic field strength**, or **magnetising force**

$$H = \frac{NI}{l} \text{ At/m}$$

where l = mean length of flux path in metres.

Hence, mmf $= NI = Hl$ At.

17 **For air, or any non-magnetic medium**, the ratio of magnetic flux density to magnetising force is a constant, i.e.

$$\frac{B}{H} = \text{a constant.}$$

This constant is μ_0, the permeability of free space (or the magnetic space constant) and is equal to $4\pi \times 10^{-7}$ H/m.

Hence, for a non-magnetic medium, $\dfrac{B}{H} = \mu_0$

18 For ferromagnetic mediums:

$$\frac{B}{H} = \mu_0 \mu_r,$$

where μ_r is the relative permeability, and is defined as

$$\frac{\text{flux density in material}}{\text{flux density in air}}.$$

Its value varies with the type of magnetic material and since μ_r is a ratio of flux densities, it has no units. From its definition, μ_r for air is 1.

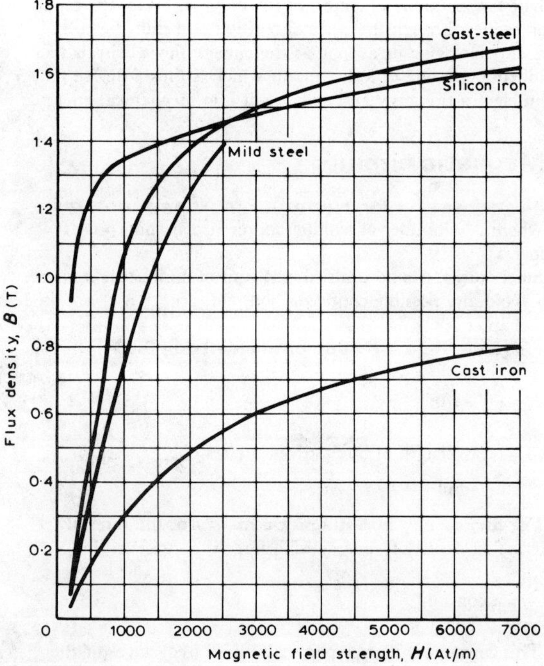

Figure 32.11

19 $\mu_0\mu_r=\mu$, called the **absolute permeability**.

20 By plotting measured values of flux density B against magnetic field strength H, a **magnetisation curve** (of $B-H$ curve) is produced. For non-magnetic materials this is a straight line. Typical curves for 4 magnetic materials are shown in *Figure 32.11*.

21 The relative permeability of a ferromagnetic material is proportional to the slope of the $B-H$ curve and thus varies with the magnetic field strength. The approximate range of values of relative permeability μ_r for some common magnetic materials are:

Cast iron $\mu_r=100-250$;
Silicon iron $\mu_r=1000-5000$;
Mumetal $\mu_r=200-5000$;
Mild steel $\mu_r=200-800$
Cast steel $\mu_r=300-900$
Stalloy $\mu_r=500-6000$

22 The 'magnetic resistance' of a magnetic circuit to the presence of magnetic flux is called **reluctance**. The symbol for reluctance is S (or R_m).

23 Reluctance

$$S=\frac{\text{mmf}}{\Phi}=\frac{NI}{\Phi}=\frac{Hl}{BA}=\frac{l}{\frac{B}{H}A}=\frac{l}{\mu_0\mu_r A}$$

24 The unit of reluctance is 1/H (or H^{-1}) or At/Wb.

25 For a series magnetic circuit having n parts, the total reluctance S is given by: $S=S_1+S_2+\ldots+S_n$. (This is similar to resistors connected in series in an electrical circuit.)

Comparison between electrical and magnetic quantities:

26

Electric circuit	*Magnetic circuit*
emf E (V) current I (A) resistance R (Ω) $I=\frac{E}{R}$ $R=\frac{\rho l}{A}$	mmf F_m (At) flux Φ (Wb) reluctance S (H^{-1}) $\Phi=\frac{\text{mmf}}{S}$ $S=\frac{l}{\mu_0\mu_r A}$

27 Ferromagnetic materials have a low reluctance and can be used as **magnetic screens** to prevent magnetic fields affecting materials within the screen.

28 **Hysteresis** is the 'lagging' effect of flux density B whenever there are changes in the magnetic field strength H.

29 When an initially unmagnetised ferromagnetic material is subjected to a varying magnetic field strength H, the flux density B produced in the material varies as shown in *Figure 32.12*, the arrows incating the direction of the cycle. *Figure 32.12* is known as the hysteresis loop.

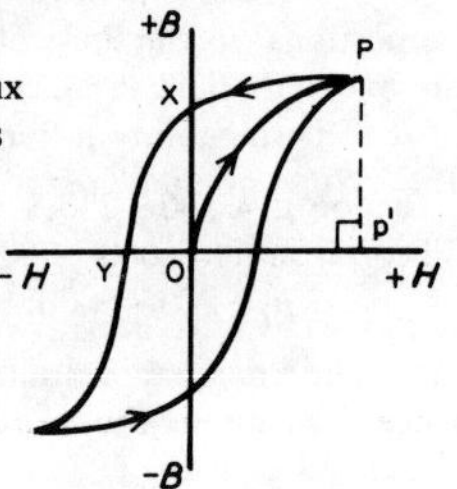

Figure 32.12

OX = **residual flux density** or **remanence**
OY = **coercive force**
PP = **saturation flux density**

30 Hysteresis results in a dissipation of energy which appears as a heating of the magnetic material. The **energy loss** associated with hysteresis is proportional to the areas of the hysteresis loop.

31 The **area** of a hysteresis loop varies with the type of material. The area, and thus the energy loss, is much greater for hard materials than for soft materials.

33 Electromagnetic induction and inductance

1 When a conductor is moved across a magnetic field, an electromotive force (emf) is produced in the conductor. If the conductor forms part of a closed circuit then the emf produced causes an electric current to flow round the circuit. Hence an emf (and thus current), is 'induced' in the conductor as a result of its movement across the magnetic field. This effect is known as **'electromagnetic induction'**.

2 **Faraday's laws of electromagnetic induction** state:

> (i) *'An induced emf is set up whenever the magnetic field linking that circuit changes.'*
> (ii) *'The magnitude of the induced emf in any circuit is proportional to the rate of change of the magnetic flux linking the circuit.'*

3 **Lenz's law** states:

> *'The direction of an induced emf is always such as to oppose the effec producing it.'*

4 An alternative method to Lenz's law of determining relative directions is given by **Fleming's Right-hand rule** (often called the geneRator rule) which states:

> *'Let the thumb, first finger and second finger on the right hand be extended such that they are all at right angles to each other, as shown in Figure 33.1. If the first finger points in the direction of the magnetic field the thumb points in the direction of motion of the conductor relative to the magnetic field, then the second finger will point in the direction of the induced emf.'*

Summarising:

First finger	–	Field
ThuMb	–	Motion
SEcond finger	–	Emf

5 In a **generator**, conductors forming an electric circuit are made to move through a magnetic field. By Faraday's law an emf is induced in the conductors and thus a source of emf is created. A

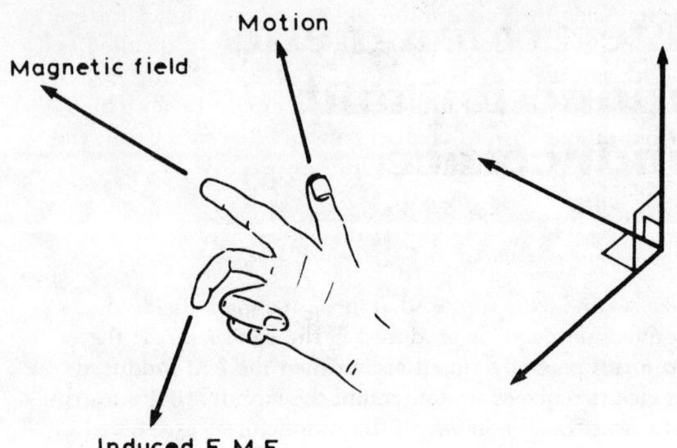

Figure 33.1

generator converts mechanical energy into electrical energy. (The action of a simple a.c. generator is described in para 2, page 228.)

6 The induced emf E set up between the ends of the conductor shown in *Figure 33.2* is given by: $\boldsymbol{E = Blv}$ **volts,** where B, the flux density is measured in teslas, l, the length of conductor in the magnetic field is measured in metres, and v, the conductor velocity, is measured in metres per second. If the conductor moves at an angle $\theta°$ to the magnetic field (instead of at 90° as assumed above) then $\boldsymbol{E = Blv \sin \theta}$.

7 **Inductance** is the name given to the property of a circuit whereby there is an emf induced into the circuit by the change of flux linkages produced by a current change.

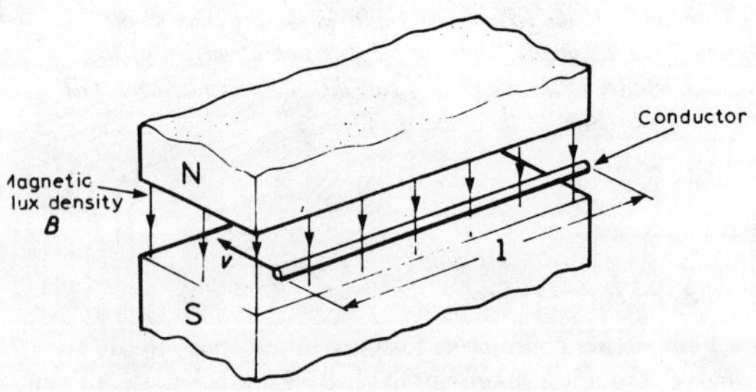

Figure 33.2

(i) When the emf is induced in the same circuit as that in which the current is changing, the property is called **self inductance**, L;
(ii) When the emf is induced in a circuit by a change of flux due to current changing in an adjacent circuit, the property is called **mutual inductance**, M.

8 The unit of inductance is the **henry, H.**

'A circuit has an inductance of one henry when an emf of one volt is induced in it by a current changing at the rate of one ampere per second.'

9 A component called an **inductor** is used when the property of inductance is required in a circuit. The basic form of an inductor is simply a coil of wire. Factors which affect the inductance of an inductor include:

(i) the number of turns of wire – the more turns the higher the inductance;
(ii) the cross-sectional area of the coil of wire – the greater the cross-sectional area the higher the inductance;
(iii) the presence of a magnetic core – when the coil is wound on an iron core the same current sets up a more concentrated magnetic field and the inductance is increased;
(iv) the way the turns are arranged a short thick coil of wire has a higher inductance than a long thin one.

10 Two examples of practical inductors are shown in *Figure 33.3* and the standard electrical circuit diagram symbol for air-cored and iron-cored inductors are shown in *Figure 33.4*.

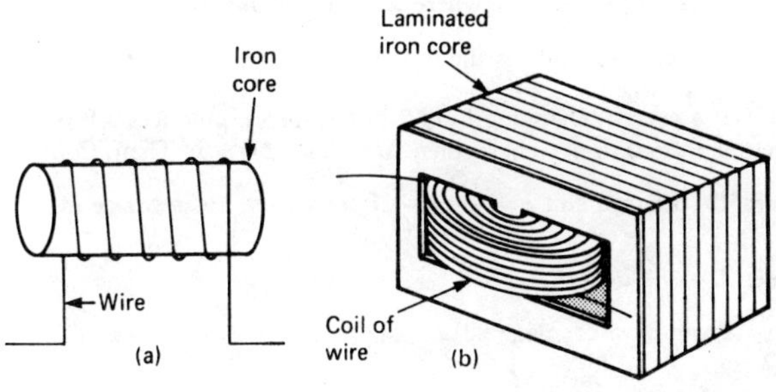

Figure 33.3

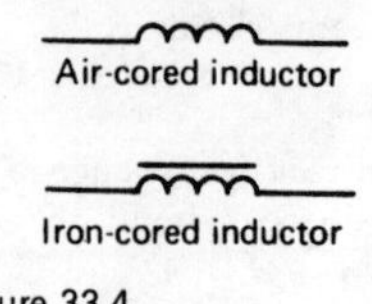

Figure 33.4

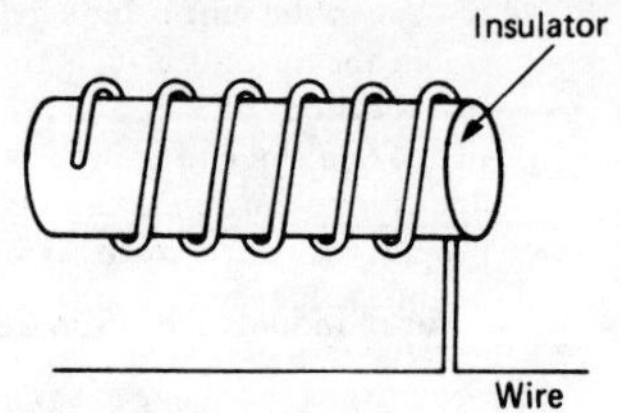

Figure 33.5

An iron-cored inductor is often called a **choke** since, when used in a.c. circuits it has a choking effect, limiting the current flowing through it.

11 Inductance is often undesirable in a circuit. To reduce inductance to a minimum the wire may be bent back on itself as shown in *Figure 33.5* so that the magnetising effect of one conductor is neutralised by that of the adjacent conductor. The wire may be coiled around an insulator as shown without increasing the inductance. Standard resistors may be non-inductively wound in this manner.

12 An inductor possesses an ability to store energy. The **energy stored**, W, in the magnetic field of an inductor is given by:

$$\mathbf{W=\frac{1}{2}LI^2\ joules}$$

13 (i) Induced emf in a coil of N turns, $E=N\left(\frac{\Delta\Phi}{t}\right)$ volts, where $\Delta\Phi$ is the change in flux, in Webers, and t is the time taken for the flux to change, in seconds.

(ii) Induced emf in a coil of inductance L henrys $E=L\left(\frac{\Delta I}{t}\right)$ volts, where ΔI is the change in current, in amperes, and t is the time taken for the current to change, in seconds.

14 If a current changing from 0 to I amperes, produces a flux change from 0 to Φ webers, then $\Delta I=I$ and $\Delta\Phi=\Phi$. Then, from para 13, induced emf $E=\frac{N\Phi}{t}=\frac{LI}{t}$, from which, **inductance of coil,** $\mathbf{L=\frac{N\Phi}{I}}$ **henrys.**

15 From para 23, page 221.

reluctance $S=\frac{\text{mmf}}{\text{flux}}=\frac{NI}{\Phi}$, from which $\Phi=\frac{NI}{S}$.

Hence, inductance of coil,

$$L=\frac{N\Phi}{I}=\frac{N}{I}\left(\frac{NI}{S}\right)=\frac{N^2}{S}, \text{ i.e. } L\alpha N^2.$$

16 Mutually induced emf in the second coil, $\boldsymbol{E_2=N\left(\frac{\Delta I_1}{t}\right)}$ **volts,** where M is the mutual inductance between two coils, in henrys, ΔI_1 is the change in current in the first coil, in amperes, and t is the time the current takes to change in the first coil, in seconds. A transformer is a device which uses the phenomenon of mutual inductance to change the value of alternating voltages.

34 Alternating currents and voltages

1 Electricity is produced by generators at power stations and then distributed by a vast network of transmission lines (called the National Grid system) to industry and for domestic use. It is easier and cheaper to generate alternating current (a.c.) than direct current (d.c.) and a.c. is more conveniently distributed than d.c. since its voltage can be readily altered using transformers. Whenever d.c. is needed in preference to a.c. devices called rectifiers are used for conversion (see paragraphs 19 to 23).

2 Let a single turn coil be free to rotate at constant angular velocity ω symmetrically between the poles of a magnet system as shown in *Figure 34.1*. An emf is generated in the coil (from

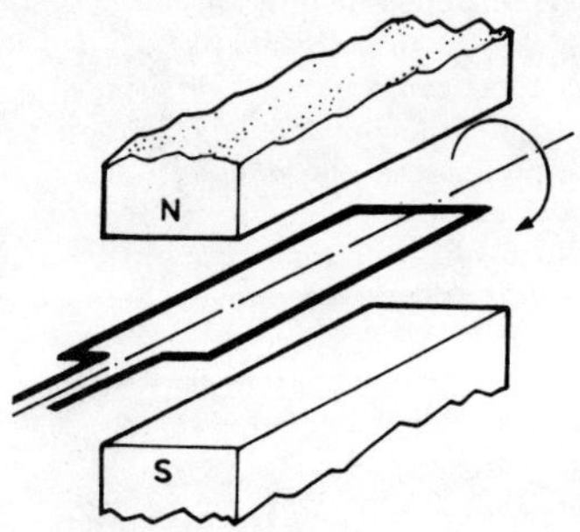

Figure 34.1

Faraday's law) which varies in magnitude and reverses its direction at regular intervals. The reason for this is shown in *Figure 34.2*. In positions (a), (e) and (i) the conductors of the loop are effectively moving along the magnetic field, no flux is cut and hence no emf is induced. In position (c) maximum flux is cut and hence maximum emf is induced. In position (g), maximum flux is cut and hence maximum emf is again induced. However, using Fleming's right-

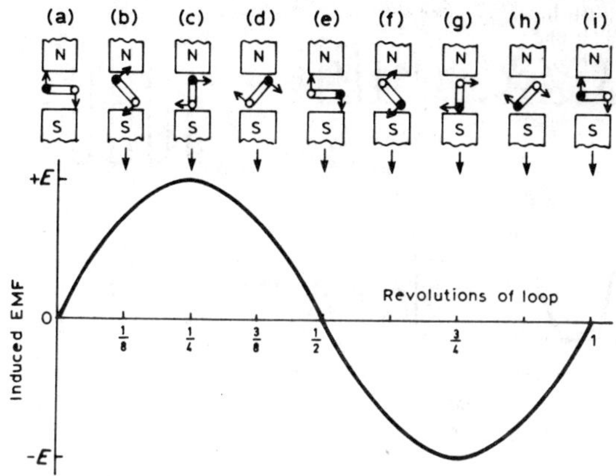

Figure 34.2

hand rule, the induced emf is in the opposite direction to that in position (c) and is thus shown as $-E$. In positions (b), (d), (f) and (h) some flux is cut and hence some emf is induced. If all such positions of the coil are considered, in one revolution of the coil, one cycle of alternating emf is produced as shown. This is the principle of operation of the **a.c. generator** (i.e. the **alternator**).

3 If values of quantities which vary with time t are plotted to a base of time, the resulting graph is called a **waveform**. Some typical waveforms are shown in *Figure 34.3*. Waveforms (a) and (b) are **unidirectional waveforms,** for, although they vary considerably with time, they flow in one direction only (i.e. they do not cross the time axis and become negative). Waveforms (c) to (g) are called **alternating waveforms** since their quantities are continually changing in direction (i.e. alternately positive and negative).

4 A waveform of the type shown in *Figure 34.3* is called a **sine wave**. It is the shape of the waveform of emf produced by an alternator and thus the mains electricity supply is of 'sinusoidal' form.

5 One complete series of values is called a **cycle** (i.e. from O to P in *Figure 34.3(g)*.

6 The time taken for an alternating quantity to complete one cycle is called the **period** or the **periodic time**, T, of the waveform.

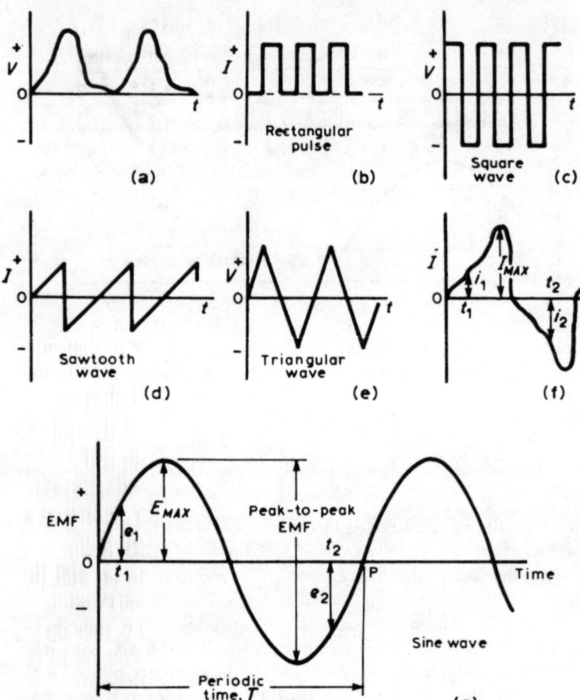

Figure 34.3

7 The number of cycles completed in one second is called the **frequency**, f, of the supply and is measured in hertz, Hz. The standard frequency of the electricity supply in Great Britain is 50 Hz.

$$T=\frac{1}{f} \text{ or } f=\frac{1}{T}.$$

8 **Instantaneous values** are the values of the alternating quantities at any instant of time. They are represented by small letters, i, v, e etc. (see *Figure 34.3(f)* and *(g)*).
9 The largest value reached in a half cycle is called the **peak value** or the **maximum value** or the **crest value** or the **ampli-**

tude of the waveform. Such values are represented by V_{MAX}, I_{MAX} etc. (see *Figure 34.3(f)* and *(g)*). A **peak-to-peak value** of emf is shown in *Figure 34.3(g)* and is the difference between the maximum and minimum values in a cycle.

10 The **average or mean value** of a symmetrical alternating quantity, (such as a sinewave), is the average value measured over a half cycle, (since over a complete cycle the average value is zero).

$$\textbf{Average or mean value} = \frac{\textbf{area under the curve}}{\textbf{length of base}}$$

The area under the curve is found by approximate methods such as the trapezoidal rule, the mid-ordinate rule or Simpson's rule. Average values are represented by V_{AV}, I_{AV}, etc.

For a sine wave, average value = 0.637 × maximum value $\left(\text{i.e. } \frac{2}{\pi} \times \text{maximum value}\right)$.

11 The **effective value** of an alternating current is that current which will produce the same heating effect as an equivalent direct current. The effective value is called the **root mean square (r.m.s.) value** and whenever an alternating quantity is given, it is assumed to be the r.m.s. value. For example, the domestic mains supply in Great Britain is 240 V and is assumed to mean '240 V r.m.s.'. The symbols used for r.m.s. values are *I*, *V*, *E*, etc. For a non-sinusoidal waveform as shown in *Figure 36.4*, the r.m.s. value is

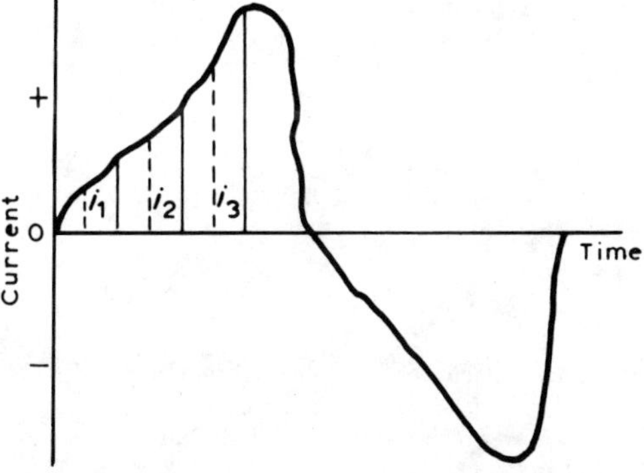

Figure 34.4

given by:

$$I=\sqrt{\left\{\frac{i_1^2+i_2^2+\ldots+i_n^2}{n}\right\}}$$

where n is the number of intervals used.

For a sine wave, r.m.s. value = 0.707 × maximum value $\left(\text{i.e. } \frac{1}{\sqrt{2}} \times \text{maximum value}\right)$.

12 (a) **Form factor** $= \frac{\text{r.m.s. value}}{\text{average value}}$.

For a sine wave, form factor = 1.11

(b) **Peak factor** $= \frac{\text{maximum value}}{\text{r.m.s. value}}$.

For a sine wave, peak factor = 1.41

The values of form and peak factors give an indication of the shape of waveforms.

13 In *Figure 34.5*, OA represents a vector that is free to rotate anticlockwise about 0 at an angular velocity of ω rad/s. A rotating

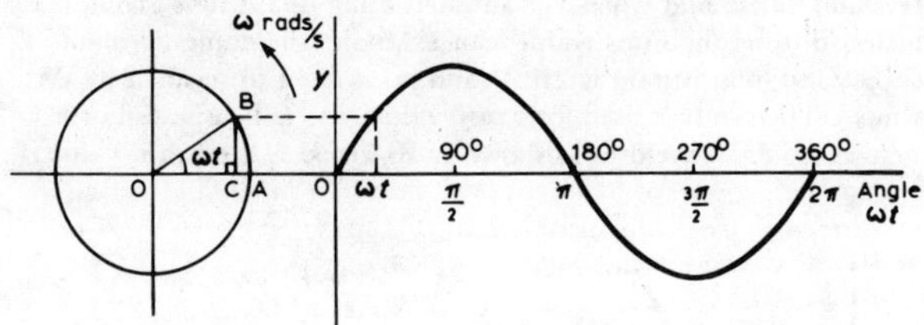

Figure 34.5

vector is known as a *phasor*. After time t seconds the vector OA has turned through an angle ωt. If the line BC is constructed perpendicular to OA as shown, then

$$\sin \omega t=\frac{BC}{OB}, \text{ i.e. } BC=OB \sin \omega t.$$

If all such vertical components are projected on to a graph of y against angle ωt (in radians), a sine curve results of maximum value OA. Any quantity which varies sinusoidally can thus be represented as a phasor.

14 A sine curve may not always start at 0°. To show this a

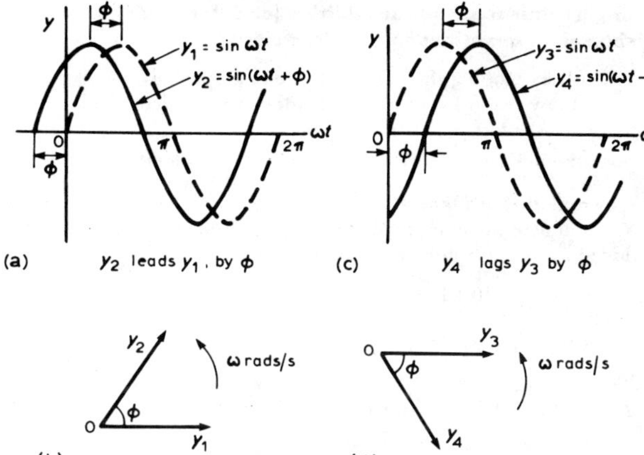

Figure 34.6

periodic function is represented by $y = \sin(\omega t \pm \phi)$, where ϕ is a phase (or angle) difference compared with $y = \sin \omega t$. In *Figure 34.6(a)*, $y_2 = \sin(\omega t + \phi)$ starts ϕ radians earlier than $y_1 = \sin \omega t$ and is thus said to lead y_1 by ϕ radians. Phasors y_1 and y_2 are shown in *Figure 34.6(b)* at the time when $t = 0$. In *Figure 34.6(c)*, $y_4 = \sin(\omega t - \phi)$ starts ϕ radians later than $y_3 = \sin \omega t$ and is thus said to lag y_3 by ϕ radians. Phasors y_3 and y_4 are shown in *Figure 34.6(d)* at the time when $t = 0$.

15 Given the general sinusoidal voltage, $V = V_{MAX} \sin(\omega t \pm \phi)$, then

(i) Amplitude or maximum value $= V_{MAX}$;
(ii) Peak to peak value $= 2\ V_{MAX}$;
(iii) Angular velocity $= \omega$ rad/s.

(iv) Periodic time, $T = \dfrac{2\pi}{\omega}$ seconds.

(v) Frequency, $f = \dfrac{\omega}{2\pi}$ Hz (hence $\omega = 2\pi f$).

(vi) ϕ = angle of lag or lead (compared with $v = V_{MAX} \sin \omega t$).

16 The **resultant of the addition (or subtraction) of two sinusoidal quantities** may be determined either:

(a) by plotting the periodic function graphically, or
(b) by resolution of phasors by drawing or calculation.

For example, currents $i_1 = 20 \sin \omega t$ and $i_2 = 10 \sin\left(\omega t + \frac{\pi}{3}\right)$ are shown plotted in *Figure 34.7*.

To determine the resultant of $i_1 + i_2$, ordinates of i_1 and i_2 are added at, say, 15° intervals. For example:

at 30°, $i_1 + i_2 = 10 + 10 = 20$ A
at 60°, $i_1 + i_2 = 8.7 + 17.3 = 26$ A
at 150°, $i_1 + i_2 = 10 + (-5) = 5$ A, and so on.

The resultant waveform for $i_1 + i_2$ is shown by the broken line in *Figure 34.7*. It has the same period, and hence frequency, as i_1 and

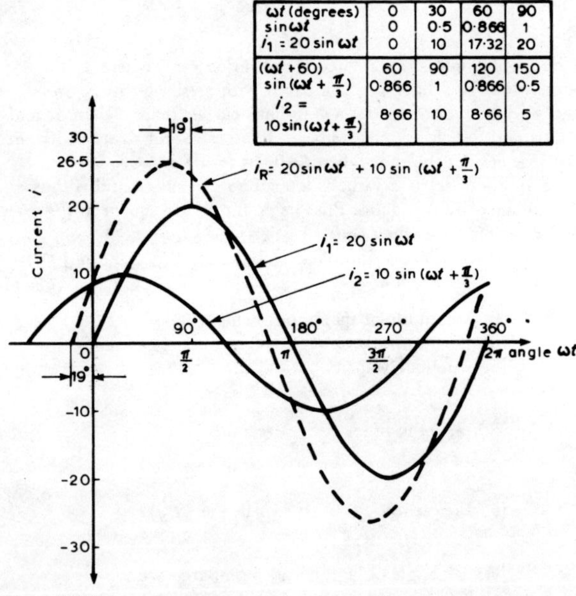

ωt (degrees)	0	30	60	90
$\sin \omega t$	0	0·5	0·866	1
$i_1 = 20 \sin \omega t$	0	10	17·32	20
$(\omega t + 60)$	60	90	120	150
$\sin(\omega t + \frac{\pi}{3})$	0·866	1	0·866	0·5
$i_2 = 10 \sin(\omega t + \frac{\pi}{3})$	8·66	10	8·66	5

Figure 34.7

i_2. The amplitude or peak value is 26.5 A. The resultant waveform leads $i_1 = 20 \sin \omega t$ by 19°, i.e., $\left(19 \times \frac{\pi}{180}\right)$ rad = 0.332 rad. Hence the sinusoidal expression for the resultant $i_1 + i_2$ is given by:

$$i_R = i_1 + i_2 = 26.5 \sin(\omega t + 0.332) \text{ A}.$$

The relative positions of i_1 and i_2 at time=0 are shown as phasors in *Figure 34.8(a)*. The phasor diagram in *Figure 13.8(b)* shows the resultant i_R, and i_R is measured as 26 A and angle ϕ as 19° (i.e. 0.33 rad) leading i_1. Hence, by drawing, $i_R = 26 \sin(\omega t + 0.33)$ A.

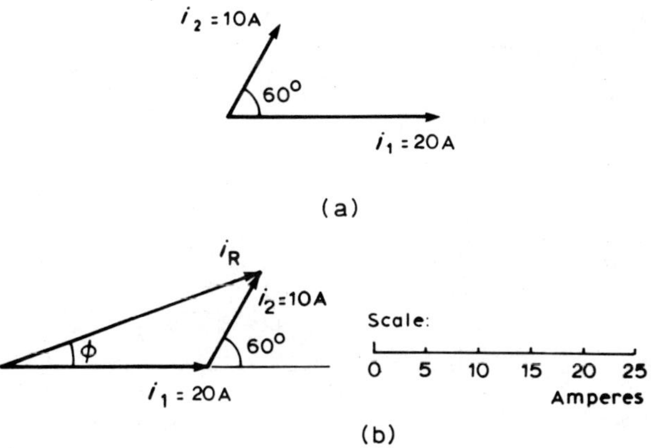

Figure 34.8

From *Figure 34.8(b)*, by the cosine rule:

$$i_R^2 = 20^2 + 10^2 - 2(20)(10) \cos 120°,$$

from which, $i_R = 26.46$ A. By the sine rule:

$$\frac{10}{\sin \phi} = \frac{26.46}{\sin 120°}$$

from which,

$$\phi = 19°\ 10' \text{ (i.e., 0.333 rad).}$$

Hence, by calculation,

$$i_R = 26.46 \sin(\omega t + 0.333) \text{ A}.$$

17 When a sinusoidal voltage is applied to a purely resistive circuit of resistance R, the voltage and current waveforms are in phase and $I = \frac{V}{R}$ (exactly as in d.c. circuit). V and I are r.m.s. values.

18 For an a.c. resistive circuit, power $P = VI = I^2R = \frac{V^2}{R}$ watts (exactly as in a d.c. circuit). V and I are r.m.s. values.

19 The process of obtaining unidirectional currents and voltages from alternating currents and voltages is called **rectification**. Automatic switching in circuits is carried out by devices called diodes (see page 268).

20 Using a single diode, as shown in *Figure 34.9*, **half-wave rectification** is obtained. When P is sufficiently positive with respect to Q, diode D is switched on and current i flows.

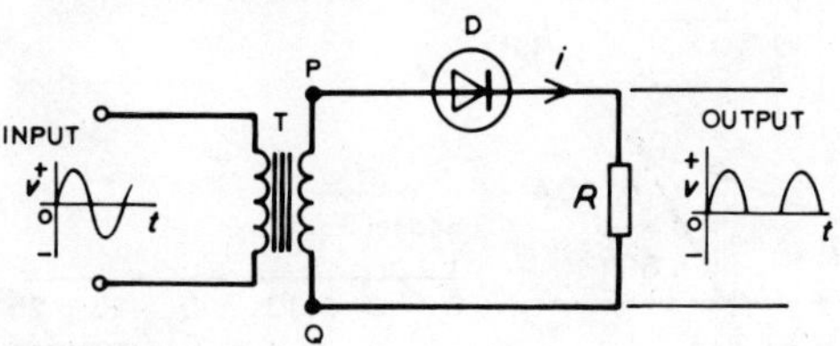

Figure 34.9

When P is negative with respect to Q, diode D is switched off. Transformer T isolates the equipment from direct connection with the mains supply and enables the mains voltage to be changed.

21 Two diodes may be used as shown in *Figure 34.10* to be obtain **full wave rectification**. A centre-tapped transformer T is used. When P is sufficiently positive with respective to Q, diode D_1 conducts and current flows (shown by the broken line in *Figure 34.8*). When S is positive with respect to Q, diode D_2 conducts and current flows (shown by continuous line in *Figure 34.10*). The current flowing in R is in the same direction for both half cycles of the input. The output waveform is thus as shown in *Figure 13.10*.

22 Four diodes may be used in a **bridge rectifier circuit**, as shown in *Figure 34.11* to obtain **full wave rectification**. As for the

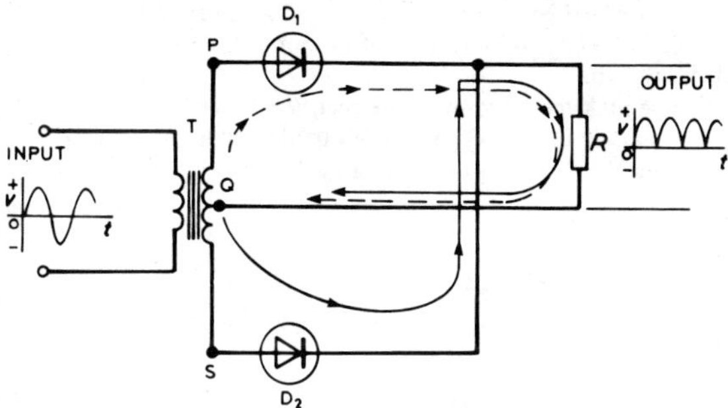

Figure 34.10

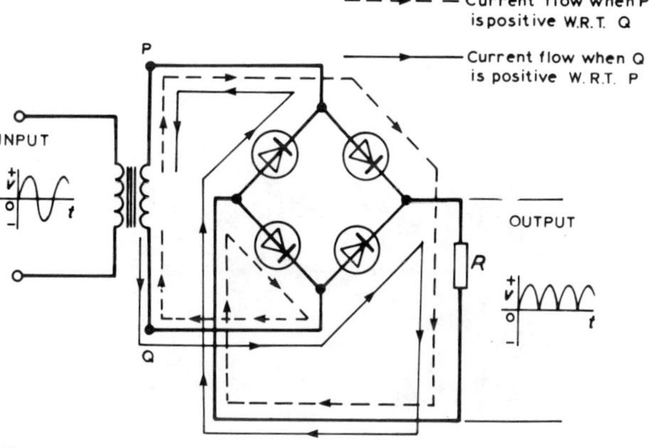

Figure 34.11

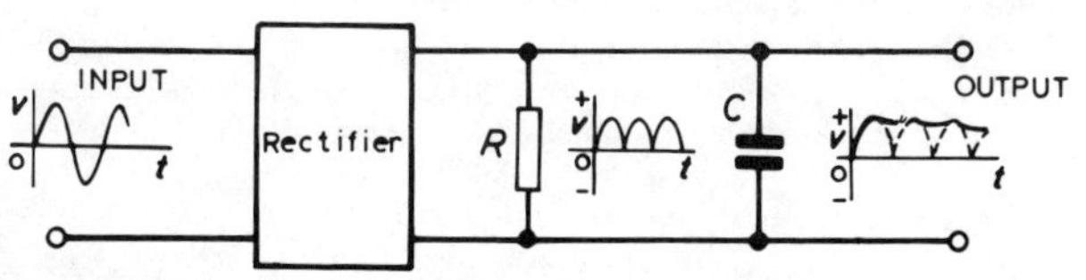

Figure 34.12

rectifier shown in *Figure 34.10*, the current flowing in R is in the same direction for both half cycles of the input giving the output waveform shown.

23 To **smooth the output** of the rectifiers described above, capacitors having a large capacitance may be connected across the load resistors R. The effect of this is shown on the output in *Figure 13.12*.

35 Single phase series a.c. circuits

1 In a **purely resistive a.c. circuit**, the current I_R and applied voltage V_R are in phase, see *Figure 35.1(a)*.
2 In a **purely inductive a.c. circuit**, the current I_L **lags** the applied voltage V_L by 90° (i.e. $\pi/2$ rads), see *Figure 35.1(b)*.
3 In a **purely capacitive a.c. circuit**, the current I_C **leads** the applied voltage V_C by 90° (i.e. $\pi/2$ rads), see *Figure 35.1(c)*.
4 In a purely inductive circuit the opposition to the flow of alternating current is called the **inductive reactance**, X_L.

$$X_L=\frac{V_L}{I_L}=2\pi fL \textbf{ ohms},$$

vhere f is the supply frequency in hertz and L is the inductance in ıenry's.
X_L is proportional to f as shown in *Figure 35.2(a)*.
5 In a purely capacitive circuit the opposition to the flow of alternating current is called the **capacitive reactance**, X_C.

$$X_C=\frac{V_C}{I_C}=\frac{1}{2\pi fC} \textbf{ ohms},$$

where C is the capacitance in farads. X_C varies with f as shown in *Figure 35.2(b)*.
6 In an a.c. series circuit containing inductance L and resistance R, the applied voltage V is the phasor sum of V_R and V_L (see *Figure 35.3(a)*) **and thus the current I lags the applied voltage V by an angle lying between 0° and 90°** (depending on the values of V_R and V_L), shown as angle ϕ. In any a.c. series circuit the current is common to each component and is thus taken as the reference phasor.
7 In an a.c. series circuit containing capacitance C and resistance R, the applied voltage V is the phasor sum of V_R and V_C (see *Figure 35.3(b)*) **and thus the current I leads the applied voltage V by an angle lying between 0° and 90°** (depending on the values of V_R and V_C), shown as angle α.

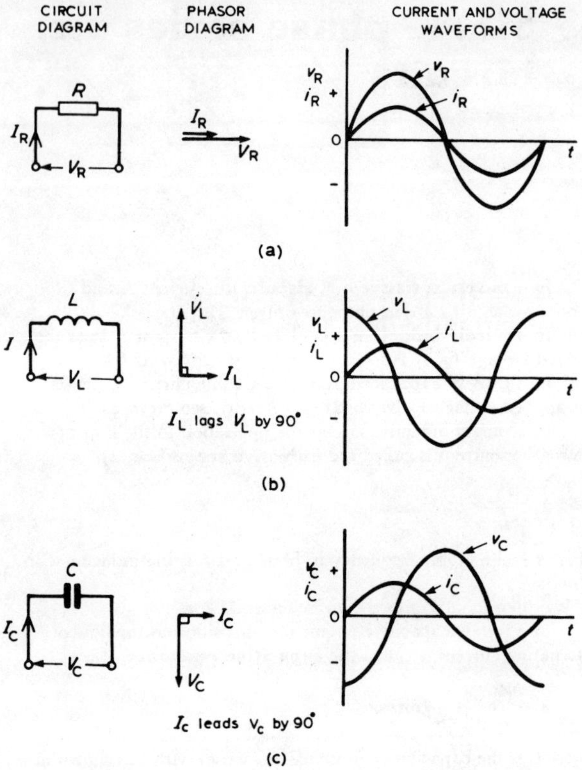

CIRCUIT DIAGRAM
PHASOR DIAGRAM
CURRENT AND VOLTAGE WAVEFORMS
R
I_R
V_R
v_R
i_R
+
0
-
t
(a)
L
I
V_L
I_L
v_L
i_L
I_L lags V_L by 90°
(b)
C
I_C
V_C
v_C
i_C
I_c leads V_c by 90°
(c)

Figure 35.1

X_L (Ω)
0
f (Hz)
(a)
X_C (Ω)
0
f (Hz)
(b)

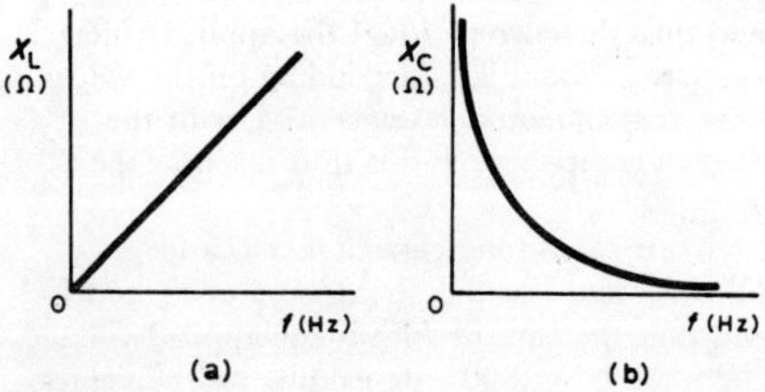

Figure 35.2

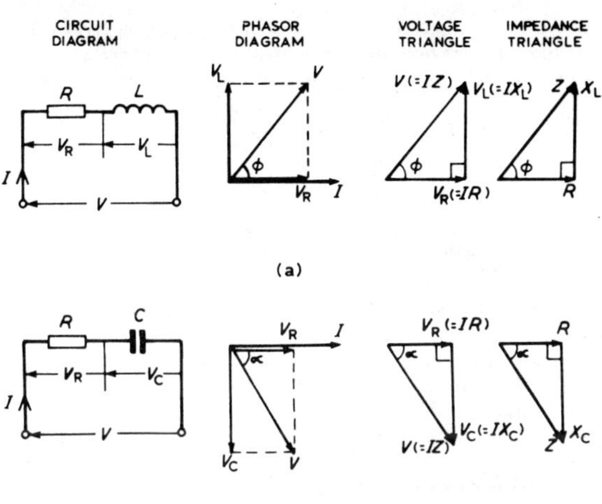

Figure 35.3

8 In an a.c. circuit, the ratio

$$\frac{\text{applied voltage } V}{\text{current } I}$$

is called the **impedance** Z,

i.e. $\boldsymbol{Z=\frac{V}{I}}$ **ohms.**

9 From the phasor diagrams of *Figure 35.3*, the '**voltage triangles**' are derived.

(a) For the *R-L* circuits:

$V=\surd(V_R^2+V_L^2)$ by Pythagoras' theorem), and

$\tan \phi=\frac{V_L}{V_R}$ (by trigonometric ratios)

(b) For the *R-C* circuit:

$V=\surd(V_R^2+V_C^2)$, and

$\tan \alpha=\frac{V_C}{V_R}$

10 If each side of the voltage triangles in *Figure 35.3* is divided by current I then the '**impedance triangles**' are derived.

(a) For the *R-L* circuit: $Z=\sqrt{(R^2+X_L^2)}$

$$\tan\phi=\frac{X_L}{R},\ \sin\phi=\frac{X_L}{Z}\text{ and }\cos\phi=\frac{R}{Z}$$

(b) For the *R-C* circuit: $Z=\sqrt{(R^2+X_C^2)}$

$$\tan\alpha=\frac{X_C}{R},\ \sin\alpha=\frac{X_C}{Z}\text{ and }\cos\alpha=\frac{R}{Z}$$

11 In an a.c. series circuit containing resistance R, inductance L and capacitance C, the applied voltage V is the phasor sum of V_R, V_L and V_C (see *Figure 35.4*). V_L and V_C are anti-phase and there are three phasor diagrams possible — each depending on the relative values of V_L and V_C.

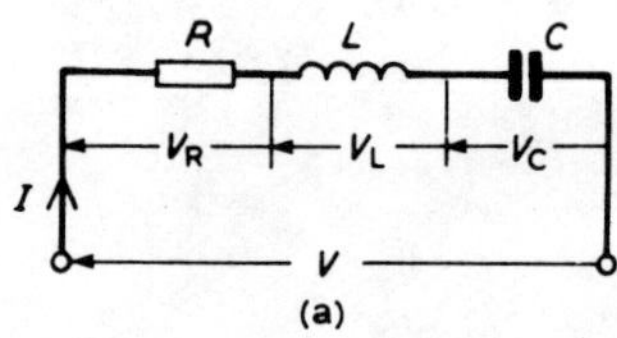

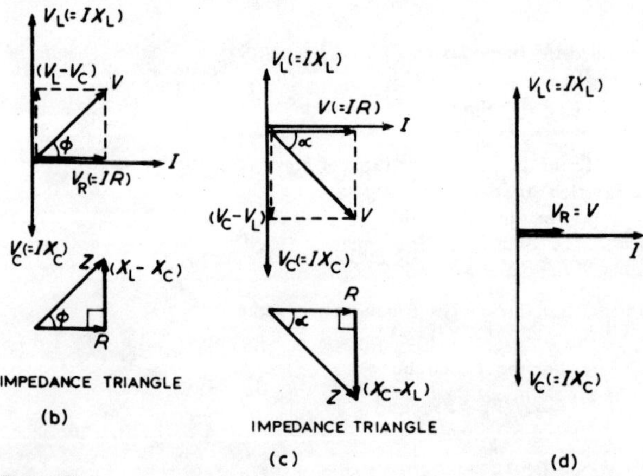

Figure 35.4

12 When $X_L > X_C$ (see *Figure 35.4(b)*):

$$Z = \sqrt{[R^2 + (X_L - X_C)^2]}$$

$$\text{and } \tan\phi = \frac{(X_L \mp X_C}{R}$$

13 When $X_C > X_L$ (see *Figure 35.4(c)*):

$$Z = R_2\sqrt{[R^2 + (X_C - X_L)^2}$$

$$\text{and } \tan\alpha = \frac{(X_C - X_L)}{R}$$

14 When $X_L = X_C$ (see *Figure 35.3(d)*), the applied voltage V and the current I are in phase. This effect is called **series resonance**. At resonance:

(i) $V_L = V_C$

(ii) $Z = R$ (i.e. the minimum circuit impedance possible in an *L-C-R* circuit).

(iii) $I = \frac{V}{R}$ (i.e. the maximum current possible in an *L-C-R* circuit).

(iv) Since $X_L = X_C$, then $2\pi f_r L = \frac{1}{2\pi f_r C}$,

from which, $f_r = \frac{1}{2\pi\sqrt{(LC)}}$ Hz, where f_r is the resonant frequency.

15 At resonance, if R is small compared with X_L and X_C, it is possible for V_L and V_C to have voltages many times greater than the supply voltage V (see *Figure 35.4(d)*).

Voltage magnification at resonance

$$= \frac{\text{voltage across } L \text{ (or } C)}{\text{supply voltage } V}$$

This ratio is a measure of the quality of a circuit (as a resonator or tuning device) and is called the **Q-factor**.

Hence **Q-factor** $= \frac{V_L}{V} = \frac{IX_L}{IR} = \frac{X_L}{R} = \mathbf{\frac{2\pi f_r L}{R}}$

$\left(\text{Alternatively, Q-factor} = \frac{V_C}{V} = \frac{IX_C}{IR} = \frac{X_C}{R} = \frac{1}{2\pi f_r CR}\right)$

At resonance $f_r = \frac{1}{2\pi\sqrt{(LC)}}$ i.e. $2\pi f_r = \frac{1}{\sqrt{(LC)}}$

Hence **Q-factor** $=\frac{2\pi fL}{R}=\frac{1}{\sqrt{(LC)}}\left(\frac{L}{R}\right)=\frac{\mathbf{1}}{\boldsymbol{R}}\sqrt{\left(\frac{\boldsymbol{L}}{\boldsymbol{C}}\right)}$

16 (a) The advantages of a high Q-factor in a series high frequency circuit is that such circuits need to be 'selective' to signals at the resonant frequency and must be less selective to other frequencies.

(b) The disadvantage of a high Q-factor in a series power circuit is that it can lead to dangerously high voltages across the insulation and may lead to electrical breakdown.

17 For series-connected impedances the total circuit impedance can be represented as a single *L-C-R* circuit by combining all values of resistance together, all values of inductance together and all values of capacitance together (remembering that for series connected capacitors $1/C=1/C_1+1/C_2+\cdots$). For example, the circuit of *Figure 35.5(a)* showing three impedances has an equivalent circuit of *Figure 35.5(b)*.

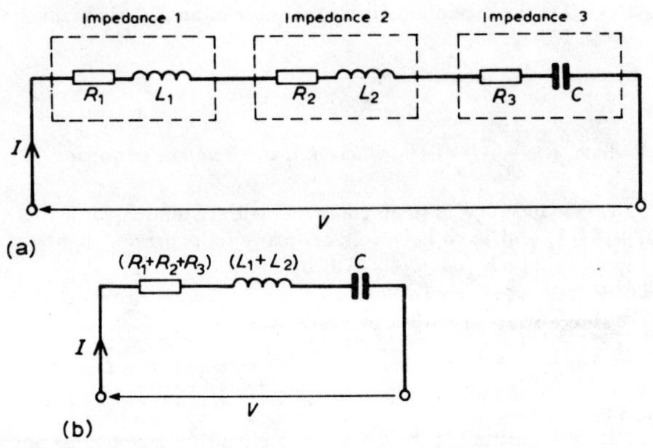

Figure 35.5

18 (a) For a purely resistive a.c. circuit, the average power dissipated, P, is given by: $P=VI=I^2R=V^2/R$ watts (V and I being r.m.s. values). (See *Figure 35.6(a)*).

(b) For a purely inductive a.c. circuit, the average power is zero. See *Figure 35.6(b)*.

(c) For a purely capacitive a.c. circuit, the average power is zero. See *Figure 35.6(c)*.

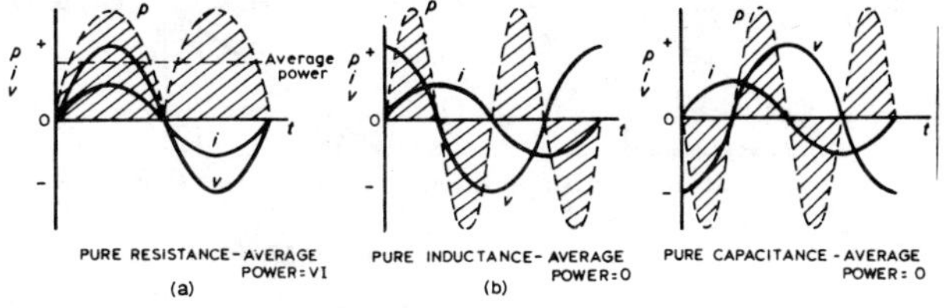

Figure 35.6

In *Figure 35.6(a)–(c)*, the value of power at any instant is given by the product of the voltage and current at that instant, i.e. the instantaneous power, $p = vi$, as shown by the broken lines.

19 *Figure 35.7* shows current and voltage waveforms for an *R-L* circuit where the current lags the voltage by angle ϕ. The waveform for power (where $p = vi$) is shown by the broken line, and its shape, and hence average power, depends on the value of angle ϕ.

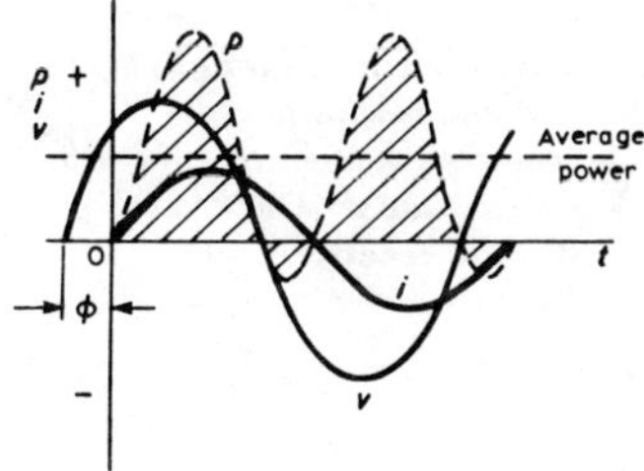

Figure 35.7

For an *R-L*, *R-C* or *L-C-R* series a.c. circuit, the average power P is given by:

$P = VI \cos \phi$ **watts** or $P = I^2R$ **watts** (V and I being r.m.s. values).

20 *Figure 35.8(a)* shows a phasor diagram in which the current I lags the applied voltage V by angle ϕ. The horizontal component of V is $V \cos \phi$ and the vertical component of V is $V \sin \phi$. If each of the voltage phasors is multiplied by I, *Figure 35.8(b)* is obtained and is known as the '**power triangle**'.

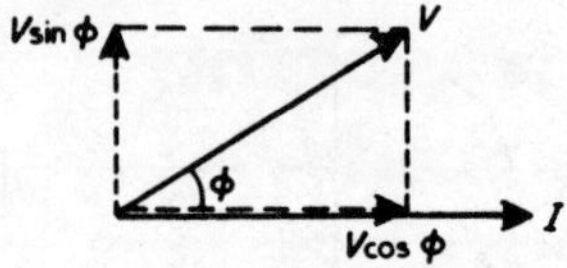

(a) PHASOR DIAGRAM

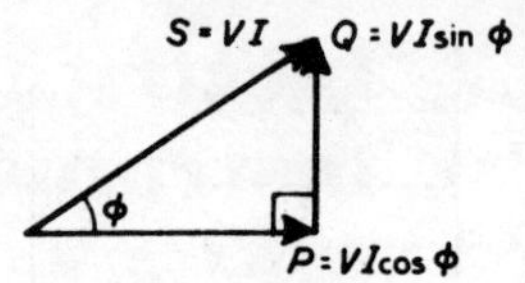

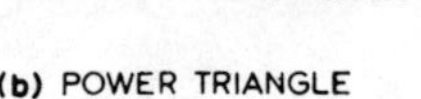

(b) POWER TRIANGLE

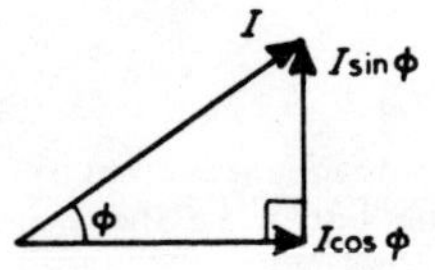

(c) CURRENT TRIANGLE

Figure 35.8

Apparent power, $S=VI$ voltamperes (VA)
True or active power, $P=VI\cos\phi$ watts (W)
Reactive power, $Q=VI\sin\phi$ reactive voltamperes (VAr)

21 If each of the phasors of the power triangle of *Figure 35.8(b)* is divided by V, *Figure 35.8(c)* is obtained and is known as the '**current triangle**'. The horizontal component of current, $I\cos\phi$, is called the **active** or the **in-phase component**. The vertical component of current, $I\sin\phi$, is called the **reactive** or the **quadrature component**.

22 $$\text{Power factor}=\frac{\text{True power } P}{\text{Apparent power } S}$$

For sinusoidal voltages and currents,

$$\text{power factor}=\frac{P}{S}=\frac{VI\cos\phi}{VI}$$

i.e. $\mathbf{p.f.}=\boldsymbol{\cos\phi}=\dfrac{\boldsymbol{R}}{\boldsymbol{Z}}$ (from *Figure 35.3(a)*).

(The relationships stated in paras 20 to 22 are also true when current I leads voltage V.)

36 Electrical measuring instruments and measurements

1 Tests and measurements are important in designing, evaluating, maintaining and servicing electrical circuits and equipment. In order to detect electrical quantities such as current, voltage, resistance or power, it is necessary to transform an electrical quantity or condition into a visible indication. This is done with the aid of instruments (or meters) that indicate the magnitude of quantities either by the position of a pointer moving over a graduated scale (called an analogue instrument) or in the form of a decimal number (called a digital instrument).

2 All analogue electrical indicating instruments require three essential devices. These are:

(a) **A deflecting or operating device** A mechanical force is produced by the current or voltage which causes the pointer to deflect from its zero position.

(b) **A controlling device** The controlling force acts in opposition to the deflecting force and ensures that the deflection shown on the meter is always the same for a given measured quantity. It also prevents the pointer always going to the maximum deflection. There are two main types of controlling device — spring control and gravity control.

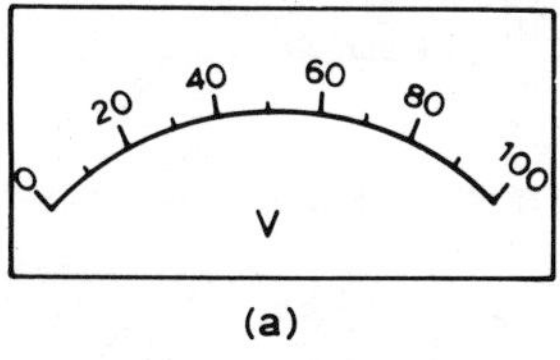

(a)

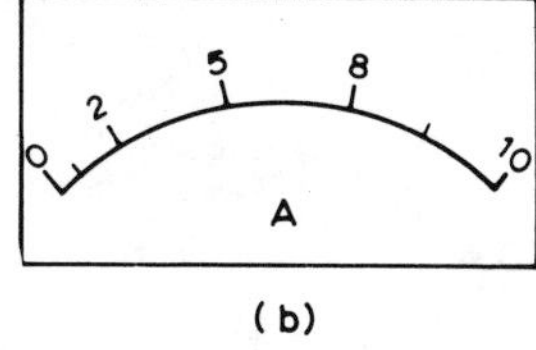

(b)

Figure 36.1

4 Comparison of the moving coil, moving iron and moving coil rectifier instruments

Type of instrument	*Moving coil*	*Moving iron*	*Moving coil rectifier*
Suitable for measuring:	Direct current and voltage.	Direct and alternating current and voltage (reading in r.m.s. value).	Alternating current and voltage (reads average value but scale is adjusted to give r.m.s. value for sinusoidal waveforms).
Scale	Linear	Non-linear	Linear
Method of control	Hairsprings	Hairsprings	Hairsprings
Method of damping	Eddy current	Air	Eddy current
Frequency limits	—	20 Hz–200 Hz	20 Hz–100 Hz

Type of instrument	*Moving coil*	*Moving iron*	*Moving coil rectifier*
Advantages	1. Linear scale. 2. High sensitivity. 3. Well shielded from stray magnetic fields. 4. Lower power consumption.	1. Robust construction. 2. Relatively cheap. 3. Measures d.c. and a.c. 4. In frequency range 20–100 Hz reads r.m.s. correctly, regardless of supply waveform.	1. Linear scale. 2. High sensitivity. 3. Well shielded from stray magnetic fields. 4. Lower power consumption. 5. Good frequency range.
Disadvantages	1. Only suitable for d.c. 2. More expensive than moving iron type. 3. Easily damaged.	1. Non-linear scale. 2. Affected by stray magnetic fields. 3. Hysteresis errors in d.c. circuits. 4. Liable to temperature errors. 5. Due to the inductance of the solenoid, readings can be affected by variation of frequency.	1. More expensive than moving iron type. 2. Errors caused when supply is non-sinusoidal.

(c) **A damping device** The damping force ensures that the pointer comes to rest in its final position quickly and without undue oscillation. There are three main types of damping used — eddy-current damping, air-friction damping and fluid-friction damping.

3 There are basically two types of scale — linear and non-linear. A **linear scale** is shown in *Figure 36.1(a)* where each of the divisions or graduations are evenly spaced. The voltmeter shown has a range 0–100 V, i.e. a full scale deflection (FSD) of 100 V.

A **non-linear scale** is shown in *Figure 36.1(b)*. The scale is cramped at the beginning and the graduations are uneven throughout the range. The ammeter shown has a FSD of 10 A.

Principle of operation of a moving coil instrument

5 A moving coil instrument operates on the motor principle. When a conductor carrying current is placed in a magnetic field, a force F is exerted on the conductor, given by $F=BIl$. If the flux density B is made constant (by using permanent magnets) and the conductor is a fixed length (say, a coil) then the force will depend only on the current flowing in the conductor.

In a moving coil instrument, a coil is placed centrally in the gap between shaped pole pieces as shown by the front elevation in *Figure 36.2(a)*. The coil is supported by steel pivots, resting in jewel bearings, on a cylindrical iron core. Current is led into and out of the coil by two phosphor bronze spiral hairsprings which are wound in opposite directions to minimise the effect of temperature change and to limit the coil swing (i.e. to control the movement) and return the movement to zero position when no current flows.

Current flowing in the coil produces forces as shown in *Figure 36.2(b)*, the directions being obtained by Fleming's left-hand rule. The two forces, F_A and F_B, produce a torque which will move the coil in a clockwise direction, i.e. move the pointer from left to right. Since force is proportional to current the scale is linear.

When the aluminium frame, on which the coil is wound, is rotated between the poles of the magnet, small currents (called eddy currents) are induced into the frame, and this provides automatically the necessary damping of the system due to the reluctance of the former to move within the magnetic field.

The moving coil instrument will only measure direct current or voltage and the terminals are marked positive and negative to

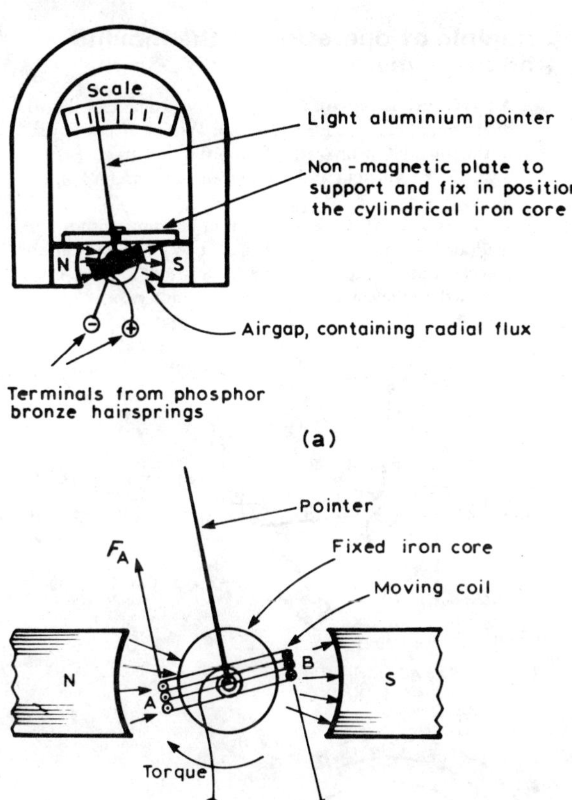

Figure 36.2

ensure that the current passes through the coil in the correct direction to deflect the pointer 'up the scale'.

The range of this sensitive instrument is extended by using shunts and multipliers. (See para. 10.)

Principle of operation of the moving iron instrument

6 (a) An **attraction type** of moving iron instrument is shown diagrammatically in *Figure 36.3(a)*. When current flows in the solenoid, a pivoted soft iron disc is attracted towards the solenoid and the movement causes a pointer to move across a scale.

(b) In the **repulsion type** moving iron instrument shown diagrammatically in *Figure 36.3(b)*, two pieces of iron are placed inside the solenoid, one being fixed, and the other attached to the spindle carrying the pointer.

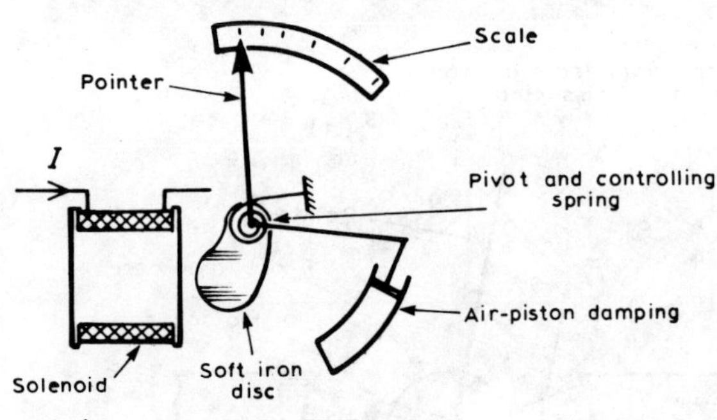

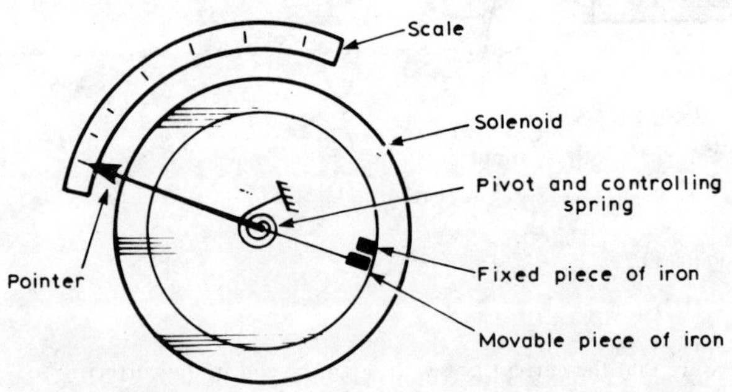

Figure 36.3

When current passes through the solenoid, the two pieces of iron are magnetised in the same direction and therefore repel each other. The pointer thus moves across the scale. The force moving the pointer is, in each type, proportional to I^2. Because of this the direction of current does not matter and the moving iron instrument can be used on d.c. or a.c. The scale, however, is non-linear.

7 A moving coil instrument, which measures only d.c., may be used in conjunction with a bridge rectifier circuit as shown in *Figure 36.4* to provide an indication of alternating currents and voltages. The average value of the full wave rectified current is

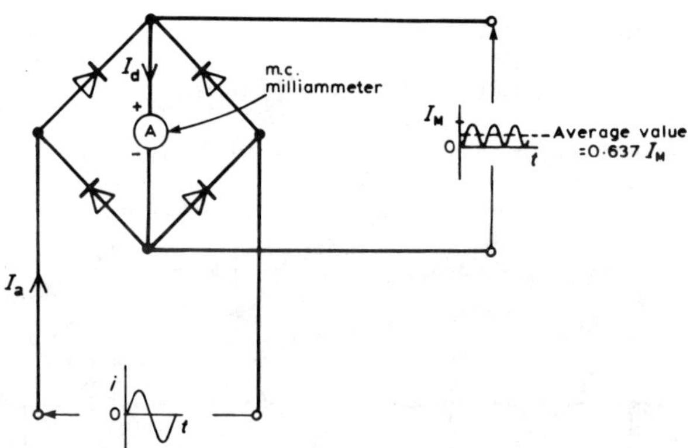

Figure 36.4

$0.637I_M$. However, a meter being used to measure a.c. is usually calibrated in r.m.s. values. For sinusoidal quantities the indication is

$$\frac{0.707I_M}{0.637I_M}, \text{ i.e. } 1.11 \text{ times the mean value.}$$

Rectifier instruments have scales calibrated in r.m.s. quantities and it is assumed by the manufacturer that the a.c. is sinusoidal.

8 An **ammeter**, which measures current, has a low resistance (ideally zero) and must be connected in series with the circuit.

9 A **voltmeter**, which measures p.d., has a high resistance

(ideally infinite) and must be connected in parallel with the part of the circuit whose p.d. is required.

Shunts and multipliers

10 There is no difference between the basic instrument used to measure current and voltage since both use a milliammeter as their basic part. This is a sensitive instrument which gives FSD for currents of only a few milliamperes. When an ammeter is required to measure currents of larger magnitude, a proportion of the current is diverted through a low value resistance connected in parallel with the meter. Such a diverting resistance is called a **shunt**.

From *Figure 36.5(a)*, $V_{PQ} = V_{RS}$.

Hence, $I_a r_a = I_S R_S$

Thus the value of the shunt, $\boldsymbol{R_S = \frac{I_a r_a}{I_S}}$ **ohms**

The milliammeter is converted into a voltmeter by connecting a high resistance (called a **multiplier**) in series with it as shown in *Figure 36.5(b)*. From *Figure 36.5(b)*, $V = V_a + V_M = Ir_a + IR_M$

Thus the value of the multiplier, $\boldsymbol{R_M = \frac{V - Ir_a}{I}}$ **ohms**

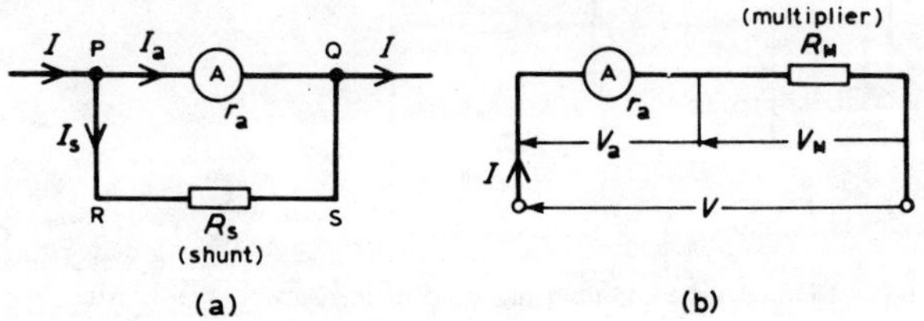

Figure 36.5

For example, let a m.c. instrument have a FSD of 20 mA and a resistance of 25 Ω. To enable the instrument to be used as a 0–10 A ammeter, a shunt resistance R_S needs to be connected in parallel with the instrument. From *Figure 36.5(a)*.

$I = 10$ A, $I_S = I - I_a = 10 - 0.020 = 9.98$ A.

Hence the value of R_S is given by:

$$R_S = \frac{I_a r_a}{I_S} = \frac{(0.020)(25)}{9.98} = \mathbf{50.10\ m\Omega}.$$

To enable the instrument to be used as a 0 to 100 V voltmeter, a multiplier R_M needs to be connected in series with the instrument, the value of R_M being given by:

$$R_M = \frac{V - Ir_a}{I} = \frac{100 - (0.020)(25)}{0.020} = \mathbf{4.975\ k\Omega}$$

11 An **ohmmeter** is an instrument for measuring electrical resistance. A simple ohmmeter circuit is shown in *Figure 36.6(a)*. Unlike the ammeter or voltmeter, the ohmmeter circuit does not receive the energy necessary for its operation from the circuit under test. In the ohmmeter this energy is supplied by a self-contained source of voltage, such as a battery. Initially, terminals XX are short-circuited and R adjusted to give FSD on the milliammeter. If current I is at a maximum value and voltage E is constant, then resistance $R = E/I$ is at a minimum value. Thus FSD on the milliammeter is made zero on the resistance scale. When terminals XX are open circuited no current flows and R $(= E/O)$ is infinity, ∞. The milliammeter can thus be calibrated directly in ohms. A cramped (non-linear) scale results and is 'back to front' (as shown in *Figure 36.6(b)*). When calibrated, an unknown resistance is

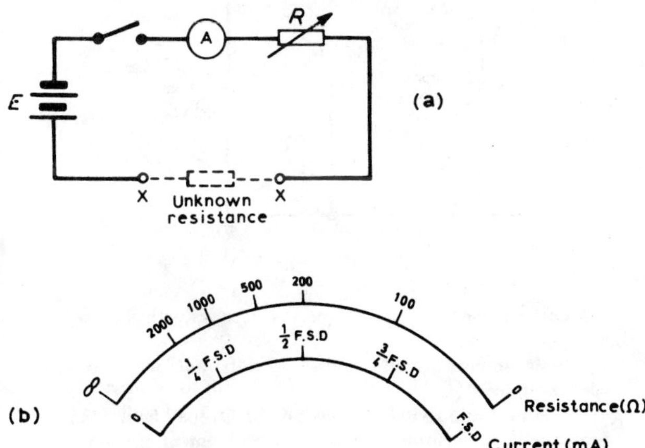

Figure 36.6

placed between terminals XX and its value determined from the position of the pointer on the scale. An ohmmeter designed for measuring low values of resistance is called a **continuity tester**. An ohmmeter designed for measuring high values of resistance (i.e. megohms) is called an **insulation resistance tester** (or **'Megger'**).

12 Instruments are manufactured that combine a moving coil meter with a number of shunts and series multipliers, to provide a range of readings on a single scale graduated to read current and voltage. If a battery is incorporated into the instrument then resistance can also be measured.

Such instruments are called **multimeters** or **universal instruments** or **multirange instruments**. An **Avometer** is a typical example. A particular range may be selected either by the use of separate terminals or by a selector switch. Only one measurement can be performed at one time. Often such instruments can be used in a.c. as well as d.c. circuits when a rectifier is incorporated in the instrument.

13 A **wattmeter** is an instrument for measuring electrical power in a circuit. *Figure 36.7* shows typical connections of a wattmeter used for measuring power supplied to a load. The instrument has two coils:

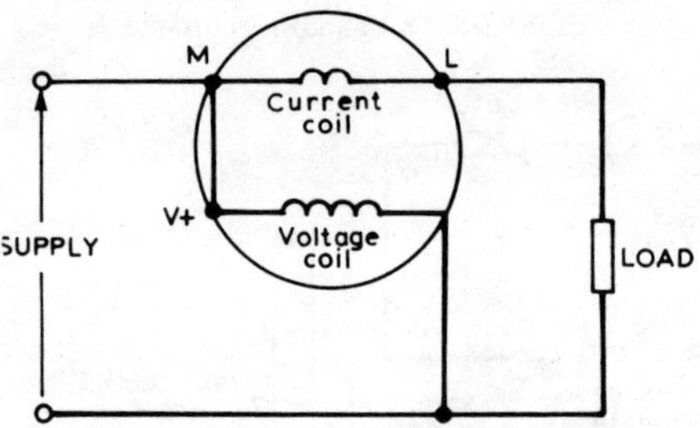

Figure 36.7

(i) a **current coil**, which is connected in series with the load (like an ammeter), and

(ii) a **voltage coil**, which is connected in parallel with the load (like a voltmeter).

14 **The cathode ray oscilloscope (CRO)** may be used in the observation of waveforms and for the measurement of voltage, current, frequency, phase and periodic time.

(i) With **direct voltage measurements**, only the Y amplifier 'volts/cm' switch on the CRO is used. With no voltage applied to the Y plates the position of the spot trace on the screen is noted. When a direct voltage is applied to the Y plates the new position of the spot trace is an indication of the magnitude of the voltage. **For example, in *Figure 36.8(a)*, with no voltage applied to the Y** plates, the spot trace is in the centre of the screen (initial position) and then the spot trace moves 2.5 cm to the final position shown, on application of a d.c. voltage. With the 'V/cm' switch on 10 V/cm the magnitude of the direct voltage is 2.5 cm × 10 V/cm, i.e. 25 V.

(ii) With **alternating voltage measurements**, let a sinusoidal waveform be displayed on a CRO screen as shown in *Figure 36.8(b)*. If the 'variable' switch is on, say, 5 ms/cm then the

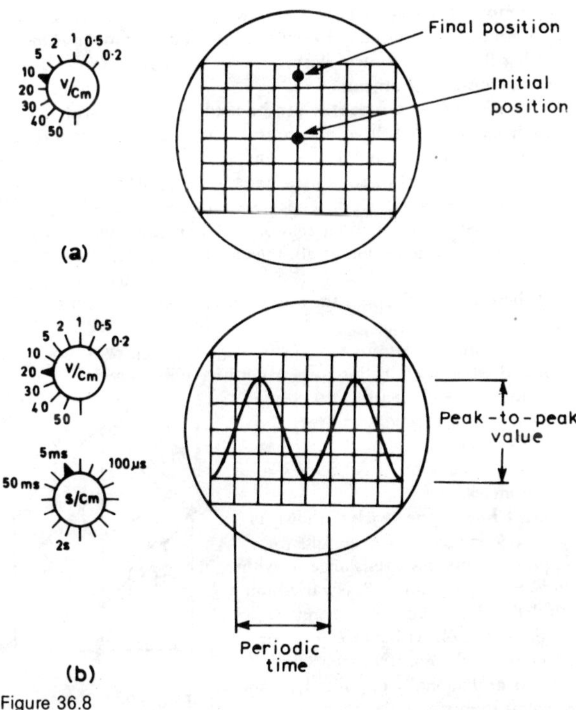

Figure 36.8

periodic time T of the sine wave is 5 ms/cm × 4 cm, i.e. 20 ms or 0.02 s.

Since frequency $f=\frac{1}{T}$,

frequency$=\frac{1}{0.02}=$ **50 Hz**.

If the 'volts/cm' switch is on, say 20 V/cm then the **amplitude** or **peak value** of the sine wave shown is 20 V/cm × 2 cm, i.e. 40 V.

Since r.m.s. voltage$=\frac{\text{peak voltage}}{\sqrt{2}}$

r.m.s. voltage$=\frac{40}{\sqrt{2}}=$ **28.28 V**.

Double beam oscilloscopes are useful whenever two signals are to be compared simultaneously. The CRO demands reasonable skill in adjustment and use. However its greatest advantage is in observing the shape of a waveform — a feature not possessed by other measuring instruments.

15 An **electronic voltmeter** can be used to measure with accuracy e.m.f. or p.d. from millivolts to kilovolts by incorporating in its design amplifiers and attentuators.

16 A **null method of measurement** is a simple, accurate and widely used method which depends on an instrument reading being adjusted to read zero current only. The method assumes:
(i) if there is any deflection at all, then some current is flowing, and
(ii) if there is no deflection, then no current flows (i.e. a null condition).
Hence it is unnecessary for a meter sensing current flow to be calibrated when used in this way. A sensitive milliammeter or microammeter with centre zero position setting is called a **galvanometer**. Two examples where the method is used are in the Wheatstone bridge and the d.c. potentiometer.

17 A **Wheatstone bridge**, shown in *Figure 36.9*, is used in d.c. circuits to compare an unknown resistance R_x with others of known values. R_3 is varied until zero deflection is obtained on the galvanometer, G. At balance (i.e. zero deflection on the galvanometer) the products of diagonally opposite resistances are equal to one another,

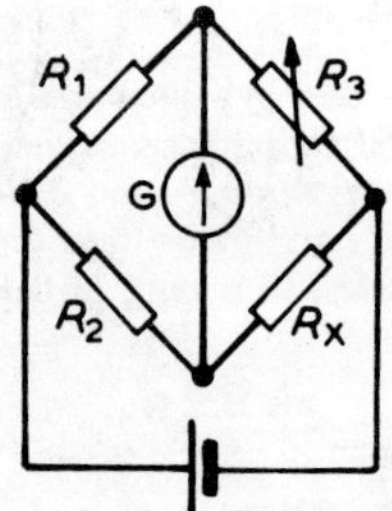

Figure 36.9

i.e. $R_1R_x=R_2R_3$

from which, $$\boxed{R_x=\frac{R_2R_3}{R_1}\ \text{ohms.}}$$

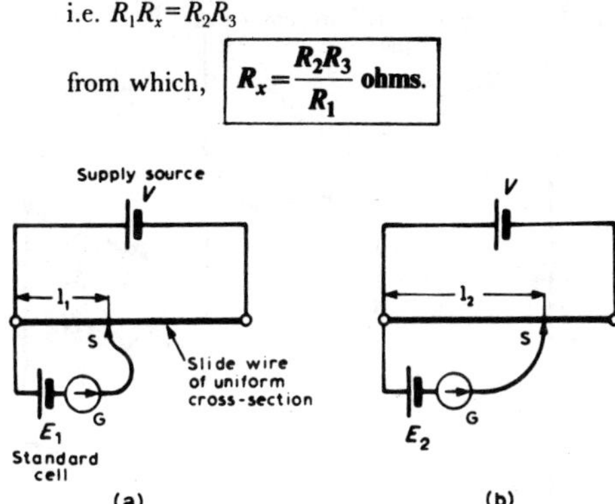

Figure 36.10

18 The **d.c. potentiometer** is a null-balance instrument used for determining values of e.m.f.'s and p.d.'s by comparison with a known e.m.f. or p.d. In *Figure 36.10(a)*, using a standard cell of known e.m.f. E_1, the slider S is moved along the slide wire until balance is obtained (i.e. the galvanometer deflection is zero), shown as length l_1. The standard cell is now replaced by a cell of unknown e.m.f., E_2 (see *Figure 36.10(b)*) and again balance is obtained (shown as l_2). Since $E_1 \propto l_1$ and $E_2 \propto l_2$.

then $\frac{E_1}{E_2}=\frac{l_1}{l_2}$ and $$\boxed{E_2=E_1\left(\frac{l_2}{l_1}\right)\ \text{volts.}}$$

A.C. bridges

19 A Wheatstone bridge type circuit, shown in *Figure 36.11*, may be used in a.c. circuits to determine unknown values of inductance and capacitance, as well as resistance. When the potential differences across Z_3 and Z_x (or across Z_1 and Z_2) are equal to magnitude and phase, then the current flowing through the galvanometer, G, is zero. At balance, $Z_1Z_x=Z_2Z_3$, from which,

$$Z_x=\frac{Z_2Z_3}{Z_1}\ \Omega$$

20 There are many forms of a.c. bridge, and these include: the Maxwell, Hay, Owen and Heaviside bridges for measuring

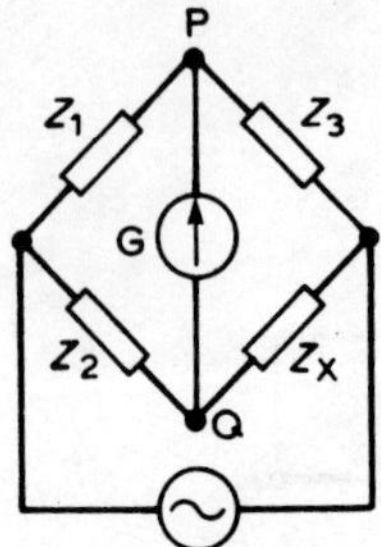

Figure 36.11

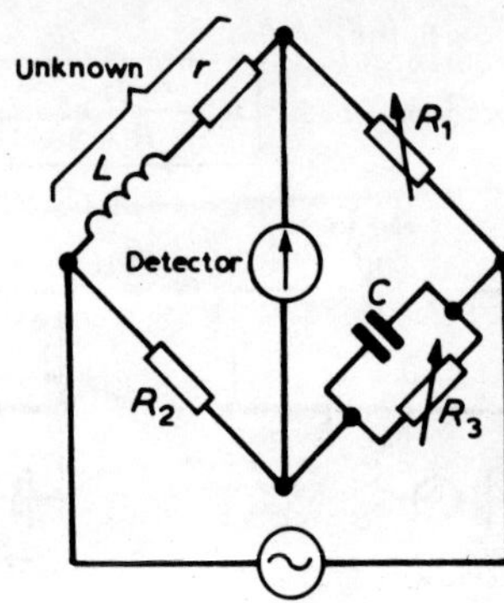

Figure 36.12

inductance, and the De Sauty, Schering and Wien bridges for measuring capacitance. A commercial or universal bridge is one which can be used to measure resistance, inductance or capacitance.

21 **Maxwell's bridge** for measuring the inductance L and resistance r of an inductor is shown in *Figure 36.12*.

At balance, $Z_1Z_2 = Z_3Z_4$

Using complex quantities,

$$Z_1 = R_1,\ Z_2 = R_2$$

$$Z_3 = \frac{R_3(-jX_C)}{(R_3 - jX_C)}\left(\text{i.e. } \frac{\text{product}}{\text{sum}}\right) \text{ and } Z_4 = r + jX_L$$

Hence, $R_1R_2 = \dfrac{R_3(-jX_C)}{(R_3 - jX_C)}(r + jX_L)$

i.e. $R_1R_2(R_3 - jX_C) = (-jR_3X_C)(r + jX_L)$

$$R_1R_2R_3 - jR_1R_2X_C = -jrR_3X_C - j^2R_3X_CX_L$$

i.e. $R_1R_2R_3 - jR_1R_2X_C = -jrR_3X_C + R_3X_CX_L$ (since $j^2 = -1$)

Equating the real parts gives: $R_1R_2R_3 = R_3X_CX_L$

from which, $X_L = \dfrac{R_1R_2}{X_C}$

i.e. $2\pi fL=\dfrac{R_1R_2}{\dfrac{1}{2\pi fC}}=R_1R_2(2\pi fC)$

Hence **inductance, $L=R_1R_2C$ henry** (1)

Equating the imaginary parts gives:

$$-R_1R_2X_C=-rR_3X_C$$

from which, **resistance $r=\dfrac{R_1R_2}{R_3}$ ohms** (2)

From equation (1), $R_2=\dfrac{L}{R_1C}$ and from

equation (2), $R_3=\dfrac{R_1}{r}R_2$. Hence $R_3=\dfrac{R_1}{r}\left(\dfrac{L}{R_1C}\right)=\dfrac{L}{Cr}$

If the frequency is constant then

$$R_3\propto\frac{L}{r}\propto w\frac{L}{r}\propto \text{Q-factor.}$$

Thus the bridge can be adjusted to give a direct indication of Q-factor.

22 The **Q-factor** for a series *L-C-R* circuit is the voltage magnification at resonance,

i.e. $\text{Q-factor}=\dfrac{\text{voltage across capacitor}}{\text{supply voltage}}=\dfrac{V_c}{V}$

(see Chapter 35, para. 15).

The simplified circuit of a **Q-meter**, used for measuring Q-factor, is shown in *Figure 36.13*. Current from a variable frequency oscillator flowing through a very low resistance r develops a variable frequency voltage, V_r, which is applied to a series *L-R-C* circuit. The frequency is then varied until resonance causes voltage V_c to reach a maximum value. At resonance V_r and V_c are noted.

Then $\text{Q-factor}=\dfrac{V_c}{V_r}=\dfrac{V_c}{Ir}$

In a practical Q-meter, V_r is maintained constant and the electronic voltmeter can be calibrated to indicate the Q-factor directly. If a variable capacitor C is used and the oscillator is set at a given frequency, then C can be adjusted to give resonance. In this way inductance L may be calculated using

$$f_r=\frac{1}{2\pi\sqrt{(LC)}}$$

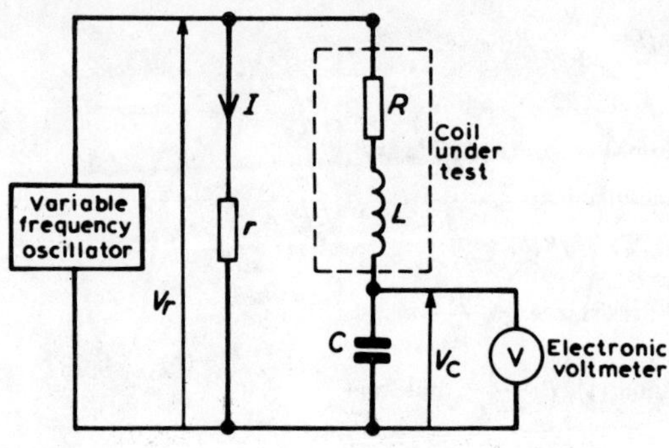

Figure 36.13

Since $Q = \dfrac{2\pi fL}{R}$, then R may be calculated.

Q-meters operate at various frequencies and instruments exist with frequency ranges from 1 kHz to 50 MHz. Errors in measurement can exist with Q-meters since the coil has an effective parallel self capacitance due to capacitance between turns. The accuracy of a Q-meter is approximately $\pm 5\%$.

Waveform harmonics

23 (i) Let an instantaneous voltage v be represented by $v = V_M \sin 2\pi ft$ volts. This is a waveform which varies sinusoidally with time t, has a frequency f, and a maximum value, V_M. Alternating voltages are usually assumed to have wave-shapes which are sinusoidal where only one frequency is present. If the waveform is not sinusoidal it is called a **complex wave**, and, whatever its shape, it may be split up mathematically into components called the **fundamental** and a number of **harmonics**. This process is called **harmonic analysis**. The fundamental (or first harmonic) is sinusoidal and has the supply frequency, f, the other harmonics are also sine waves having frequencies which are integer multiples of f. Thus, if the supply frequency is 50 Hz, then the third harmonic frequency is 150 Hz, the fifth 250 Hz, and so on.
(ii) A complex waveform comprising the sum of the fundamental and a third harmonic of about half the amplitude of the

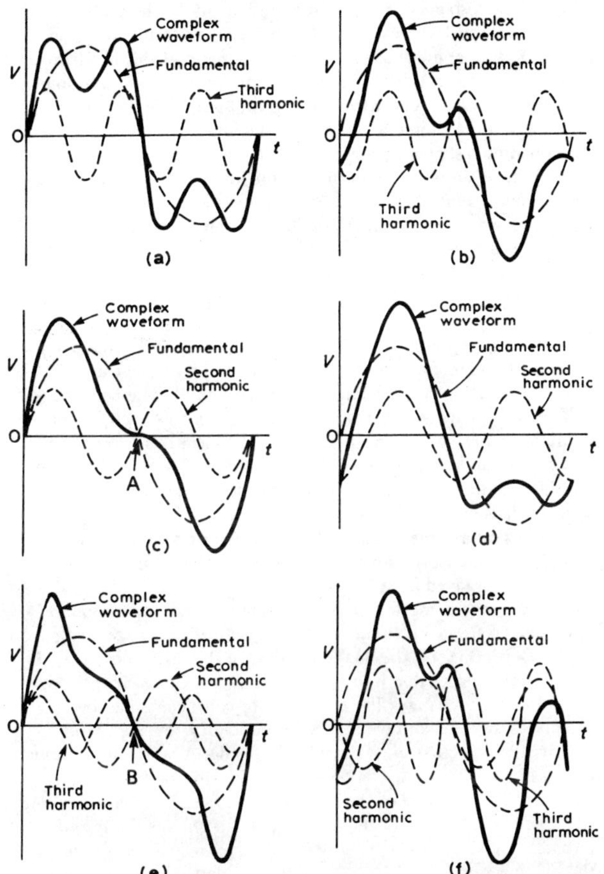

Figure 36.14

fundamental is shown in *Figure 36.14(a)*, both waveforms being initially in phase with each other. If further odd harmonic waveforms of the appropriate amplitudes are added, a good approximation to a square wave results.

In *Figure 36.14(b)* the third harmonic is shown having an

initial phase displacement from the fundamental. The positive and negative half cycles of each of the complex waveforms shown in *Figure 36.14(a)* and *(b)* are identical in shape, and this is a feature of waveforms containing the fundamental and only odd harmonics.
(iii) A complex waveform comprising the sum of the fundamental and a second harmonic of about half the amplitude of the fundamental is shown in *Figure 36.14(c)*, each waveform being initially in phase with each other. If further even harmonics of appropriate amplitudes are added a good approximation to a triangular wave results. In *Figure 36.14(c)* the negative cycle appears as a mirror image of the positive cycle about point A. In *Figure 36.14(d)* the second harmonic is shown with an initial phase displacement from the fundamental and the positive and negative half cycles are dissimilar.
(iv) A complex waveform comprising the sum of the fundamental, a second harmonic and a third harmonic is shown in *Figure 36.14(e)*, each waveform being initially 'in-phase'. The negative half cycle appears as a mirror image of the positive cycle about point B. In *Figure 36.14(f)*, a complex waveform comprising the sum of the fundamental, a second harmonic and a third harmonic are shown with initial phase displacement. The positive and negative half cycles are seen to be dissimilar.

The features mentioned relative to *Figures 36.16(a)* to *(f)* make it possible to recognise the harmonics present in a complex waveform displayed on a CRO.

24 Some measuring instruments depend for their operation on power taken from the circuit in which measurements are being made. Depending on the 'loading' effect of the instrument (i.e. the current taken to enable it to operate), the prevailing circuit conditions may change. The resistance of voltmeters may be calculated since each have a stated **sensitivity** (or '**figure of merit**'), often stated in 'kΩ per volt' of FSD. A voltmeter should have as high a resistance as possible (– ideally infinite). In a.c. circuits the impedance of the instrument varies with frequency and thus the loading effect of the instrument can change. Electronic measuring instruments have advantages over instruments such as the moving iron or moving coil meters, in that they have a much higher input resistance (some as high as 1000 MΩ) and can handle a much wider range of frequency (from d.c. up to MHz).

25 Instruments for a.c. measurements are generally calibrated with a sinusoidal alternating waveform to indicate r.m.s. values when a sinusoidal signal is applied to the instrument. Some instruments, such as the moving iron and electrodynamic instruments, give a true r.m.s. indication. With other instruments the indication is either scaled up from the mean value (such as

with the rectifier moving coil instrument) or scaled down from the peak value. Sometimes quantities to be measured have complex waveforms (see para. 23), and whenever a quantity is non-sinusoidal, errors in instrument readings can occur if the instrument has been calibrated for sine waves only. Such waveform errors can be largely eliminated by using electronic instruments.

26 **Errors** are always introduced when using instruments to measure electrical quantities. Besides possible errors introduced by the operator or by the instrument disturbing the circuit, errors are caused by the limitations of the instrument used.

The calibration accuracy of an instrument depends on the precision with which it is constructed. Every instrument has a margin of error which is expressed as a percentage of the instruments full scale deflection. For example, an instrument may have an accuracy of $\pm 2\%$ of FSD. Thus, if a voltmeter has a FSD of 100 V and it indicates say 60 V, then the actual voltage measured may be anywhere between 60 V $\pm$ (2% of 100 V), i.e. 60 ± 2 V, i.e. between 58 V and 62 V. As a percentage of the voltmeter reading this error is $\pm 2/60 \times 100\%$, i.e. $\pm 3.33\%$. Hence the accuracy can be expressed as 60 V $\pm 3.33\%$. It follows that an instrument having a 2% FSD accuracy can give relatively large errors when operating at conditions well below FSD.

When more than one instrument is used in a circuit then a cumulative error results. For example, if the current flowing through and the p.d. across a resistor is measured, then the percentage error in the ammeter is added to the percentage error in the voltmeter when determining the maximum possible error in the measured value of resistance.

Decibel units

27 In electronic systems, the ratio of two similar quantities measured at different points in the system, are often expressed in logarithmic units. By definition, if the ratio of two powers P_1 and P_2 is to be expressed in decibel (dB) units then the number of decibels, X, is given by:

$$\boldsymbol{X = 10 \lg\left(\frac{P_2}{P_1}\right) \text{ dB}}$$

Thus, when the power ratio, $\frac{P_2}{P_1} = 1$,

then the decibel power ratio $= 10 \lg 1 = 0$

when the power ratio, $\frac{P_2}{P_1}=100$,

then the decibel power ratio $=10 \lg 100=+20$

(i.e. a power gain),

and when the power ratio, $\frac{P_2}{P_1}=\frac{1}{10}$,

then the decibel power ratio $=10 \lg \frac{1}{10}=-10$,

(i.e. a power loss or attentuation).

28 Logarithmic units may also be used for voltage and current ratios. Power P, is given by $P=I^2R$ or $P=V^2/R$.

Substituting in equation (1) gives:

$$X=10 \lg \left(\frac{I_2^2R_2}{I_1^2}\right) \text{ dB or } X=10 \lg \left(\frac{V_2^2/R_2}{V_1^2/R_1}\right) \text{ dB}$$

If $R_1=R_2$, then $X=10 \lg \left(\frac{I_2^2}{I_1^2}\right)$ dB or $X=10 \lg \left(\frac{V_2^2}{V_1^2}\right)$ dB

$$X=20 \lg \left(\frac{I_2}{I_1}\right) \text{ dB or } X=20 \lg \left(\frac{V_2}{V_1}\right) \text{ dB}$$

(from the laws of logarithms).

Thus if the current input to a system is 5 mA and the current output is 20 mA, the decibel current ratio

$=20 \lg \left(\frac{I_2}{I_1}\right)=20 \lg \left(\frac{20}{5}\right)=20 \lg 4=$ **12 dB gain**.

29 From equation (1), X decibels is a logarithmic ratio of two similar quantities and is not an absolute unit of measurement. It is therefore necessary to state a **reference level** to measure a number of decibels above or below that reference. The most widely used reference level for power is 1 mW, and when power levels are expressed in decibels, above or below the 1 mW reference level, the unit given to the new power level is dBm.

A voltmeter can be re-scaled to indicate the power level directly in decibels. The scale is generally calibrated by taking a reference level of 0 dB when a power of 1 mW is dissipated in a 600 Ω resistor (this being the natural impedance of a simple transmission line). The reference voltage V is then obtained from

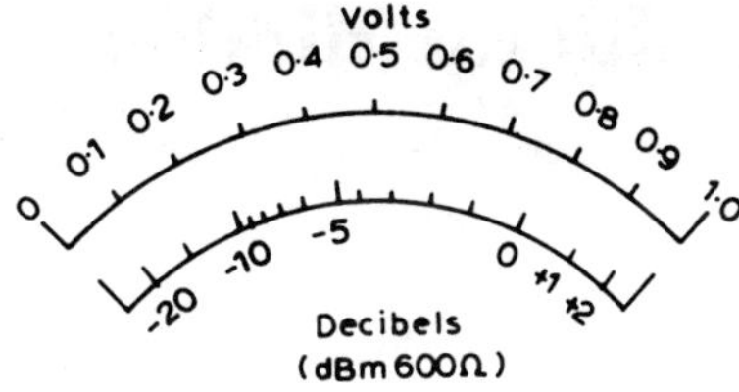

Figure 36.15

$$P=\frac{V^2}{R}, \text{ i.e. } 1\times10^{-1}=\frac{V^2}{600} \text{ from which, } V=0.775 \text{ volts.}$$

In general, the number of dBm, $X=20\ \lg\left(\frac{V}{0.775}\right)$

Thus $V=0.20$ V corresponds to $20\ \lg\left(\frac{0.20}{0.775}\right)=-11.77$ dBm and

$V=0.90$ V corresponds to $20\ \lg\left(\frac{0.90}{0.775}\right)=+1.3$ dBm, and so on.

A typical **decibelmeter**, or **dB meter scale** is shown in *Figure 36.15.* Errors are introduced with dB metres when the circuit impedance is not 600 Ω.

37 Semiconductor diodes

1 Materials may be classified as conductors, semiconductors or insulators. The classification depends on the value of resistivity of the material. Good conductors are usually metals and have resistivities in the order of 10^{-7} to 10^{-8} ohm metres. Semiconductors have resistivities in the order of 10^{-3} to 3×10^{3} ohm metres. The resistivities of insulators are in the order of 10^{4} to 10^{14} ohms. Some typical approximate values at normal room temperatures are:

Conductors:	Aluminium	$2.7 \times 10^{-8}\,\Omega m$	
	Brass (70 Cu/30 Zn)	$8 \times 10^{-8}\,\Omega m$	
	Copper (pure annealed)	$1.7 \times 10^{-8}\,\Omega m$	
	Steel (mild)	$15 \times 10^{-8}\,\Omega m$	
Semiconductors:	Silicon	$2.3 \times 10^{3}\,\Omega m$	at 27°C
	Germanium	$0.45\,\Omega m$	at 27°C
Insulators:	Glass $\geqslant 10^{10}\,\Omega m$		
	Mica $\geqslant 10^{11}\,\Omega m$		
	P.V.C. $\geqslant 10^{13}\,\Omega m$		
	Rubber (pure) 10^{12} to $10^{14}\,\Omega m$.		

2 In general, over a limited range of temperatures, the resistance of a conductor increases with temperature increase. The resistance of insulators remains approximately constant with variation of temperature. The resistance of semiconductor materials decreases as the temperature increases. For a specimen of each of these materials, having the same resistance (and thus completely different dimensions), at, say, 15°C, the variation for a small increase in temperature to t°C is as shown in *Figure 37.1*.

3 The most important semiconductors used in the electronics industry are **silicon** and **germanium**. As the temperature of these materials is raised above room temperature, the resistivity is reduced and ultimately a point is reached where they effectively become conductors. For this reason, silicon should not operate at a working temperature in excess of 150°C to 200°C, depending on its purity, and germanium should not operate at a working temperature in excess of 75°C to 90°C, depending on its purity. As the temperature of a semiconductor is reduced below normal room

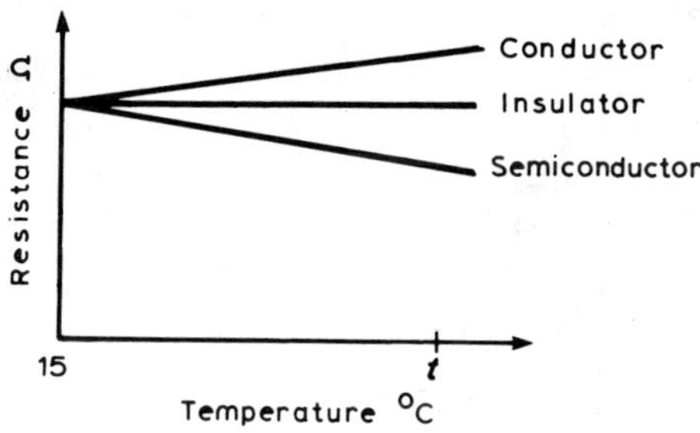

Figure 37.1

temperature, the resistivity increases until, at very low temperatures, the semiconductor becomes an insulator.

4 Adding extremely small amounts of impurities to pure semiconductors in a controlled manner is called **doping**. Antimony, arsenic and phosphorus are called *n*-type impurities and form an ***n*-type material** when any of these impurities are added to silicon or germanium. The amount of impurity added usually varies from 1 part impurity in 10^5 parts semiconductor material to 1 part impurity to 10^8 parts semiconductor material, depending on the resistivity required. Indium, aluminium and boron are called *p*-type impurities and form a ***p*-type material** when any of these impurities are added to a semiconductor.

5 In semiconductor materials, there are very few charge carriers per unit volume free to conduct. This is because the 'four electron structure' in the outer shell of the atoms (called valency electrons), form strong covalent bonds with neighbouring atoms, resulting in a tetrahedral structure with the electrons held fairly rigidly in place. A two-dimensional diagram depicting this is shown for germanium in *Figure 37.2*.

Arsenic, antimony and phosphorus have five valency electrons and when a semiconductor is doped with one of these substances, some impurity atoms are incorporated in the tetrahedral structure. The 'fifth' valency electron is not rigidly bonded and is free to conduct, the impurity atom donating a charge carrier. A two-dimensional diagram depicting this is shown in *Figure 37.3*, in which a phosphorus atom has replaced one of the germanium atoms. The resulting material is called *n*-type material, and contains free electrons.

Indium, aluminium and boron have three valency electrons and when a semiconductor is doped with one of these substances,

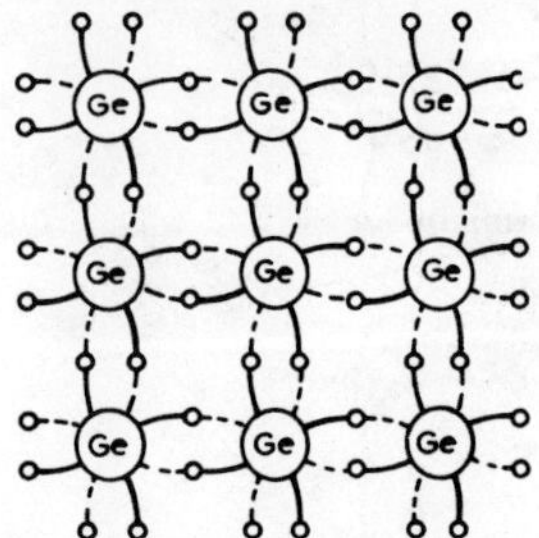

Figure 37.2

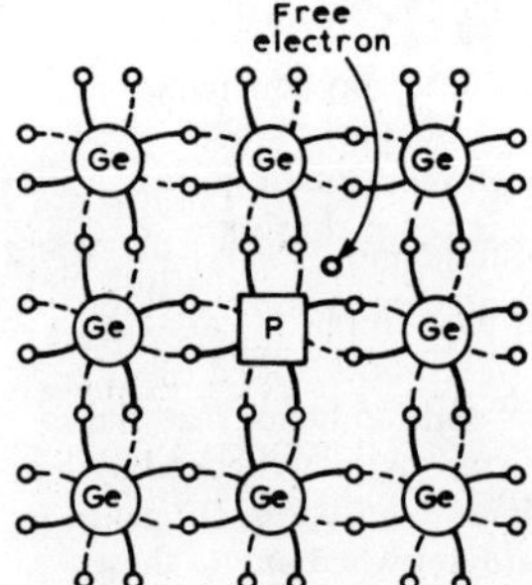

Figure 37.3

some of the semiconductor atoms are replaced by impurity atoms. One of the four bonds associated with the semiconductor material is deficient by one electron and this deficiency is called a **hole**. Holes give rise to conduction when a potential difference exists across the semiconductor material due to movement of electrons from one hole to another, as shown in *Figure 37.4*. In this figure, an electron moves from A to B, giving the appearance that the hole moves from B to A. Then electron C moves to A, giving the appearance that the hole moves to C, and so on. The resulting material is *p*-type material containing holes.

6 A **p-n junction** is a piece of semiconductor material in which part of the material is *p*-type and part is *n*-type. In order to examine the charge situation, assume that separate blocks of *p*-type and *n*-type materials are pushed together. Also assume that a hole is a positive charge carrier and that an electron is a negative charge carrier. At the junction, the donated electrons in the *n*-type

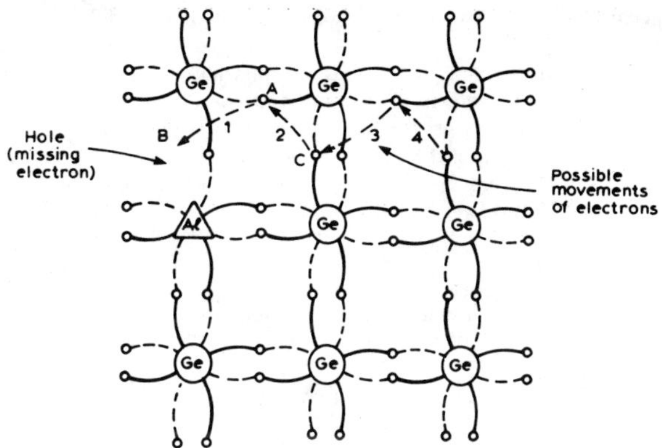

Figure 37.4

material, called **majority carriers**, diffuse into the *p*-type material (diffusion is from an area of high density to an area of lower density) and the acceptor holes in the *p*-type material diffuse into the *n*-type material as shown by the arrows in *Figure 37.5*. Because the *n*-type material has lost electrons, it acquires a positive potential with respect to the *p*-type material and thus tends to

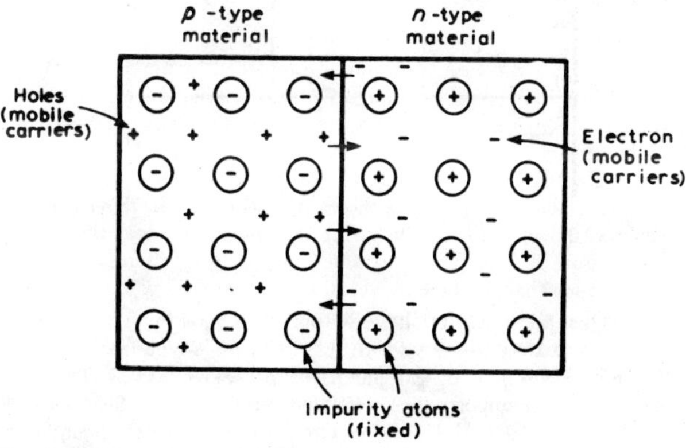

Figure 37.5

prevent further movement of electrons. The *p*-type material has gained electrons and becomes negatively charged with respect to the *n*-type material and hence tends to retain holes. Thus after a short while, the movement of electrons and holes stops due to the potential difference across the junction, called the **contact potential**. The area in the region of the junction becomes depleted of holes and electrons due to electron-hole recombinations, and is called a **depletion layer**, as shown in *Figure 37.6*.

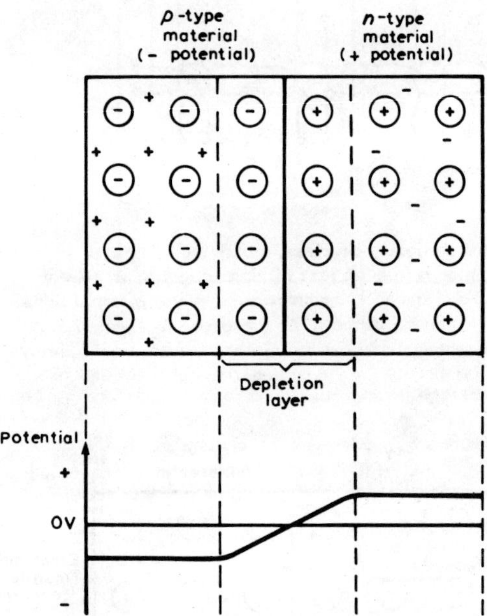

Figure 37.6

7 When an external voltage is applied to a *p-n* junction making the *p*-type material positive with respect to the *n*-type material, as shown in *Figure 37.7*, the *p-n* junction is **forward biased**. The applied voltage opposes the contact potential, and, in effect, closes the depletion layer. Holes and electrons can now cross the junction and a current flows. An increase in the applied voltage above that

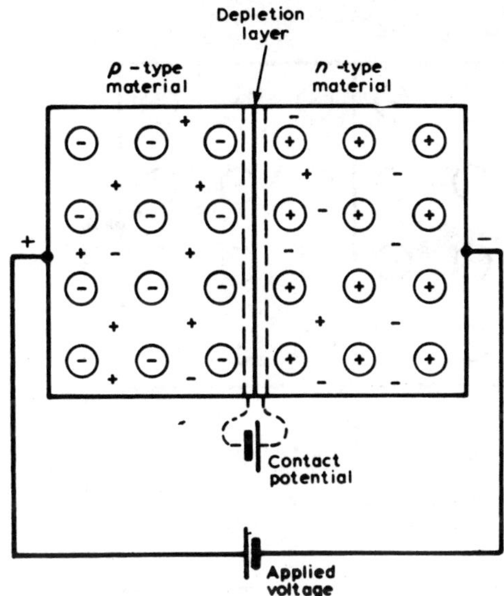

Figure 37.7

required to narrow the depletion layer (about 0.2 V for germanium and 0.6 V for silicon), results in a rapid rise in the current flow. Graphs depicting the current-voltage relationship for forward biased *p-n* junctions, for both germanium and silicon, called the forward characteristics, are shown in *Figure 37.8*.

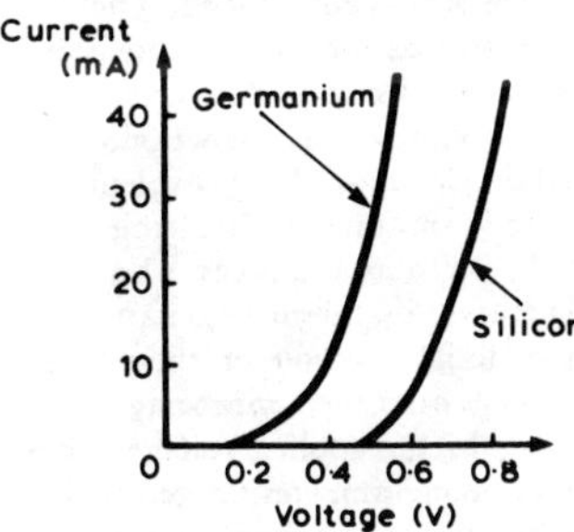

Figure 37.8

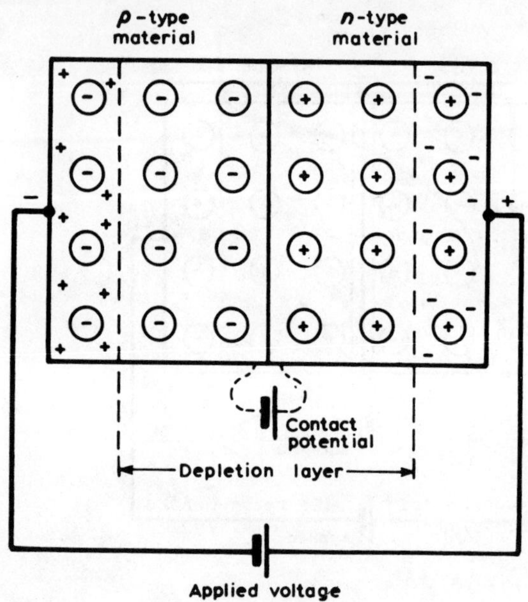

Figure 37.9

When an external voltage is applied to a *p-n* junction making the *p*-type material negative with respect to the *n*-type material, as shown in *Figure 37.9*, the *p-n* junction is **reverse biased**. The applied voltage is now in the same sense as the contact potential and opposes the movement of holes and electrons, due to opening up the depletion layer. Thus, in theory, no current flows. However, at normal room temperature, certain electrons in the covalent bond lattice acquire sufficient energy from the heat available to leave the lattice, generating mobile electrons and holes. This process is called electron-hole generation by thermal excitation.

The electrons in the *p*-type material and holes in the *n*-type material caused by thermal excitation, are called **minority carriers** and these will be attracted by the applied voltage. Thus, in practice, a small current of a few micro-amperes for germanium and less than one micro-ampere for silicon, at normal room temperature, flows under reverse bias conditions. Typical reverse

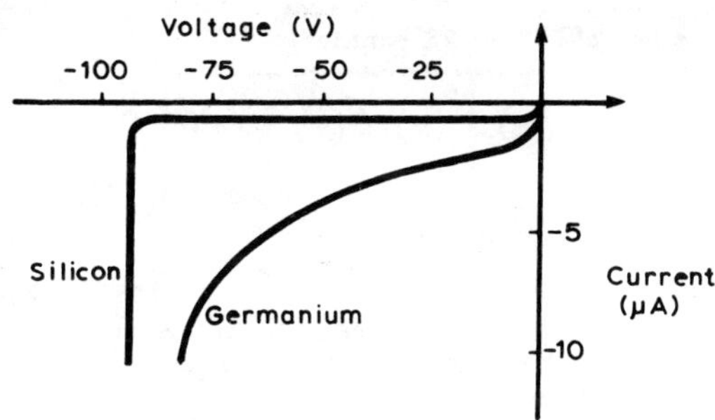

Figure 37.10

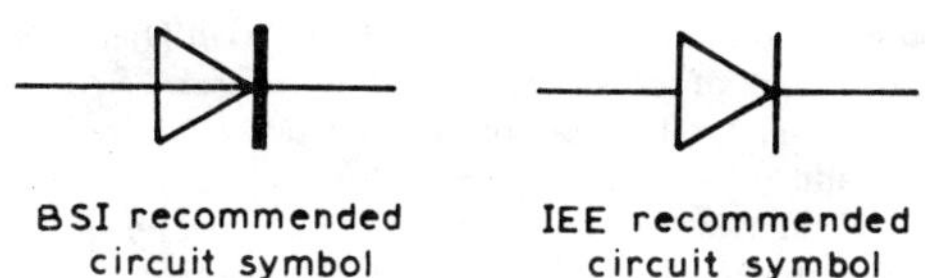

Figure 37.11

characteristics are shown in *Figure 37.10* for both germanium and silicon.

8 A **semiconductor diode** is a device having a *p-n* junction, mounted in a container, suitable for conducting and dissipating the heat generated in operation and having connecting leads. Its operating characteristics are as shown in *Figures 37.8* and *37.10*. Two circuit symbols for semiconductor diodes are in common use and are shown in *Figure 37.11*.

38 Transistors

1 The bipolar junction transistor consists of three regions of semiconductor material. One type is called a *p-n-p* transistor, in which two regions of *p*-type material sandwich a very thin layer of *n*-type material. A second type is called a *n-p-n* transistor, in which two regions of *n*-type material sandwich a very thin layer of *p*-type material. Both of these types of transistors consist of two *p-n* junctions placed very close to one another in a back-to-back arrangement on a single piece of semiconductor material. Diagrams depicting these two types of transistors are shown in ***Figure 38.1***.

The two *p*-type material regions of the *p-n-p* transistor are called the emitter and collector and the *n*-type material is called the base. Similarly, the two *n*-type material regions of the *n-p-n* transistor are called the emitter and collector and the *p*-type material region is called the base, as shown in ***Figure 38.1***.

2 Transistors have three connecting leads and in operation an electrical input to one pair of connections, say the emitter and base

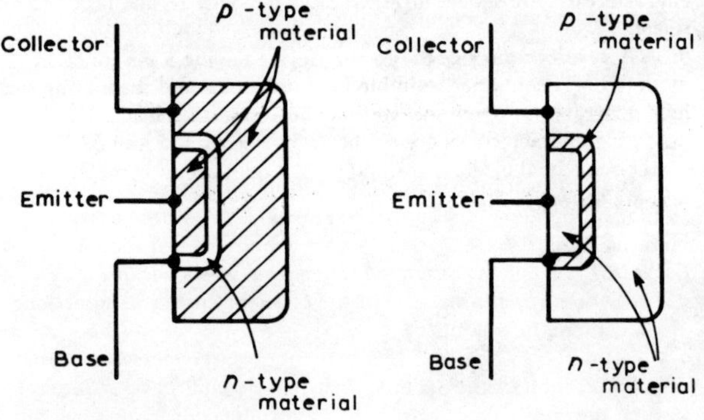

Figure 38.1

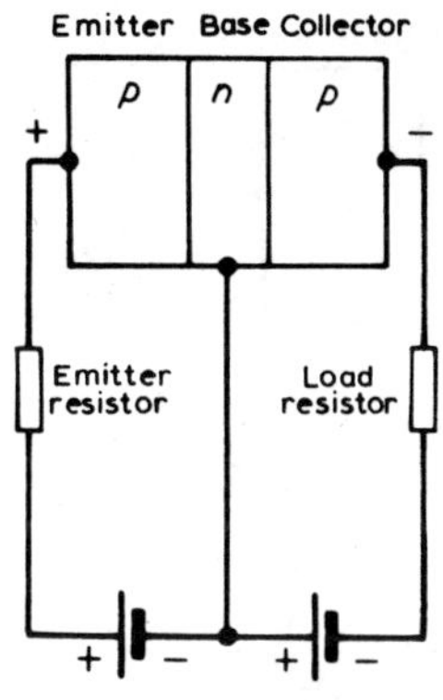

(a) *p-n-p* transistor

(b) *n-p-n* transistor

Figure 38.2

connections can control the output from another pair, say the collector and emitter connections. This type of operation is achieved by appropriately biasing the two internal *p-n* junctions. When batteries and resistors are connected to a *p-n-p* transistor, as shown in *Figure 38.2(a)*, the base-emitter junction is forward biased and the base-collector junction is reverse biased. Similarly, an *n-p-n* transistor has its base-emitter junction forward biased and its base-collector junction reverse biased when the batteries are connected as shown in *Figure 38.2(b)*.

3 For a silicon *p-n-p* transistor, biased as shown in *Figure 38.2(a)*, if the base-emitter junction is considered on its own, it is forward baised and a current flows. This is depicted in *Figure 38.3(a)*. For example, if R_E is 1000Ω, the battery is 4.5 V and the voltage drop across the junction is taken as 0.7 V, the current flowing is given by (4.5−0.7)/1000=3.8 mA. When the base-collector junction is considered on its own, as shown in *Figure 38.3(b)*, it is reverse biased and the collector current is something less than one microampere.

However, when both external circuits are connected to the transistor, most of the 3.8 mA of current flowing in the emitter, which previously flowed from the base connection, now flows out through the collector connection due to transistor action.

4 In a *p-n-p* transistor, connected as shown in *Figure 38.2(a)*, transistor action is accounted for as follows:

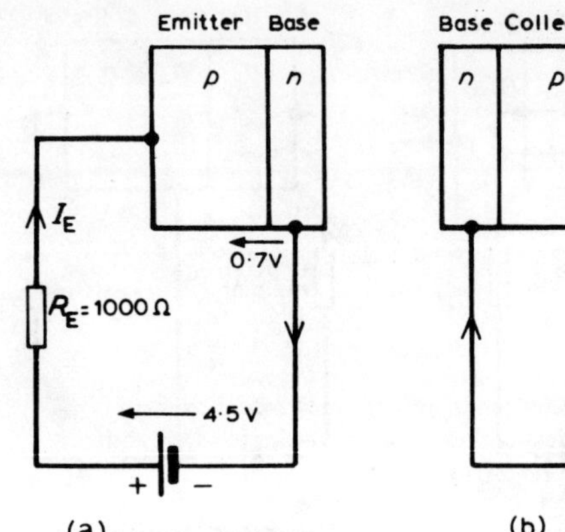

Figure 38.3

(a) The majority carriers in the emitter *p*-type material are holes.
(b) The base-emitter junction is forward biased to the majority carriers and the holes cross the junction and appear in the base region.
(c) The base region is very thin and is only lightly doped with electrons so although some electron-hole pairs are formed, many holes are left in the base region.
(d) The base-collector junction is reverse biased to electrons in the base region and holes in the collector region, but forward biased to holes in the base region. These holes are attracted by the negative potential at the collector terminal.
(e) A large proportion of the holes in the base region cross the base-collector junction into the collector region, creating a collector current. Conventional current flow is in the direction of hole movement.

The transistor action is shown diagrammatically in *Figure 38.4*. For transistors having very thin base regions, up to $99\frac{1}{2}\%$ of the holes leaving the emitter cross the base collector junction.

5 In an *n-p-n* transistor, connected as shown in *Figure 38.2(b)*, transistor action is accounted for as follows:

(a) The majority carriers in the *n*-type emitter material are electrons.

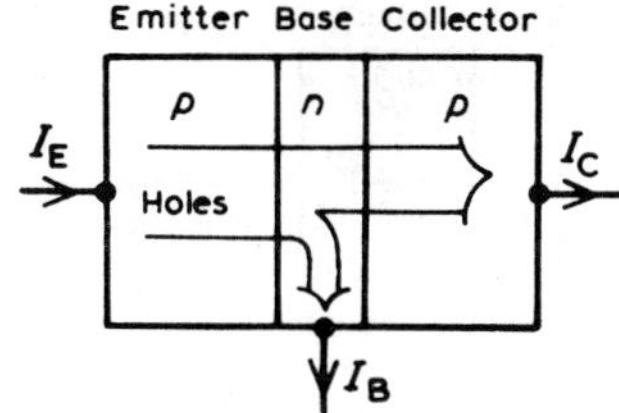

Figure 38.4

(b) The base-emitter junction is forward biased to these majority carriers and electrons cross the junction and appear in the base region.
(c) The base region is very thin and only lightly doped with holes, so some recombination with holes occurs but many electrons are left in the base region.
(d) The base-collector junction is reverse biased to holes in the base region and electrons in the collector region, but is forward biased to electrons in the base region. These electrons are attracted by the positive potential at the collector terminal.
(e) A large proportion of the electrons in the base region cross the base-collector junction into the collector region, creating a collector current.

The transistor action is shown diagrammatically in *Figure 38.5*. As stated in para. 4, conventional current flow is taken to be in the direction of hole flow, that is, in the opposite direction to electron flow, hence the directions of the conventional current flow are as shown in *Figure 38.5*.

5 For a *p-n-p* transistor, the base-collector junction is reverse biased for majority carriers. However, a small leakage current, I_{CBO}

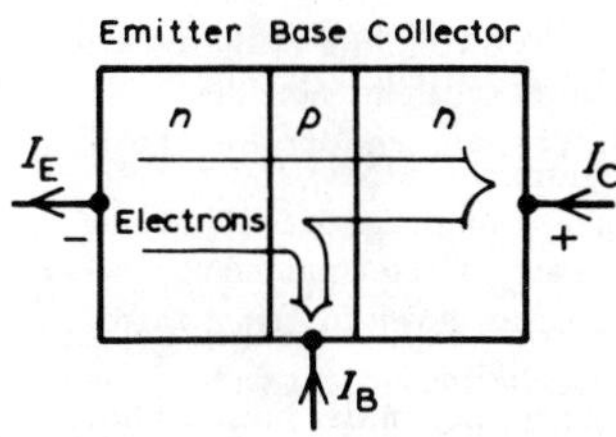

Figure 38.5

flows from the base to the collector due to thermally generated minority carriers (electrons in the collector and holes in the base), being present.

The base-collector junction is forward biased to these minority carriers. If a proportion, α, (having a value of up to 0.995 in modern transistors), of the holes passing into the base from the emitter, pass through the base-collector junction, then the various currents flowing in a *p-n-p* transistor are as shown in *Figure 38.6*.

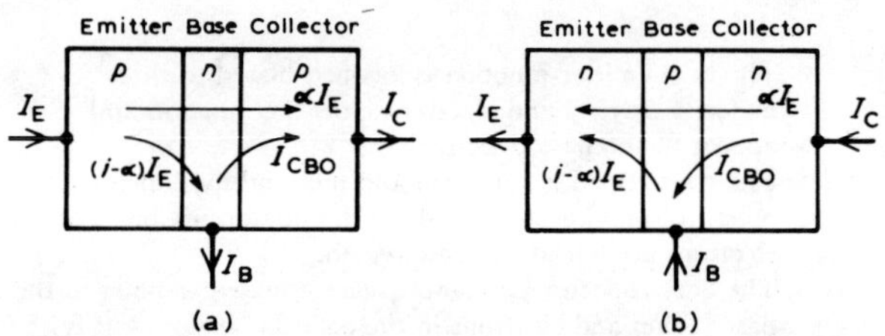

Figure 38.6

Similarly, for a *n-p-n* transistor, the base-collector junction is reverse biased for majority carriers, but a small leakage current, I_{CBO}, flows from the collector to the base due to thermally generated minority carriers (holes in the collector and electrons in the base), being present. The base-collector junction is forward biased to these minority carriers. If a proportion, α, of the electrons passing through the base-emitter junction also pass through the base-collector junction, then the currents flowing in an *n-p-n* transistor are as shown in *Figure 38.6*.

7 Symbols are used to represent *p-n-p* and *n-p-n* transistors in circuit diagrams and two of these in common use are shown in *Figure 38.7*. The arrow head drawn on the emitter of the symbol is in the direction of conventional emitter current (hole flow). The potentials marked at the collector, base and emitter are typical values for a silicon transistor having a potential difference of 6 V between its collector and its emitter.

The voltage of 0.6 V across the base and emitter is that required to reduce the potential barrier and if it is raised slightly to, say, 0.62 V, it is likely that the collector current will double to about 2 mA. Thus a small change of voltage between the emitter and the base can give a relatively large change of current in the

BSI SYMBOLS IEE SYMBOLS

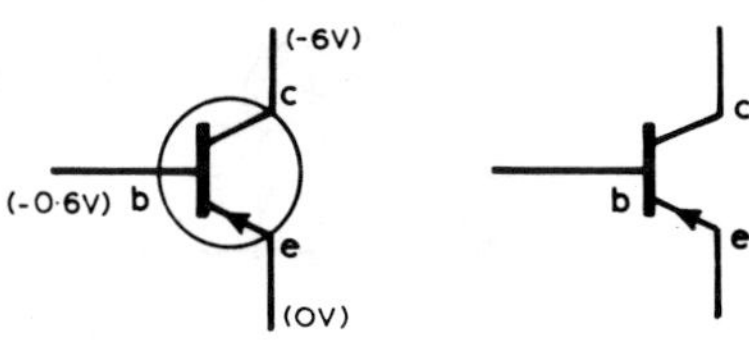

p-n-p transistor

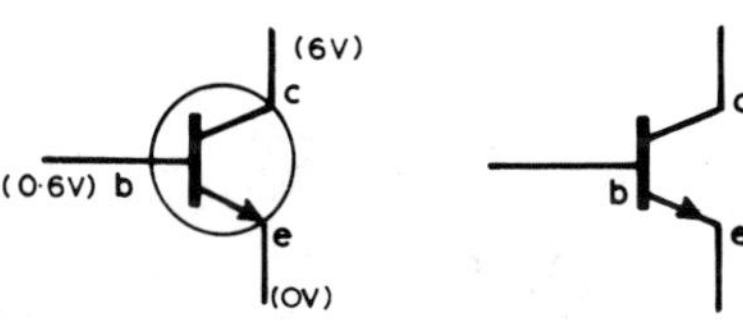

n-p-n transistor

Figure 38.7

emitter circuit. Because of this, transistors can be used as amplifiers.

8 There are three ways of connecting a transistor, depending on the use to which it is being put. The ways are classified by the electrode which is common to both the input and the output. They are called

(a) common-base configuration, shown in *Figure 38.8(a)*,
(b) common-emitter configuration shown in *Figure 38.8(b)*, and
(c) common-collector configuration, shown in *Figure 38.8(c)*.

These configurations are for an *n-p-n* transistor. The current flows shown are all reversed for a *p-n-p* transistor.

9 The effect of changing one or more of the various voltages and currents associated with a transistor circuit can be shown graphically and these graphs are called the characteristics of the transistor. As there are five variables (collector, base and emitter currents and voltages across the collector and base and emitter and base) and also three configurations, many characteristics are possible. Some of the possible characteristics are given below.

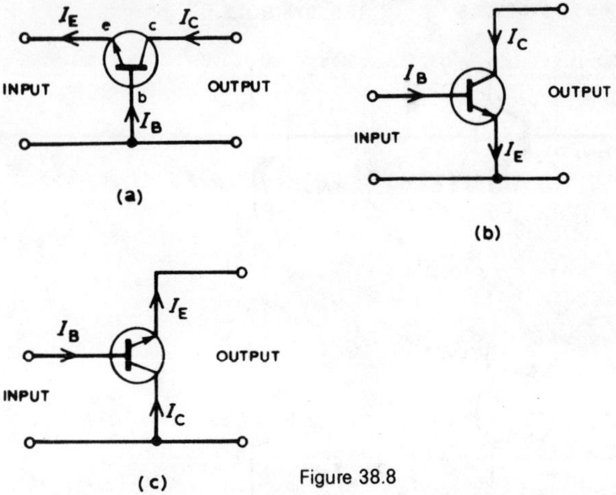

Figure 38.8

(a **Common-base configuration**

(i) **Input characteristic**. With reference to *Figure 38.8(a)*, the input to a common-base transistor is the emitter current, I_E, and can be varied by altering the base-emitter voltage V_{EB}. The base-emitter junction is essentially a forward biased junction diode, so as V_{EB} is varied, the current flowing is similar to that for a junction diode, as shown in *Figure 38.9* for a silicon transistor. *Figure 38.9* is called the input characteristic for an *n-p-n* transistor having common-base configuration. The variation of the collector-base voltage C_{CB} has little effect on the characteristic. A similar characteristic can be obtained for a *p-n-p* transistor, these having reversed polarities.

(ii) **Output characteristics** The value of the collector current I_C is very largely determined by the emitter current, I_E. For a given value of I_E, the collector-base voltage, V_{CB}, can be varied and has little effect on the value of I_C. If V_{CB} is made slightly negative, the collector no longer attracts the majority carriers leaving the emitter and I_C falls rapidly to zero. A family of curves for various values of I_E are possible and some of these are shown in

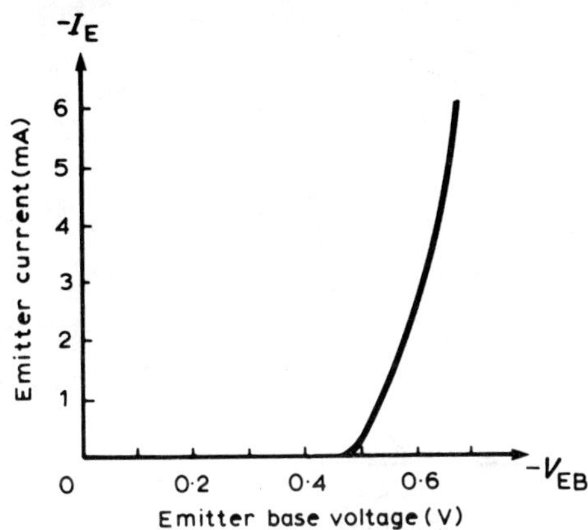

Figure 38.9

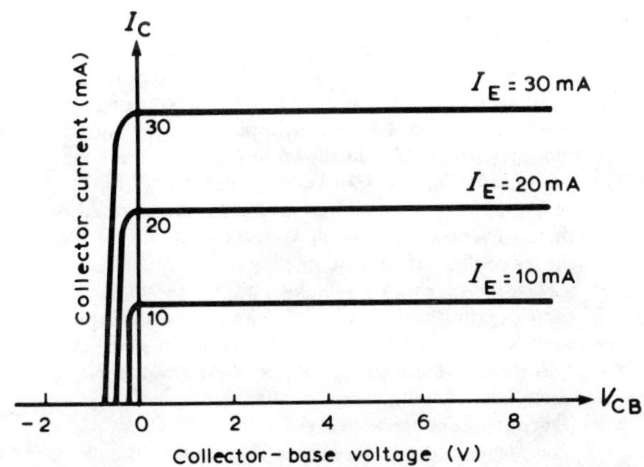

Figure 38.10

Figure 38.10. *Figure 38.10* is called the output characteristic from an *n-p-n* transistor having common-base configuration. Similar characteristics can be obtained for a *p-n-p* transistor, these having reversed polarities.

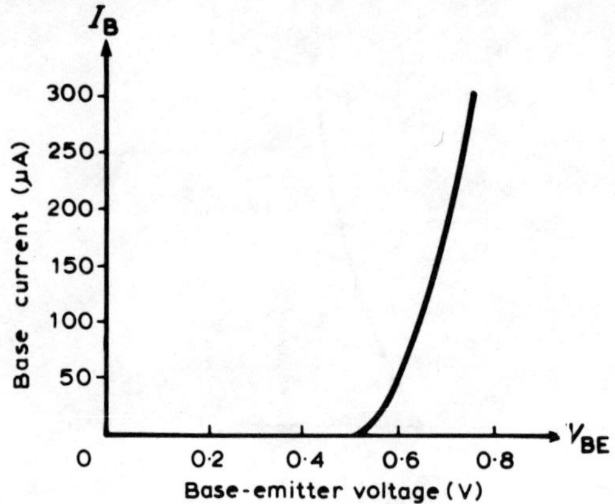

Figure 38.11

(b Common-emitter configuration

(i) **Input characteristic** In a common-emitter configuration (see *Figure 38.8(b)*), the base current is now the input current. As V_{EB} is varied, the characteristic obtained is similar in shape to the input characteristic for a common-base configuration shown in *Figure 38.9*, but the values of current are far less. With reference to *Figure 38.6(a)*, as long as the junctions are biased as described, the three currents, I_E, I_C and I_B keep the ratio $1:\alpha:(1-\alpha)$ whichever configuration is adopted.

Thus the base current changes are much smaller than the corresponding emitter current changes and the input characteristic for an *n-p-n* transistor is as shown in *Figure 38.11*. A similar characteristic can be obtained for a *p-n-p* transistor, these having reversed polarities.

(ii) **Output characteristics** A family of curves can be obtained depending on the value of base current I_B and some of these for an *n-p-n* transistor are shown in *Figure 38.12*. A similar set of characteristics can be obtained for a *p-n-p* transistor, these having reversed polarities. These characteristics differ from the common-base output characteristics in the following ways:

The collector current reduces to zero without having to reverse the collector voltage, and the characteristics slope

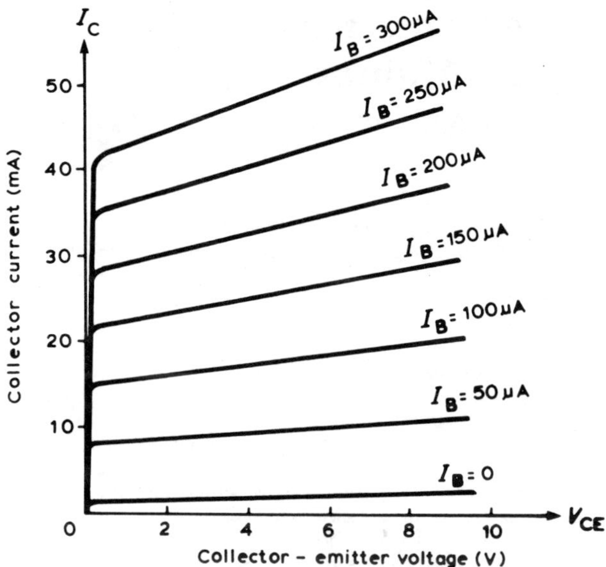

Figure 38.12

upwards indicating a lower output resistance (usually kilohms for a common-emitter configuration compared with megohms for a common-base configuration).

39 Atoms, molecules, compounds and mixtures

1 **Atoms** have been defined as the **smallest parts of matter** which cannot be changed by **chemical means**. In nature, the inert gases neon, argon, krypton and xenon together with helium exist in nature as single uncombined atoms. A few other elements like gold and silver do not exist as single atoms but are not combined with any other element. The **majority of elements** are found **in combination** with other elements.

2 **Molecules** is the collective term given to substances which contain at least **two atoms chemically combined together in a simple ratio**. Examples of **molecules containing one element** are hydrogen H_2, nitrogen N_2, oxygen O_2, fluorine F_2, phosphorus P_4, and sulphur S_8. This state of matter can be changed by chemical means during chemical reactions.

3 **Compounds** are substances which **contain different elements combined in a simple ratio** according to the **combining powers** or **valencies** of the constituents. Some examples of compounds which are commonly found in everyday life are, table salt or sodium chloride ($NaCl$), sugar or sucrose ($C_{12}H_{22}O_{11}$), battery acid or sulphuric acid (H_2SO_4), chalk or calcium carbonate ($CaCO_3$) and camping gas or propane (C_3H_8). **The smallest amount of a compound which can exist** and can be changed by chemical methods **is called a molecule**.

4 **Mixtures** are **combinations of compounds** which **can be separated by physical methods**. These methods include (i) **filtering**, (ii) **distillation**, (iii) **evaporation**, (iv) **decanting** (v) **separating funnels**, (vi) **magnetism** and (vii) **fractional crystallisation**.

(i) **Filtration** can be used to separate a **mixture of liquid and solid**, or by making use of the different solubilities of the compounds in the mixture, by dissolving one substance, but not the other. For example, in a mixture of sand and salt, the salt dissolves in water leaving the sand undissolved. A typical arrangement for filtration is shown in *Figure 39.1(i)*.

(ii) **Distillation** can also be used to separate a **mixture of a solid and a liquid**, the liquid being boiled off leaving the solid behind. More commonly **fractional distillation** is used to

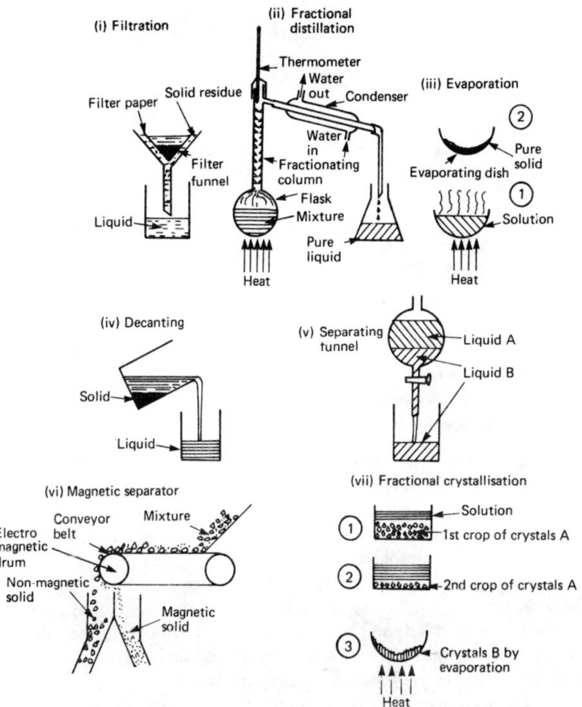

Figure 39.1 Apparatus for separating mixtures of chemicals

separate **mixtures of miscible liquids** as happens for the separation of the different boiling point fractions of crude oil. A typical arrangement is shown in *Figure 39.1(ii)*.

(iii) **Evaporation** can be used to separate the **components of a solution** by allowing the liquid solvent to be removed leaving behind the solid, as shown in *Figure 39.1(iii)*.

(iv) **Decanting** is a very simple way of separating **a solid with a liquid**, achieved by carefully pouring off the liquid layer (see *Figure 39.1(iv)*.

(v) **Separating funnels** are pieces of apparatus designed to **separate two immiscible liquids** of different density. The liquids from **two layers** which can be run off into different

Table 39.1 The properties of compounds and mixtures

Compounds	*Mixtures*
1. The substance has a constant composition.	The composition of the mixture can be varied by mass.
2. The substance can only be separated into its components by chemical methods.	The substances can be separated by physical methods.
3. There can be considerable amounts of heat energy released when the substance is formed.	There is little or no heat energy released when the substances are mixed.
4. The properties of a compound are quite different to the properties of the constituent elements.	The properties are those of each of the substances forming the mixture.

containers. A typical separating funnel is shown in *Figure 39.1(v)*.
(vi) **Magnetism** can only be utilised to separate magnetic materials from non-magnetic materials. This technique is used in separating crushed metallic ores or for retrieving iron based articles for recycling. A suitable arrangement is shown in *Figure 39.1(vi)*.
(vii) **Fractional crystallisation** can be used to separate **two solids**, which though both soluble in the same solvent, have quite **different solubilities**. The method involves dissolving the solids and taking '**crops**' of crystals as they are formed. Finally, the soluble substance is obtained by evaporation.

5 The differences between compounds and mixtures are summarised in *Table 39.1*.

40 The laws of chemical combination

Law of conservation of matter

1 The law states that '**matter cannot be created or destroyed in a chemical reaction**'. This law is based on the fact that atoms are not composed or destroyed by a chemical reaction. For example, when 1.0 g of powdered iron is heated with 0.4 g of sulphur, 1.1 g of iron(II) sulphide is produced and no sulphur remains. The total mass of elements used is $1+0.4=1.4$ g but the mass of iron sulphide is only 1.1 g. This means that $1.4-1.1=0.3$ g of iron remain unchanged in the reaction.

Law of constant composition

2 The law states that '**a compound always contains the same elements combined together in the same proportion by mass**'. This is because the **chemical composition of a compound cannot change**; they are all identical molecules. For example, when three specimens of copper(II) oxide are prepared by different methods and then reduced to metallic copper by heating in hydrogen the results obtained are:

Sample A 2.65 g of copper oxide gave 2.12 g of copper
Sample B 2.26 g of copper oxide gave 1.81 g of copper
Sample C 3.18 g of copper oxide gave 2.54 g of copper.

By finding the **ratio of** the **mass of copper** to **the mass of oxygen** for the three samples, the **constancy of the ratio** can be determined in the following way,

	Sample A	**Sample B**	**Sample C**
Copper oxide	2.65	2.26	3.18
Copper	2.12	1.81	2.54
Oxygen	0.53	0.45	0.64
$\frac{\text{Mass of copper}}{\text{Mass of oxygen}}$	4	4.02	3.97

These results show that the elements are approximately combined in the same proportions by mass.

Law of multiple proportions

3 The law states that **'if two elements A and B combine together to form more than one compound, then the different masses of A which combine with a fixed mass of B are in a simple ratio'**. This follows from the principle that **in compounds, atoms combine in a simple ratio**. For example, when 2.07 g of lead reacts with oxygen to form lead oxide the amount of lead oxide formed is 2.13 g, whereas when lead dioxide is formed the mass is 2.39 g.
The ratio of lead:oxygen in lead oxide is 2.07:0.16=**12.94**
The ratio of lead:oxygen in lead dioxide is 2.07:0.32=**6.47**
The ratio of these 2 ratios is 12.94:6.47=2, which is **a simple ratio** in accordance with the law stated.

Law of chemical combination by volume

4 **Gay-Lussac's law of volumes** states **'when gases react, the volume of gases are in a simple ratio to each other and to the volume of products, provided all the volumes are measured at the same temperature and pressure'**. Under constant conditions of temperature and pressure, some examples are,
(i) **1 volume of hydrogen + 1 volume of chlorine forms 2 volumes of hydrogen chloride.**
(ii) **2 volumes of hydrogen + 1 volume of oxygen forms 2 volumes of steam (above 373 K).**
(iii) **2 volumes of ammonia forms 1 volume of nitrogen and 3 volumes of hydrogen.**
(iv) When **40 cm^3 of hydrogen** is mixed with **30 cm^3 of chlorine** and the mixture is sparked in an **eudiometer** (an instrument for measuring volumes of gases) **10 cm^3 of hydrogen gas remains** after the hydrogen chloride gas has been absorbed from the mixture. By **Gay-Lussac's law**:

1 volume of chlorine + 1 volume of hydrogen forms 2 volumes of hydrogen chloride.

or **30 cm^3 of chlorine + 30 cm^3 of hydrogen** forms **60 cm^3 of hydrogen chloride**. This means that **40 cm^3–30 cm^3 of hydrogen are** unchanged, that is, **10 cm^3**.
5 **Avagadro's Hypothesis** states that **'equal volumes of all gases under the same conditions of temperature and pressure contain equal numbers of molecules'**. Under constant conditions of temperature and pressure the reaction

between hydrogen and chlorine to form hydrogen chloride can be written as:

1 volume of hydrogen + 1 volume of chlorine → 2 volumes of hydrogen chloride

By Avagadro's hypothesis

1 molecule of hydrogen + 1 molecule of chlorine → 2 molecules of hydrogen chloride

similarly

1 volume of ammonia → 1 volume of nitrogen + 3 volumes of hydrogen.

6 The various laws can be combined. Thus under conditions of constant temperature and pressure, 40 cm^3 of carbon monoxide (CO), requires 20 cm^3 of oxygen (O_2) for complete combustion to form 40 cm^3 of carbon dioxide (CO_2). Using Gay-Lussac's law

40 cm^3 of CO + 20 cm^3 of O_2 → 40 cm^3 of CO_2
By taking 20 cm^3 as 1 volume this becomes
2 volumes of CO + 1 volume of O_2 → 2 volumes of CO_2

From Avagadro's hypothesis:

2 molecules of CO + 1 molecules of O_2 forms 2 molecules of CO_2 or $2CO + O_2 \rightarrow 2CO_2$, which is the reacting ratio between carbon monoxide and oxygen.

41 Relative atomic masses, molecular masses and the 'mole' concept

1 When the **laws of chemical combination** were formulated it was necessary to decide on **an arbitary value** for the mass of one element so that comparative reacting masses could be made.

Because **hydrogen gas** was used in many of the early experiments and was the lightest element known, it was given the **value of unity**. The relative masses of elements **combining with 1 g of hydrogen** were then found. Because hydrogen reacts readily with only a few elements it was **replaced by oxygen** as an arbitary standard with a **value of mass of 16** compared to hydrogens value of 1. This gave a different set of relative masses for the elements. The inadequacy of oxygen as an arbitary standard led to the selection of **the carbon-12 isotope of carbon as the arbitary standard**. The **relative atomic mass of an element is defined as the weight in grams of the number of atoms of the element contained in 12.00 g of carbon-12. Alternatively, the relative atomic mass of an element is given by**

$$\frac{\text{the mass of one atom of the element}}{1/12 \text{ of the mass of one atom of carbon-12}}$$

2 The **number of atoms** contained in the relative atomic mass of an element has been shown by **Avagardro to be 6.02 × 10^{23}**. This number is called the **Avagadro number** or the **Avagadro constant**.

An alternative definition for the relative atomic mass of an element is the mass of 6.02×10^{23} atoms of the element. For example, the relative atomic mass of chlorine can be found by using a technique called mass spectrometry to find the **isotopes** present and their **relative abundancies** in the element. The $^{35}_{17}Cl$ isotope is present in 75.53% and the $^{37}_{17}Cl$ isotope is 24.47%.

To calculate the relative atomic mass of chlorine, the average mass of one atom of chlorine is found by considering 100 atoms of chlorine. 75.53 of these atoms each have a mass of 35 atomic mass units (a.m.u.), and 24.47 atoms each have a mass of 37 a.m.u. 75.53 atoms of ^{35}Cl have a mass of $35 \times 75.53 = 2643.55$ a.m.u. 24.47 atoms of ^{37}Cl have a mass of $37 \times 24.47 = 905.39$ a.m.u. The combined mass is $2643.55 + 905.39 = 3548.94$ a.m.u.

The average mass of 1 chlorine atom $= \dfrac{3548.94}{100} = 35.498$ a.m.u.

and the mass of 6.02×10^{23} atoms of chlorine can be defined as 35.49 g.

The 'mole' concept

3 The word '**mole**' has been adopted to **represent the Avagadro number of atoms** of an element, that is, the **relative atomic mass** of an element. Thus, one mole of sodium weighs 23.0 g or one tenth of a mole of sodium weighs 2.3 g.

4 When applied to molecules, **one mole of molecules** is the **relative molecular mass** of that molecule, which is the **summation of the individual relative atomic masses** of the constituent atoms. For example, calcium carbonate contains calcium, carbon and oxygen in the ratio 1:1, 3 (i.e. $CaCO_3$). The accurate relative atomic masses are Ca=40.1, C=12.01, O=16.00, thus the relative molecular mass is $40.1 + 12.01 + (3 \times 16.00) = 100.11$. For many purposes the relative atomic masses are rounded up to the nearest whole number except for chlorine and copper which are 35.5 and 63.5 respectively.

5 When applied to solutions, **a 1 molar, (1 M), solution** is one in which **1 mole of a solute is dissolved in a solvent** in order that the **volume of the solution is 1000 cm^3** (**1 dm^3 or 1 litre**). This means that if the concentration of the solution is known in **moles per dm^3**, the number of moles in any volume of solution can be determined. For example, to find how many moles of sodium hydroxide, NaOH, are contained in 200 cm^3 of a 2 M, (2 molar), solution:

1000 cm^3 of the solution contains 2 moles of NaOH

Thus, 1 cm^3 of the solution contains $\dfrac{2}{1000}$ moles of NaOH and

200 cm^3 of the solution contains $\dfrac{2}{1000} \times 200$ moles of NaOH. That is, the number of moles of sodium hydroxide is 0.4.

In order to find the mass of sodium hydroxide required to make 200 cm^3 of 2 M solution:

200 cm^3 of a 2 M solution requires $\frac{2}{1000} \times 200$ moles.

0.4 moles of NaOH has a mass found by the equation:

mass of NaOH = number of moles of NaOH × relative molecular mass of NaOH

$$= 0.4 \times (23 + 16 + 11)$$
$$0.4 \times 40$$
$$= 16 \text{ g}$$

That is, the mass of sodium hydroxide required is 16 g.

6 When applied to gases, the **molar volume of any gas is defined as occupying 22.4 dm^3 at a temperature of 273 K and pressure 101.3 kPa** (atmospheric pressure). Volumes of gases are easier to measure than masses. Using the molar volume definition, if the volume of a gas is known, the number of moles and hence the mass of the gas can be determined. For example, to find the number of moles of carbon dioxide gas which are contained in 100 cm^3 of the gas measured at 273 K and 101.3 kPa. Use is made of the above definition that at 101.3 kPa and 273 K, 22400 cm^3 of any gas is the volume of 1 mole of the gas.

Thus, 22400 cm^3 of CO_2 are equivalent to 1 mole of CO_2.

and 1 cm^3 of CO_2 is equivalent to $\frac{1}{22400}$ moles of CO_2.

Thus, 100 cm^3 of CO_2 are equivalent to

$$\frac{1}{2240} \times 100 \text{ moles of } CO_2$$

$$= \frac{1}{224} \text{ or } 0.00446 \text{ moles of carbon dioxide.}$$

In order to find the mass of carbon dioxide gas occupying 100 cm^3 at 273 K and 101.3 kPa, use is made of the fact that 100 cm^3 of CO_2 is equivalent to 0.00446 moles.

Mass of carbon dioxe = Number of moles of carbon dioxide

× Relative molecular mass of carbon dioxide

$$= 0.00446 \times (12 + 2 \times 16)$$
$$= 0.00466 \times 44$$
$$= 0.196 \text{ g.}$$

The mass of carbon dioxide is 0.196 g.

7 If the temperature and pressure of the gas are different from the values stated, the volume must be converted to these values using the gas laws (see Section 25).

42 Atomic structure of matter

1 **Matter** is the collective word used to describe all of the different substances that exist. There are **three types of matter — solids, liquids and gases**, all of which contain one or more basic materials called **elements**.

An **element** is a substance which **cannot be separated into anything simpler by chemical means**. The elements which have been identified are shown in their **systematic arrangement** in the **periodic table**, in *Table 42.1*. The names of the elements are abbreviated in order that **chemical formulae** can be represented in a simple format; their full names are given in *Table 42.2*, for some of the elements.

2 Elements are made up of **very small particles called atoms**. An **atom** is the **smallest part of an element which cannot be divided or destroyed by chemical means**. In atomic theory, a simplified model of an atom can be considered to resemble a miniature solar system. It consists of a **positively charged central nucleus** around which **negatively charged particles called electrons orbit** at certain **fixed distances** away from the nucleus, called **shells** (see *Figure 42.1*).

The nucleus which occupies only a **very small part of the total volume** of an atom contains two different types of particles called **protons**, which are **positively charged**, and **neutrons** which are **not charged**. The number of protons in the nucleus and the number of electrons orbiting around it **must be equal** because electrically, **atoms are neutral**. The number of protons, which **uniquely defines** an atom, is called **the atomic number (z) of the element** of which the atom is a part. An element may be defined as **a substance containing only atoms with the same atomic number**. Each element has a unique atomic number.

3 **Isotopes**. Atoms which contain the same number of protons **do not always contain the same number of neutrons**. For example, hydrogen can exist in three forms as shown in *Figure 42.2*, namely **hydrogen** which contains 1 proton and 1 electron, **deuterium** which has 1 proton 1 neutron and 1 electron and **tritium** which contains 1 proton 2 neutrons and 1 electron. These

Table 42.1 The Periodic Table

1s	^{1}H 1.008	^{2}He 1.00																			1st short period
2s	^{3}Li 6.94	^{4}Be 9.01												2p	^{5}B 10.8	^{6}C 12.01	^{7}N 14.01	^{8}O 16.00	^{9}F 19.0	^{10}Ne 20.2	2nd short period
3s	^{11}Na 23.0	^{12}Mg 24.3												3p	^{13}Al 27.0	^{14}Si 28.1	^{16}P 31.0	^{16}S 32.1	^{17}Cl 35.5	^{18}Ar 39.9	3rd short period
4s	^{19}K 39.1	^{20}Ca 40.1	3d	^{21}Sc 45.0	^{22}Ti 47.9	^{23}V 50.9	^{24}Cr 62.0	^{25}Mn 54.9	^{26}Fe 55.9	^{27}Co 58.9	^{28}Ni 58.7	^{29}Cu 63.5	^{30}Zn 65.4	4p	^{31}Ga 69.7	^{32}Ge 72.6	^{33}As 74.9	^{34}Se 79	^{35}Br 79.9	^{36}Kr 83.6	1st long period
5s	^{37}Rb 85.5	^{38}Sr 87.6	4d	^{39}Y 88.9	^{40}Zr 91.2	^{41}Nb 92.9	^{42}Mo 95.9	^{43}Tc 99	^{44}Ru 101.1	^{45}Rh 102.9	^{46}Pd 106.4	^{47}Ag 107.9	^{48}Cd 112.4	5p	^{49}In 114.8	^{50}Sn 118.7	^{51}Sb 121.8	^{52}Te 127.6	^{53}I 126.9	^{54}Xe 131.3	2nd long period
6s	^{55}Cs 132.9	^{56}Ba 137.3	5d		^{72}Hf 178.5	^{73}Ta 181.0	^{74}W 183.9	^{75}Re 186.2	^{76}Os 190.2	^{77}Ir 192.2	^{78}Pt 195.1	^{79}Au 197	^{80}Hg 200.6	6p	^{81}Tl 204.4	^{82}Pb 207.2	^{83}Bi 209	^{84}Po (210)	^{85}At (210)	^{86}Rn (222)	3rd long period
7s	^{87}Fr (223)	^{88}Ra (226)																			

s-block *d*-block *p*-block

4f	^{57}La 138.9	^{58}Ce 140.1	^{59}Pr 140.9	^{60}Nd 144.2	^{61}Pm 147	^{62}Sm 150.4	^{63}Eu 152	^{64}Gd 157.3	^{65}Tb 158.4	^{66}Dy 162.5	^{67}Ho 164.9	^{68}Er 167.3	^{69}Tm 168.4	^{70}Yb 173	^{77}Lu 175
5f	^{89}Ac 227	^{90}Th 232	^{91}Pa 231	^{92}U 238	^{93}Np 237	^{94}Pu 242	^{95}Am 243	^{96}Cm 247	^{97}Bk 245	^{98}Cf 257	^{99}Es 254	^{100}Fm 253	^{101}Md 256	^{102}No 254	^{103}Lr 257

f-block

Table 42.2 The electronic structures of the elements in shells of electrons and orbitals of electrons

Element	*At No (Z)*	*Shell Structure*						*Orbital structure*											
		K	*L*	*M*	*N*	*O*	*P*	1*s*	2*s*	2*p*	3*s*	3*p*	4*s*	3*d*	4*p*	5*s*	4*d*	5*p*	6*s*
Hydrogen	1	1						1											
Helium	2	2						2											
Lithium	3	2	1					2	1										
Beryllium	4	2	2					2	2										
Boron	5	2	3					2	2	1									
Carbon	6	2	4					2	2	2									
Nitrogen	7	2	5					2	2	3									
Oxygen	8	2	6					2	2	4									
Fluorine	9	2	7					2	2	5									
Neon	10	2	8					2	2	6									
Sodium	11	2	8	1				2	2	6	1								
Magnesium	12	2	8	2				2	2	6	2								
Aluminium	13	2	8	3				2	2	6	2	1							
Silicon	14	2	8	4				2	2	6	2	2							
Phosphorus	15	2	8	5				2	2	6	2	3							
Sulphur	16	2	8	6				2	2	6	2	4							
Chlorine	17	2	8	7				2	2	6	2	5							
Argon	18	2	8	8				2	2	6	2	6							
Potassium	19	2	8	8	1			2	2	6	2	6	1						
Calcium	20	2	8	8	2			2	2	6	2	6	2						
Scandium	21	2	8	9	2			2	2	6	2	6	2	1					
Titanium	22	2	8	10	2			2	2	6	2	6	2	2					
Vanadium	23	2	8	11	2			2	2	6	2	6	2	3					
Chromium	24	2	8	13	1			2	2	6	2	6	1	5					
Manganese	25	2	8	13	2			2	2	6	2	6	2	5					
Iron	26	2	8	14	2			2	2	6	2	6	2	6					
Cobalt	27	2	8	15	2			2	2	6	2	6	2	7					
Nickel	28	2	8	16	2			2	2	6	2	6	2	8					
Copper	29	2	8	18	1			2	2	6	2	6	1	10					
Zinc	30	2	8	18	2			2	2	6	2	6	2	10					
Gallium	31	2	8	18	3			2	2	6	2	6	2	10	1				
Germanium	32	2	8	18	4			2	2	6	2	6	2	10	2				
Arsenic	33	2	8	18	5			2	2	6	2	6	2	10	3				
Selenium	34	2	8	18	6			2	2	6	2	6	2	10	4				
Bromine	35	2	8	18	7			2	2	6	2	6	2	10	5				
Krypton	36	2	8	18	8			2	2	6	2	6	2	10	6				
Rubidium	37	2	8	18	8	1		2	2	6	2	6	2	10	6	1			
Strontium	38	2	8	18	8	2		2	2	6	2	6	2	10	6	2			
Iodine	53	2	8	18	18	7		2	2	6	2	6	2	10	6	2	10	5	
Xenon	54	2	8	18	18	8		2	2	6	2	6	2	10	6	2	10	6	
Caesium	55	2	8	18	18	8	1	2	2	6	2	6	2	10	6	2	10	6	1
Barium	56	2	8	18	18	8	2	2	2	6	2	6	2	10	6	2	10	6	2

three forms are called isotopes of hydrogen. They all have an atomic number of 1 and are electrically neutral.

4 The arrangement of the elements into the modern periodic table shown in *Table 42.1* has been achieved as a result of many attempts to classify the elements into a meaningful pattern. Simple classification into metals and non-metals was not considered to be adequate, and many of the physical properties of the elements were

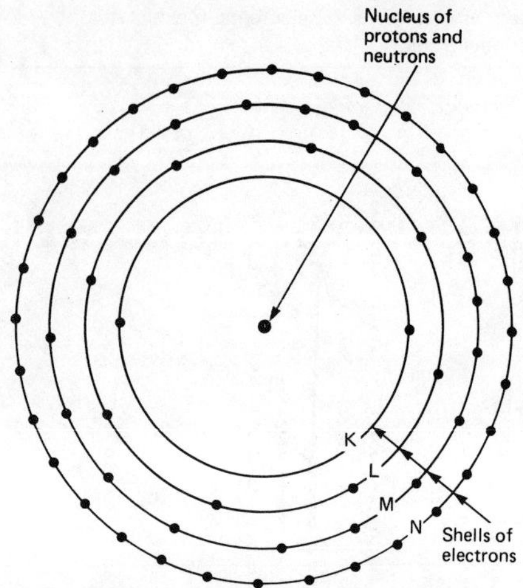

Figure 42.1 The structure of the atom

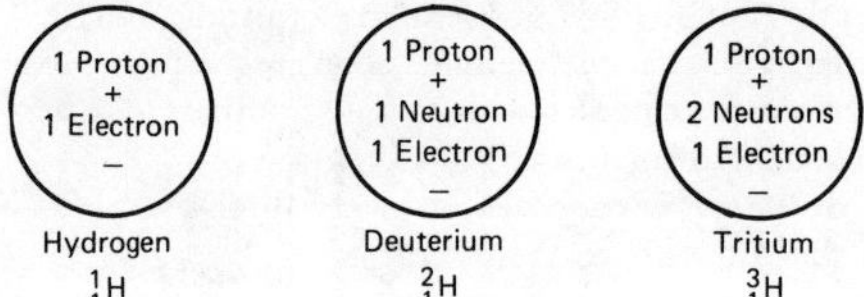

Figure 42.2 The isotopes of hydrogen

investigated for signs of **periodicity**, i.e. regular increases and decreases of physical values for similar elements known to have similar chemical properties. The method of investigation was **to plot a graph of a physical property of the element against its atomic number**.

5 It has been found that the **most useful property** of the elements to show periodicity is the **first ionisation energy (1st I.E.)** which is **the energy required to remove one mole of electrons from one mole of gaseous atoms** under standard conditions. The graph in *Figure 42.3* shows the variation of the 1st

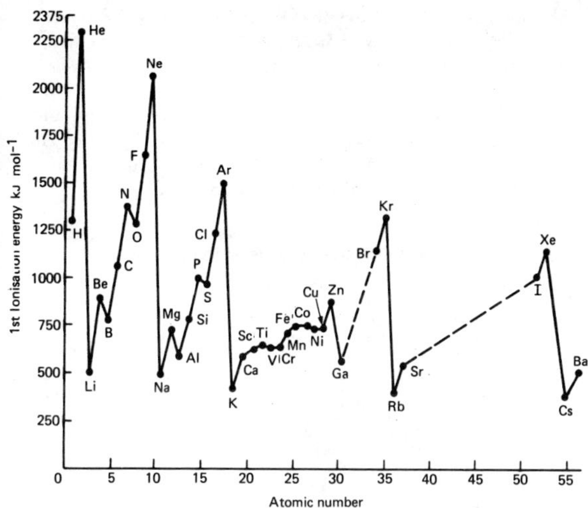

Figure 42.3 Graph of first ionisation energy of the elements against their atomic numbers

I.E. against atomic number for the first 56 elements, omitting some of the transition elements. The elements helium, neon, argon, krypton and xenon are **found at peaks** with atomic numbers 2, 10, 18 and 54. These elements which, with the exception of helium, are called the **noble or rare gases**, are very similar in chemical properties in that they are distinctly unreactive.

Similarly, the elements lithium, sodium, potassium, rubidium and caesium **occur at the troughs** of the graph with atomic numbers 3, 11, 19, 37 and 55. These elements are all similar in their properties and are called the **alkaline metals**. The distribution of the elements in this graph is such that any elements selected from similar positions on the increasing part of the peaks shows similar chemical properties. For example, the element at one place below the maximum peak value are fluorine, chlorine, bromine and iodine which occur at 9, 17, 35 and 53l and these are the **halogens** which have similar properties to each other.

6 Since the first ionisation energy of an element is an indication of how difficult energetically it is to remove an electron from the element, the elements at the peaks with the high value should be unreactive, and this is found to be true. Conversely the elements in troughs should be very reactive for the same reason, and this is also

true. This indicates that **different elements contain different types of electrons which are being removed** to find the first ionisation energy, as can be seen from the values shown in *Table 42.3.*

7 The elements with the highest 1st I.E. values are helium and the noble or rare gases, Ne, Ar, Kr and Xe. Each of these elements is followed by an element which has a low 1st I.E. value namely Li, Na, K, Rb and Cs. This **marked change** in value for

Table 42.3 Some physical properties of a selection of elements

Z = atomic number A = relative atomic mass
ρ = density at 298 K (unless shown otherwise) g cm^{-3}
$\Delta H^{\ominus}_{m}$ = molar enthalpy change on melting at T_m (latent heat of fusion) kJ mol^{-1}
R_{cov} = covalent radius of the atom nm (*van der Waal radius)
T_m = melting point temperature K *1st I E* = first ionisation energy kJ mol^{-1}

Element	Z	A	ρ	T_m	$\Delta H^{\ominus}_{m}$	R_{cov}	*1st I E*
H	1	1.0	0.07^{20K}	14	0.06	0.037	1310
He	2	4.0	0.12^{4K}	$4^{103\ atm}$	$0.02^{103\ atm}$	0.099*	2370
Li	3	6.9	0.53	454	3.01	0.123	520
Be	4	9.0	1.85	1556	11.72	0.106	900
B	5	10.8	2.55	2300	22.18	0.088	800
C	6	12.0	2.25	4000	1.87	0.077	1090
N	7	14.0	0.81^{77K}	63	0.36	0.070	1400
O	8	16.0	1.14^{90K}	54	0.22	0.066	1310
F	9	19.0	1.11^{73K}	53	0.26	0.064	1680
Ne	10	20.2	1.21^{27K}	25	0.33	0.160*	2080
Na	11	23.0	0.97	371	2.60	0.157	500
Mg	12	24.3	1.74	923	8.95	0.140	740
Al	13	27.0	2.70	932	10.75	0.126	580
Si	14	28.1	2.33	1683	46.44	0.117	790
P	15	31.0	2.20	$870^{143\ atm}$	$4.71^{143\ atm}$	0.110	1010
S	16	32.1	2.07	369	0.38	0.104	1000
Cl	17	35.5	1.56^{239K}	172	3.20	0.099	1260
Ar	18	39.9	1.40^{85K}	84	1.18	0.192*	1520
K	19	39.1	0.86	336	2.30	0.203	420
Ca	20	40.1	1.55	1123	8.66	0.174	590
Sc	21	45.0	2.99	1673	16.11	0.144	630
Ti	22	47.9	4.54	1950	15.48	0.132	660
V	23	50.9	6.11	2190	17.57	0.122	650
Cr	24	52.0	7.19	2176	13.81	0.117	650
Mn	25	54.9	7.42	1517	14.64	0.117	720
Fe	26	55.9	7.86	1812	15.36	0.116	760
Co	27	58.9	8.90	1768	15.23	0.116	760
Ni	28	58.7	8.90	1728	17.61	0.115	740
Cu	29	63.5	8.94	1356	13.05	0.135	750
Zn	30	65.4	7.13	693	7.38	0.131	910
Ga	31	69.7	5.91	303	5.59	0.126	580
Ge	32	74.9	5.32	1210	31.8	0.122	760
Br	35	79.9	3.12^{266K}	266	5.29	0.111	1140
Kr	36	83.8	2.16^{120K}	166	1.64	0.197*	1350
Rb	37	85.5	1.53	312	2.36	0.216	400
Sr	38	87.6	2.58	1043	9.20	0.191	550
I	53	126.9	4.94	387	7.87	0.128	1010
Xe	54	131.3	3.52^{64K}	161	2.29	0.217*	1170
Cs	55	132.9	1.87	302	2.13	0.235	380
Ba	56	137.3	3.50	983	7.66	0.198	500

elements differing by **one unit in atomic number** is evidence for the existence of **shells of electrons** around the central nucleus.

8 (i) Further evidence about the arrangement of electrons around the nucleus has been obtained by **removing each electron from an element in turn**, and measuring the energy for each successive electron removed. A graph of the number of the electron removed against the logarithm of the ionisation energies is shown in *Figure 42.4* for calcium.

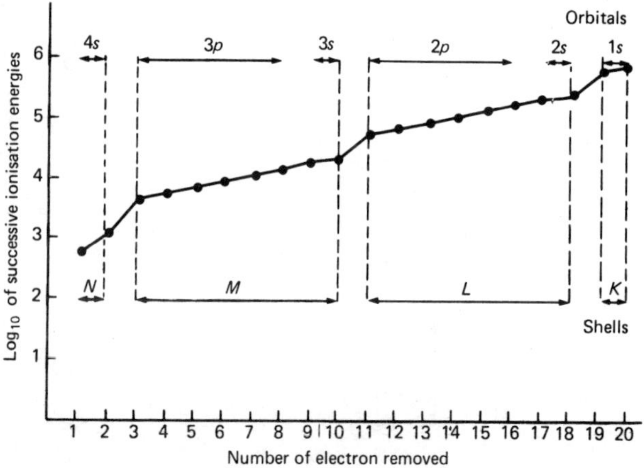

Figure 42.4 Graph of the log of successive ionisation energies of calcium against the number of the electron removed

(ii) The marked increases in removing the 3rd, 11th and 19th electrons can only be explained by assuming that **the electrons exist in sets of electrons which are particularly stable**. These sets of electrons are called shells of electrons and are assigned the letters *K*, *L*, *M*, *N* and *O* to identify them. The *K* shell contains 2 electrons, the *L*, 8 electrons, the *M* and *N* 18 electrons and the *O* shell 32 electrons.

(iii) A further sub-division is made into *s*, *p*, *d* and *f* **energy levels** at intervals of 2, 6, 10 and 14 electrons respectively. The letters *s*, *p*, *d* and *f* signify sets of orbitals in which the *s* **energy level contains one orbital**, the *p* **energy level, three orbitals**, the *d* **level five orbitals**, and the *f* **energy level seven orbitals**, in which each orbital can contain a maximum of 2 electrons.

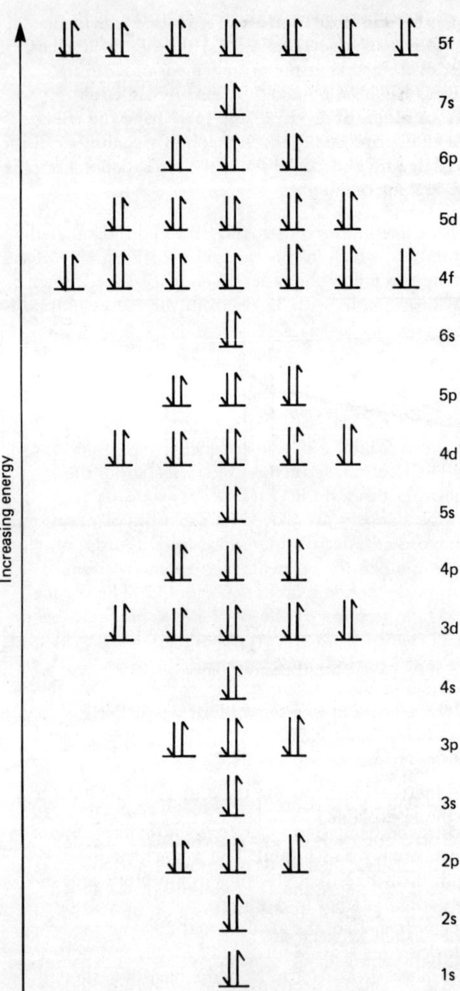

Figure 42.5 The energy levels of the orbitals

(iv) The order of the energy levels with respect to the shells of electrons is shown in *Figure 42.5*, where the 1*s* orbital is lowest and the 7*s* orbital the highest.

9 The ordering of the energy levels of electrons provides a second method of writing down the electronic structure of the elements. For example, phosphorus, which has 15 electrons, can be represented by:

P, $1s^2 2s^2 2p^6 3p^3$.

A rapid check for the correct structure is to add together the indices of the numbers, which in this case gives 2+2+6+2+3 =15. It is important to follow the order given in *Figure 42.5* such that in bromine for example, with 35 electrons the representation is

Br, $1s^2 2s^2 2p^6 3s^2 3p^6 4s^2 3d^{10} 4p^5$

and not

Br, $1s^2 2s^2 2p^6 3s^2 3p^6 3d^{10} 4s^2 4p^5$

This is because the 4*s* orbital is of a lower energy level than any of the 3*d* orbitals. The electronic structures of a selection of the elements are shown in this form in *Table 42.2.*

10 (i) When the elements are placed into **groups of elements of similar electronic structure** in vertical lines in order of increasing atomic number, the elements take up the positions which are shown in the periodic table in *Table 42.2*. The vertical lines are referred to as **groups of the periodic table**, whereas the horizontal lines of elements, which correspond to the filling of shells of electrons, are called **periods of elements.**

(ii) The classification of the elements in this way shows that the **properties of the elements are dependent upon their electronic structures**, since elements with similar electronic structures show similar properties.

11 A further sub-division of the elements has been made into ***s*-block, *p*-block and *d*-block elements.** The ***s*-block elements** are those elements whose outermost electrons (those most distant from the nucleus), **occupy an *s*-orbital.** The ***p*-block elements** are elements with their outermost electrons in ***p*-orbitals** and the ***d*-block elements** with electrons in ***d*-orbitals.**

12 The *s*-block elements are the alkaline and alkaline-earth metals, the *p*-block elements are a mixture of metals and non-metals and the *d*-block elements are the dense transition metals.

13 This classification of matter into shells of electrons and orbitals of electrons is of great use in explaining the bonding that exists within the elements and between the elements.

43 Radioactivity

1 Some materials are able to **emit particles and radiation spontaneously**. They undergo changes in their nuclei in an effort to become stabilised. Because the **nucleus changes**, the **constituent atoms change their atomic number** and hence become **a different element**. This change is called **radioactive decay** and the changes continue until finally a stable element is obtained. Such a **decay pattern** for uranium-238 is shown in *Figure 43.1*.

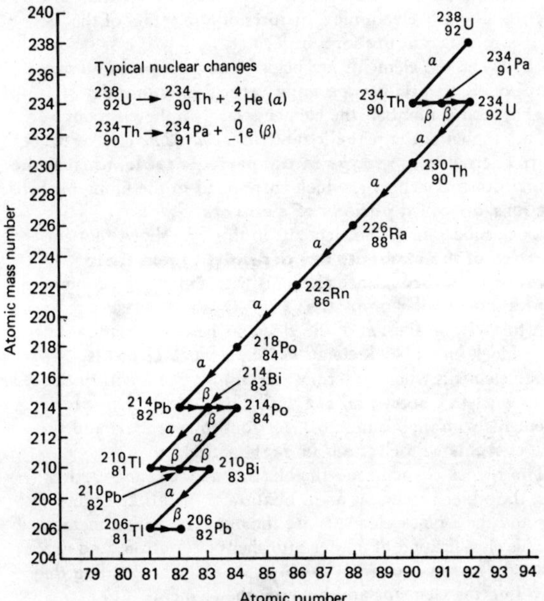

Figure 43.1 The uranium-238 decay series

2 There are three main types of emission from radioactive substances.
(i) **Alpha particles** (α), which are **helium nuclei** without any electrons, Helium nuclei contain 2 protons and 2 neutrons giving a mass number of 4 and a charge of +2. They are usually epresented in equations by the symbols $^{4}_{2}\mathbf{He}$.
(ii) **Beta particles** (β) which can be represented by the symbols $^{0}_{-1}\mathbf{e}$ and are **high velocity electrons**, that is, particles of negligible mass and a charge of − 1 units.
(iii) **Gamma radiation** (γ) which is **electromagnetic radiation with wavelengths in the region of** $\mathbf{10^{-13}}$ **m**. The properties of the particles and their penetrating powers are summarised in *Table 43.1.*
3 When an α-particle is emitted by a radioactive substance such as **uranium-238**, symbol $^{238}_{92}U$, the equation showing the change in the composition of the nucleus is

$$\underset{\text{uranium}}{^{238}_{92}U} \longrightarrow \underset{\text{thorium}}{^{234}_{90}Th} + \underset{\text{alpha particle}}{^{4}_{2}He(\alpha)}$$

From *Figure 43.1*, it can be seen that Thorium-234 is produced by the decay of uranium-238. It is important to notice that the combined mass numbers of ^{234}Th and ^{4}He when added are the same as the mass number of uranium-238. Similarly the sum of the atomic numbers, 90 and 2 is also the same as the value for uranium, i.e. 92.
4 When a β-particle is emitted by thorium the change can be represented by the equation

Table 43.1 The properties of radioactive particles

Radiation	*Mass*	*Charge*	*Deflection*	*Penetration*
α	4 a.m.u. 6.68×10^{-27} kg	+2 $+3.2 \times 10^{-19}$ C	Electric field Magnetic field	A sheet of paper
β	$\frac{1}{1836}$ a.m.u. 9.11×10^{-31} kg	−1 -1.6×10^{-19} C	Electric field Magnetic field	A few millimetres of lead
γ	0	0	None	Up to 15 cm of lead

$$^{234}_{90}\text{Th} \longrightarrow {}^{234}_{91}\text{Pa} + {}^{0}_{-1}\text{e}(\beta)$$

thorium palladium beta particle

The **decay product** has increased its atomic number by 1 unit which means a proton must have been formed during the emission of the β-particle and a new element is formed, palladium. The proton and the β-particle are formed by the **distintegration of a neutron**.

$$^{1}_{0}\text{n} \longrightarrow {}^{1}_{1}\text{H} + {}^{0}_{-1}\text{e}$$

neutron proton electron (β)

The **decay series** shown in *Figure 43.1* shows the interconversion of elements due to these nucleur changes.

5 Gamma radiation can be emitted when α- or β-particles are lost from a nucleus in order to restore the correct energy level for the newly formed isotope. Gamma radiation is uncharged electromagnetic radiation travelling at the speed of light.

6 **Nuclear stability** Some elements show no tendency to emit radioactivity. It has been found that the nuclei of these elements are composed of either equal or very nearly equal numbers, of protons and neutrons. As the **ratio of neutrons to protons increases**, the elements are found **to be more radioactive**. One theory is that the **mutual repulsion between protons in the nucleus is overcome by the neutrons present**. However, when the number of neutrons is greater than the number of protons the nucleus disintegrates spontaneously in an effort to become stable. For example, $^{238}_{92}$Uranium emits 7 α-particles, 2 β-particles and γ radiation to change eventually into $^{206}_{82}$Lead.

The graph shown in *Figure 43.2* compares the number of neutrons and protons in the elements. Those within the shaded area are the **naturally occurring stable isotopes**. However, when the number of neutrons is much greater than the number of protons, one or more neutrons disintegrate to form protons. The release of β-particles gives a ratio nearer to 1. The **elements which have mass numbers greater than 209 are unstable** to the extent that they emit α-particles. This emission reduces the number of protons by 2 and the number of neutrons by 2, which may cause further instability leading to β-particle emission.

7 **The detection of radioactivity**. Radioactivity is detected by three main techniques, i.e. **Geiger counters, scintillation counters** or **solid-state detectors**.

(i) **The Geiger counter** is an easily transportable meter which often incorporates both audio and visual detection signals. The counter is composed of a Geiger-Muller tube shown

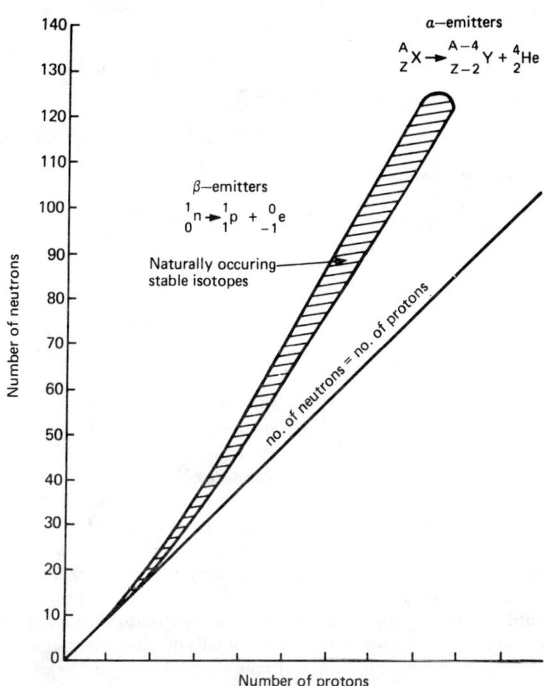

Figure 43.2 A composition of the numbers of protons and neutrons on the elements

diagrammatically in *Figure 43.3* connected to a d.c. supply and a scaler or ratemeter. When the radiation from the radioactive source enters through the mica window it causes the argon gas to ionise. These ions are accelerated toward the electrodes causing further ionisations to occur. On contact with the electrodes the ions produce a pulse which is amplified and led into a recorder. If the recorder is a scaler the total number of pulses can be found for a given time period, whereas, if a ratemeter is used, the average pulse rate is displayed as '**counts per second**' and an audio signal is given out by a loudspeaker as a series of clicks.

(ii) **Scintillation counters**. Certain molecules called **phosphors** or **fluors** produce light photons when radioactive radiation falls on them. These minute flashes of light are detected by a

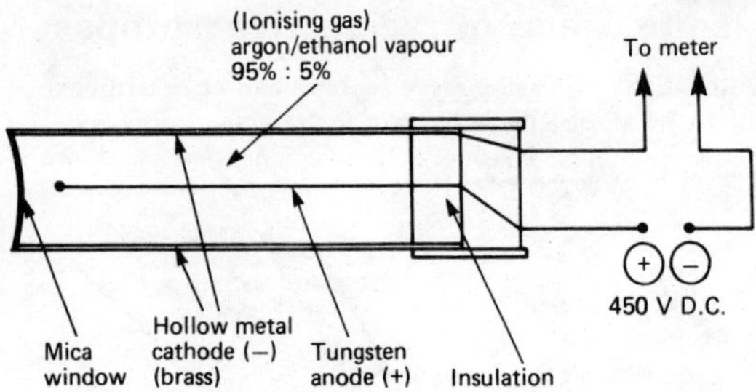

Figure 43.3 A Geiger–Müller tube

photomultiplier tube and changed into an electrical current. Liquid scintillation counting is used for low energy beta radiation counting. Some examples of phosphors are *p-terphenyl,2,5-diphenyloxazale (P.P.O.)* **a primary phosphor** which emits light at 3.7×10^{-7} m and *1,4-bis-(5-phenyloxazol-2-yl) benzene (P.O.P.O.P.)* **a secondary phosphor** which emits light at 4.3×10^{-7} m. The magnitude of the current produced by the photomultiplier tube is proportional to the incident radioactivity which can be determined by calibration with standard sources.

(iii) **Solid-state detectors**. Some materials are **semi-conductors** which readily conduct, when radioactivity falls on them. Some examples are cadmium sulphide and diamond. The **intensity of radiation** is measured by the **increase in conductivity** of the materials.

8 **The units of radioactivity**. The **curie** was first defined as **the activity of one gram of radium**, but has now been standardised as **3.7×10^{10} disintegrations per second (d.p.s.)**. Small values of activity are reported in values of

$$\textbf{a millicurie} = \frac{\textbf{one curie}}{\textbf{1000}} \text{ or } \textbf{a microcurie} = \frac{\textbf{one curie}}{\mathbf{10^6}}$$

9 **The rate of disintegration or rate of decay of radioactive istopes**. All radioactive isotopes **disintegrate naturally at a rate which cannot be speeded up or slowed down**, but which is characteristic of that isotope. If the rate of disintegration is measured at specific times it can be seen that the **time taken** for the rate of disintegration to **drop by half** is a **constant value** for that particular isotope and is called **the half life**.

Half-life values of radioactive isotopes

10 The **half-life of a radioactive isotope** can be determined graphically by measuring the activity of the sample at different times and then plotting a graph of activity against time, as shown for nitrogen-14 in *Figure 43.4*.

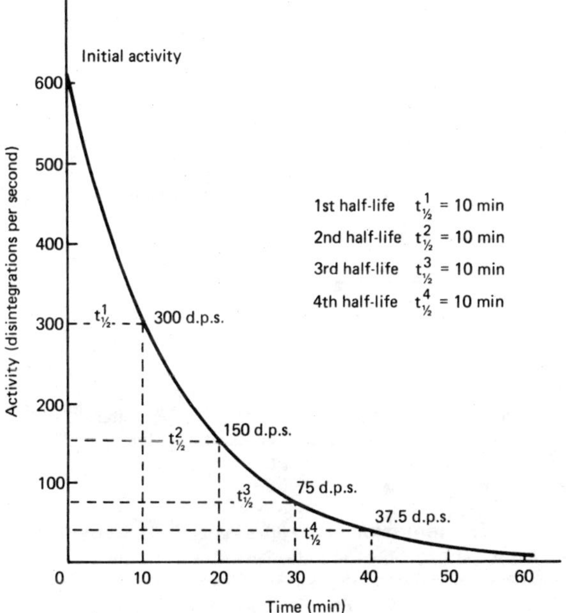

Figure 43.4 The half-life of nitrogen-14

Half-life values for different isotopes vary considerably over a wide range as can be seen from the values of some typical isotopes in *Table 43.2*. For example, when a sample of **mendelerium-256**, which has an **activity of 8000 d.p.s.** and a **half-life of 30 minutes**, has its activity determined after **4 hours** it is found to be 31.25 d.p.s. **This is because after each half-life, the activity is halved**. Thus after 30 min, the activity will be 4000 d.p.s. after

Table 43.2

Isotope	*Radiation*	*Half-life*
$^{3}_{1}H$	beta	12.1 years
$^{14}_{6}C$	beta	5730 years
$^{16}_{7}N$	beta gamma	7.14 sec.
$^{20}_{8}O$	beta gamma	14 sec.
$^{36}_{17}Cl$	beta gamma	3×10^5 years
$^{45}_{20}Ca$	beta	180 days
$^{227}_{89}Ac$	beta	13.5 years
$^{223}_{88}Ra$	alpha	11.43 days
$^{230}_{90}Th$	alpha	8×10^4 years
$^{214}_{83}Bi$	beta, alpha	19.7 min
$^{214}_{82}Pb$	beta	26.8 min
$^{211}_{82}Pb$	beta	36.1 min

1 hr (2 half-lives) 2000 d.p.s. after $1\frac{1}{2}$ hr 1000 d.p.s., after 2 hr 500 d.p.s., after $2\frac{1}{2}$ hr 250 d.p.s., after 3 hr 125 d.p.s. after $3\frac{1}{2}$ hr 62.5 d.p.s. and **finally after 4 hr an activity of 31.25 disintegrations per second.**

The disintegration law represented mathematically

11 Let N be the number of radioactive atoms at a time t; this number is proportional to the rate at which they disintegrate, $-\mathrm{d}N/\mathrm{d}t$, negative because the rate is decreasing. Mathematically this can be represented as

$$-\frac{\mathrm{d}N}{\mathrm{d}t} \propto N \text{ or}$$

$$-\frac{\mathrm{d}N}{\mathrm{d}t} = \lambda N,$$ the constant being called the **decay constant**.

This equation can be rewritten as

$$\frac{\mathrm{d}N}{N} = -\lambda\,\mathrm{d}t$$

Integrating both sides of the equation gives

$$\log_e N = -\lambda t \text{ .a constant} \quad (1)$$

If N is the number of atoms when $t=0$

Then, $\log_e N_o = -\lambda(0) + \text{a constant}$

i.e. **the constant** $= \log_e N_o$

Substituting this value into equation (1) gives

$$\log_e N - \log_e N_o = -\lambda t \text{ or}$$

$$\log_e \frac{N}{N_o} = -\lambda t \text{ or}$$

$$N = N_o e^{-\lambda t}$$

This equation **relates the number of atoms N which exist at a time t to the original number of atoms, N_o.**

A mathematical expression for the half-life

12 The **half-life of a radioactive isotope** can be related to **concentrations** by the equation

$$N_{\frac{1}{2}} = N_a e^{-\lambda t_{\frac{1}{2}}}$$

where $N_{\frac{1}{2}}$ is half the original number of atoms N_a it follows that

$$\frac{N_{\frac{1}{2}}}{N_a} = \frac{1}{2}, \text{ i.e. } \frac{1}{2} = e^{-\lambda t_{\frac{1}{2}}} \text{ or } e^{\lambda t_{\frac{1}{2}}} = 2$$

Taking natural logarithms

$$t_{\frac{1}{2}} = \frac{\log_e 2}{\lambda} \text{ or } t_{\frac{1}{2}} = \frac{0.693}{\lambda}.$$

For example, since the half-life of phosphorus 32 is 14.3 days, the radioactive decay constant can be found by using the equation

$$t_{\frac{1}{2}} = \frac{0.693}{\lambda}, \text{ i.e. } \lambda = \frac{0.693}{t_{\frac{1}{2}}}$$

By substituting in the value of $t_{\frac{1}{2}}$,

$$\lambda = \frac{0.693}{14.3} = 0.0485 \text{ day}^{-1} \text{ or } 5.61 \times 10^{-7} \text{ s}^{-1}$$

A knowledge of the decay constant for an isotope allows the activity **of that radioactive isotope to be found at any time**. For example, to find the activity (d.p.s.) of a sample of phosphorus 42 after 2 hours, given the original activity of the sample is 600 d.p.s., with a half-life of 5.61×10^{-7} s^{-1}.

Using the relationship

$$\log_e \frac{N}{N_o} = -\lambda t$$

and rearranging it to the $\log_{10}$ base gives

$$\log_{10} \frac{N_o}{N} = \frac{\lambda t}{2.303}$$

Substituting the values given into this equation gives

$$\log_{10} \frac{6000}{N} = \frac{5.61 \times 10^{-7} \times (2 \times 60 \times 60)}{2.303}$$

By taking antilogs,

$$\frac{600}{N} = \text{antilog } 1.75 \times 10^{-3} = 1.002$$

$$\text{or } N = \frac{600}{1.002} = 598.88$$

Hence after 2 hours the activity of the $^{42}_{15}P$ is **598.88 d.p.s.**

13 **Nuclear transformations** The **bombardment of the atoms** of an element with **high velocity particles** causes changes to occur within the nuclei of the atoms. Some examples of these transformations are

(i) **with α particles**

$^{14}_{7}N$	$+ \, ^{4}_{2}He \longrightarrow$	$^{17}_{8}O$	$+ \, ^{1}_{1}H$
nitrogen-14	α-particle	oxygen-17	proton

(ii) **with neutrons**

$^{16}_{8}O$	$+ \, ^{1}_{0}n \longrightarrow$	$^{13}_{6}C$	$+ \, ^{4}_{2}He$
oxygen-16	neutron	carbon-13	α-particle

(iii) **with protons**

$^{9}_{4}Be$	$+ \, ^{1}_{1}H \longrightarrow$	$^{6}_{3}Li$	$+ \, ^{4}_{2}He$
beryllium-9	proton	lithium-6	α-particle

14 **Nuclear fission** When some isotopes of elements of high atomic mass are bombarded with neutrons of appropriate energy, they **disintegrate into two nuclei of smaller atomic mass**. This is accompanied by the release of two or more high speed neutrons and a release of a large amount of energy. For example,

when Uranium-235 is treated in this way, the following changes occur:

$$^{235}_{92}U + ^{1}_{0}n \rightarrow ^{144}_{56}Ba + ^{90}_{36}Kr + 2^{1}_{0}n + \text{energy}$$

This '**splitting' of the atom** is called **fission** and the nuclei produced are called **fission products**. The 2 neutrons released during the disintegration means that there is available a source of neutrons for further nuclear disintegrations. For a particular mass of U-235 in one lump (about 30 kg) the releasing and reacting of the neutrons and nuclei becomes **a chain reaction**. This mass is called the **critical mass** and is necessary to ensure that the ratio of the number of neutrons escaping from the uranium-235 to those held within the mass is of the correct order.

15 **The atomic reactor** In an atomic energy power station, the **energy released as heat during fission** is used to produce steam to drive turbines. The nuclear reactions have to be carefully

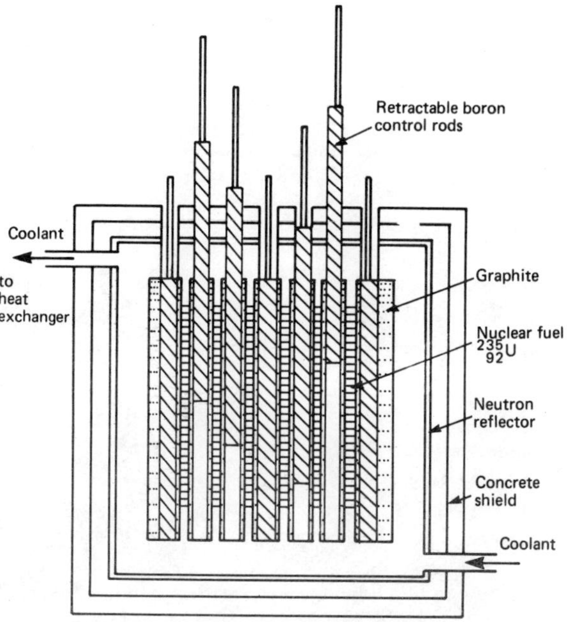

Figure 43.5 An atomic pile

controlled to prevent an explosion occurring. This can be achieved by **controlling the number of neutrons** causing the nuclear disintegrations using **neutron absorbers** such as boron or cadium. The **nuclear fuel** is contained in a **graphite block** which acts as **a moderator** in slowing down the neutrons produced by fission to a suitable velocity, to react with further atoms of nuclear fuel.

The atomic pile, which is the heart of the nuclear power station, is shown diagrammatically in *Figure 43.5*. The chain reaction is controlled by raising or lowering the boron control rods. The heat produced is absorbed by the coolant which is led into a heat exchanger to produce steam.

16 **The atomic bomb**. When two masses of $^{235}_{92}U$ whose combined mass exceed the critical mass, are kept apart and then suddenly brought together, the chain reaction which occurs is uncontrolled and a nuclear explosion occurs due to the vast release of energy.

44 Chemical bonding of the elements

1 When the elements are in their pure states they can be broadly divided into **metals** and **non-metals** by consideration of their physical properties. The way in which the atoms of the elements are held together in molecules is called **chemical bonding**.

2 (i) The type of bonding which binds the atoms of the metallic elements together is called **metallic bonding**. X-ray crystallographic studies have shown that the metal atoms are **closely packed** together in such a way that they occupy a minimum volume, and hence a minimum potential energy exists between atoms. An **ordered** arrangement of spheres can be used to represent the types of bonding that occur, and the three possible structures are shown in *Figure 44.1*.

(ii) When the layers of atoms are arranged as shown in *Figure 44.1(a)*, there is only one way of obtaining a minimum volume, which is shown in *Figure 44.1(b)*. This type of arrangement is termed **cubic close-packing**, and the structure is called a **body-centred cubic** structure, in which any single atom is touching **twelve** other atoms. For this reason the atom is said to have a **co-ordination number** of twelve.

(iii) When the layers of atoms are arranged as shown in *Figure 44.1(c)* there are two ways in which a minimum volume can be achieved. These are shown in *Figure 44.1(d)* and *1(e)*. The structure in *Figure 44.1(d)* shows the first two layers of a repeating structure, i.e. *AB AB AB*. In this structure **gaps** can be seen when looking on to the arrangement of atoms indicated by the letter *X* in the diagram. This is called a **hexagonal close-packed** structure, in which each atom touches twelve others, i.e. it has a co-ordination number of twelve.

(iv) The other possible arrangement of atoms is shown in *Figure 44.1(e)*. In this structure the gaps present in *Figure 44.1(d)* have been filled by placing the third layer of atoms over the gaps designated by *X* in that structure. This means that the repeat order of the structure is *ABC ABC ABC*. This structure is called a **face-centred cubic** structure with a co-ordination number of twelve.

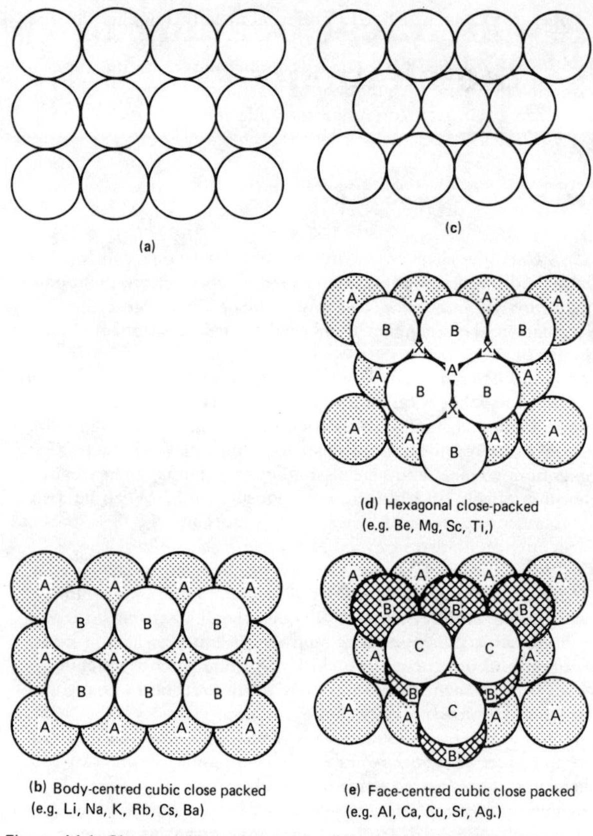

Figure 44.1 Close-packing of atoms in different metallic elements

3 The amount of **vacant space** in the three different structures is **32%** in *Figure 44.1(b)* and **26%** in *Figure 44.1(d)* and *(e)*. Examples of metals having the different types of bonding are also given in *Figure 44.1*.

4 Some of the physical properties of most metals are that they are **malleable**, **ductile**, **good conductors of heat and electricity** and in addition the majority of the metals are **dense** and have **high melting points** and **boiling points**. These

properties can be explained in terms of metallic bonding theory in the following ways.

Since the theory of metallic bonding states that the atoms of the metals are not directly bonded together, but close-packed together, it follows that the atoms within the solid can be redistributed by externally applied mechanical forces. For example, continuous hammering of a piece of metal causing it to change it's shape is a process which is termed **malleability**. Similarly, metals can be drawn into a wire by the application of a tensile force to the ends of a metal rod, this being called **ductility**.

Good electrical conductivity is explained by the presence of **non-bonded electrons** in the close-packed structures. Since the electrons within the structure are essentially **free electrons**, the passage of a stream of electrons takes place readily.

Good thermal conductivity is explained by the ready acceptance and transfer of heat energy from one atom to another when close-packed. High density in the metals is explained by the close-packing together of the atoms in either the hexagonal, or cubic close-packed structures. The high melting points and boiling points are explained by the amount of energy required to separate the atoms from each other, and due to the close packing, a large amount of energy is required.

Hence all of the properties can be explained in terms of the close-packed non-directionally bonded theory.

5 The non-metallic elements in their pure states are held together by a different type of bonding called **covalent bonding**. This involves the **sharing of two electrons** between two different atoms which holds the electrons in a distinct position relative to each other. Each element can share as many electrons as are necessary to achieve a stable electronic configuration. These stable electronic configurations are those of the **inert** or **noble gases**, neon, argon, krypton or helium. The number of bonds formed is thus dependent on the electronic configuration of the particular element and some examples are shown for simple gaseous elements in *Figure 44.2*. The number of **bonding** electrons surrounding each atom are shown by **crosses (X)** or **dots (●)**, each pair of shared electrons forming a covalent bond. In this way, one (H—H), two (O=O) or three (N≡N) covalent bonds can be formed between two atoms of the elements.

6 One of the non-metals, bromine, is a **liquid** at room temperature. As can be seen from *Figure 44.2*, each bromine molecule contains two bromine atoms joined by one covalent bond. The reason why bromine is a liquid at this temperature rather than a gas (like chlorine) can only be explained by the presence in the structure of liquid bromine of some **forces of attraction**

Element	Electronic configuration	Nearest noble gas	Number of electrons shared	Dot and cross diagram	Formula
Gases					
H_2	1	He 2·	1	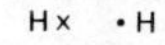 H× •H	H—H
F_2	2·7	Ne 2·8	1	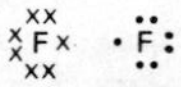 ××F× •F: (with ×× and •• pairs)	F—F
Cl_2	2·8·7	Ar 2·8·8	1	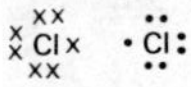 ××Cl× •Cl: (with ×× and •• pairs)	Cl—Cl
O_2	2·6	Ne 2.8	2	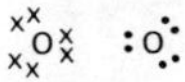 ××O×× :O:	O=O
N_2	2·5	Ne 2·8	3	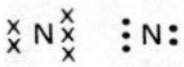 ××N××× :N:	N≡N
Liquid					
Br_2	2·8·18·7	Kr 2·8·18·8	1	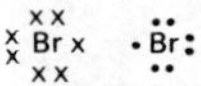 ××Br× •Br: (with ×× and •• pairs)	Br—Br

Figure 44.2 The formation of covalent bonds

Element	Molecular formula	Simplest Structure	Structure type	Melting point K	Boiling point K
Carbon (diamond)	C_n	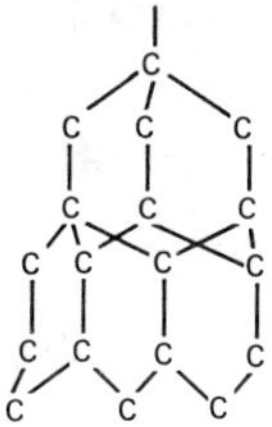	Giant structure	3850	5100
Carbon (graphite)	C_n	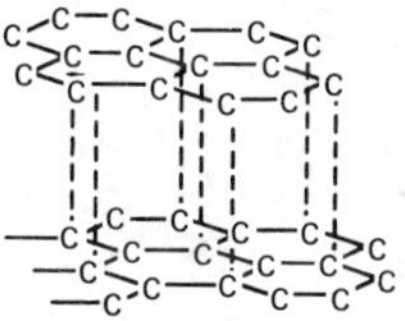	Layer structure	4000	—
Phos phorus (white)	P_4		Molecular crystal	317	554
Sulphur	S_8	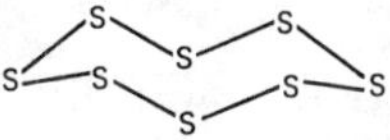	Molecular crystal	392	718
Iodine	I_2	I—I	Molecular crystal	387	456

Figure 44.3 Covalent bonding in solid non-metallic elements

between the molecules. These forces of attraction are called **van der Waal** forces, or **intermolecular forces**.

7 Some non-metallic elements are solids at room temperature and these include carbon, silicon, phosphorus, sulphur and iodine. The chemical bonding which occurs in these solid elements is either purely covalent, as in diamond (carbon) and silicon, or a

combination of covalent bonding and extensive intermolecular forces of attraction (or van der Waal forces), as shown in *Figure 44.3* by the layer structure of graphite and the molecular crystal structures of phosphorus, sulphur and iodine. The extent of the intermolecular forces can be established by the melting points of the respectively solids as shown in *Figure 44.3*.

45 Chemical bonding in compounds

1 When **two different elements** combine together to form a chemical compound the bonding between the elements can be classified into four main types. These are:

(a) **ionic bonding** which only occurs between **metals** and **non-metals**.

(b) **covalent bonding** which occurs between **different non-metals** or **non-metals** and elements of **intermediate** metallic/non-metallic character (e.g. aluminium),

(c) **dative** covalent bonding which occurs between particular molecules and in the formation of transition metal complexes, and

(d) **intermolecular** bonding which occurs between molecules which are already covalently bonded.

2 **Ionic bonds** are **permanent** electrostatic forces of attraction formed between **positively charged metallic ions** and **negatively charged non-metallic ions**.

3 Metal atoms lose electrons to attain a **stable** electronic structure identical to that of a noble gas. For example, sodium atoms have the electronic structure $2 \cdot 8 \cdot 1$, and by losing one electron becomes the sodium ion, Na^+ with an electronic structure of $2 \cdot 8$, the same as neon. The number of electrons lost by a metal atom is the number required to attain a noble gas structure and is usually 1, 2 or 3. Positive ions are called **cations** and are assigned positive because the structure will have an **excess of protons**. Some examples of positive ions are shown in *Figure 45.1*.

4 Non-metallic atoms gain electrons to form **negative ions**, called **anions**. The number of electrons gained is that number which will give the negative ion a **stable** structure. For example, oxygen has the structure $2 \cdot 6$ and by gaining 2 electrons the oxide ion is formed with the electronic structure $2 \cdot 8$, the same as neon. Other examples are given in *Figure 45.1*.

5 The combination of cations (positive) and anions (negative) into compounds, must be in a ratio which will be **electrically neutral**. For example, calcium forms Ca^{+2} ions and chlorine forms Cl^{-1} ions, and in order that calcium chlorine shall be neutral the ratio of cations to anions must be 1:2, i.e. $Ca^{+2}:2Cl^{-1}$ and the

Element	Electronic structure	Electrons removed	Ion formed	Electronic structure	Element	Electronic structure	Electrons added	Ion formed	Electronic structure
H	1	1	H^{+}	0	H	1	1	H^{-}	2·
Li	2·1	1	Li^{+}	2	N	2·5	3	N^{3-}	2·8
Na	2·8·1	1	Na^{+}	2·8	O	2·6	2	O^{2-}	2·8
Mg	2·8·2	2	Mg^{2+}	2·8	F	2·7	1	F^{-}	2·8
Al	2·8·3	3	Al^{3+}	2·8	P	2·8·5	3	P^{3-}	2·8·8
K	2·8·8·1	1	K^{+}	2·8·8	S	2·8·6	2	S^{2-}	2·8·8
Ca	2·8·8·2	2	Ca^{2+}	2·8·8	Cl	2·8·7	1	Cl^{-}	2·8·8
Rb	2·8·18·1	1	Rb^{+}	2·8·18	Br	2·18·7	1	Br^{-}	2·8·18·8
Cs	2·8·18·18·1	1	Cs^{+}	2·8·18·18	I	2·8·18·18·7	1	I^{-}	2·8·18·18·8

Figure 45.1 The formation of ions from elements

Cation	Anion	Ratio	Formula	Name
Li^+	F^-	1:1	LiF	Lithium fluoride
Na^+	O^{2-}	2:1	Na_2O	Sodium oxide
Mg^{2+}	Cl^-	1:2	$MgCl_2$	Magnesium chloride
Ca^{2+}	O^{2-}	1:1	CaO	Calcium oxide
K^+	S^{2-}	2:1	K_2S	Potassium sulphide
Mg^{2+}	S^{2-}	1:1	MgS	Magnesium sulphide
Al^{3+}	Cl^-	1:3	$AlCl_3$	Aluminium chloride
Al^{3+}	O^{2-}	2:3	Al_2O_3	Aluminium oxide

Figure 45.2

formula is represented as $CaCl_2$. Other examples of compound formation are given in *Figure 45.2.*

6 In the solid state, the external appearance of ionic compounds shows clearly that different **crystal shapes** are possible for different compounds. The external shape of the crystal does not give any information about the internal structure of the crystal lattice.

7 The composition of inorganic solids has been shown by **X-ray crystallographic** methods devised by Sir William Bragg and his son Sir William L. Bragg to exist in regular arrangements of particles. They showed that the diffraction of light through a diffraction grating produces a series of bands of intensified light interspersed with bands of darkness. This was interpreted using the wave theory of light, as the **reinforcement** of two waves in phase (see *Figure 45.3(ii)*), to explain the intensified bands, and the **cancellation** of light as a result of two wave forms out of phase (see *Figure 45.3(iii)*).

The ability of a diffraction grating to produce diffraction patterns depends on the wavelength of the type of light used. Since visible light is one type of **electromagnetic radiation**, other forms of radiation should be capable of diffraction if a suitable diffraction grating can be found.

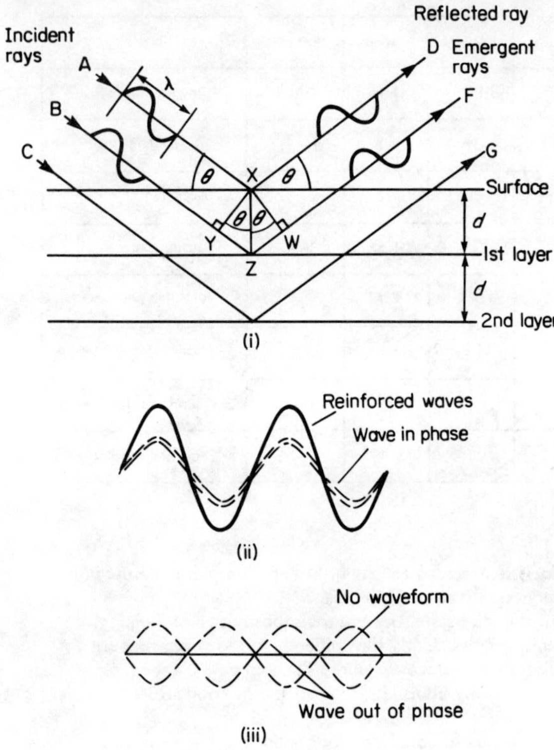

Figure 45.3

Sir William Bragg and his son discovered that crystal lattices were able to form diffraction patterns using **X-rays**. The existence of diffraction patterns was very good evidence in support of the theory that in solid state the particles are arranged in a regular pattern which could be resolved into layers depending on the way in which the crystal lattice is viewed.

If a set of planes are considered as shown in *Figure 45.3(i)*, the wave labelled AXD which is incident onto the surface of the crystal at an angle θ, strikes the surface particle at X and is reflected at an angle of θ. The second wave BZF meets the surface at a position where no particle is present and passes through the lattice structure until it strikes a particle at Z and is diffracted back through the crystal lattice surface. The waves are in phase when they are incident to the lattice surface but the wave BZF must travel a longer path than the wave AXD. For the emergent waves to be in phase, the extra distance travelled by the wave BSF must be equal to a whole number of wavelengths. Bragg found that by varying the incident angle θ, regions of intense radiation could be detected, followed by regions where no radiation could be detected. For a fixed wavelength of X-radiation, λ, Bragg related the **extra distance** travelled by the second wave as:

$$\mathbf{YZ+ZW}=n\lambda \text{ where } n=1, 2, \text{etc.}$$

Using trigonometry

$$\mathbf{YZ+ZW}=2d \sin \theta \text{ hence } \mathbf{YZ+ZW}=2d \sin \theta=n\lambda$$

This equation is called the Bragg equation and for varying values of n and θ the **interplanar distance** d can be found. A knowledge of these interplanar distances correspond to the distance between ions or atoms (single or in groups) and the lattice crystal can be discovered by constructing models of them.

9 The application of the Bragg equation can be made using the following data. When a crystal of sodium chloride is held in a certain orientation and X-radiation of wavelength 0.0597 nm is allowed to fall upon it, regions of radiation can be detected when angle of incident radiation is 12.25° and 25.1°. To find the interplanar distance use must be made of the theory that in order that a region of radiation is detected, the Bragg equation must be obeyed. The equation is

$$2d \sin \theta=n\lambda$$

where d is the interplanar distance, θ is the angle of incidence n is an integer and λ is the wavelength.

Assuming $n=1$, and using $\theta=12.25°$ and $\lambda=0.0597$ nm and substituting these values into the equation gives

$$2d \sin 12.25°=1\times 0.0597 \text{ nm}$$

Rearranging this expression gives

$$d=\frac{1\times 0.0597}{2\times \sin 12.25}=\frac{0.0597}{2\times 0.2122}=0.1407 \text{ nm}$$

Hence the interplanar distance is 0.1407 nm.

For the second value of $\theta=25.1°$, the substitution of the appropriate values into the Bragg equation assuming that $n=2$

gives

$$2d \sin 25.1 = 2 \times 0.0597 \text{ nm}$$

Rearranging this expression gives

$$d = \frac{2 \times 0.0597}{2 \times \sin 25.1} = \frac{0.1194}{2 \times 4242} = 0.1407$$

Again the interplanar distance is 0.1407 nm.

Using this interplanar distance together with values found using **different orientations** of the sodium chloride crystal allow a **model** to be constructed to **represent the solid structure**.

9 The structures of the ionic solids can be simply classified into the types **AB** and $\mathbf{AB_2}$ where **A** is the metal ion and **B** is the non-metal ion.

10 **Ionic solids** of the type **AB** can be sub-divided again into two main types, the **sodium chloride** oe **rock salt structure** and the **caesium chloride structure**. These are the **lattice** structures of the group 1 halides which are shown in *Figure 45.4(a)*

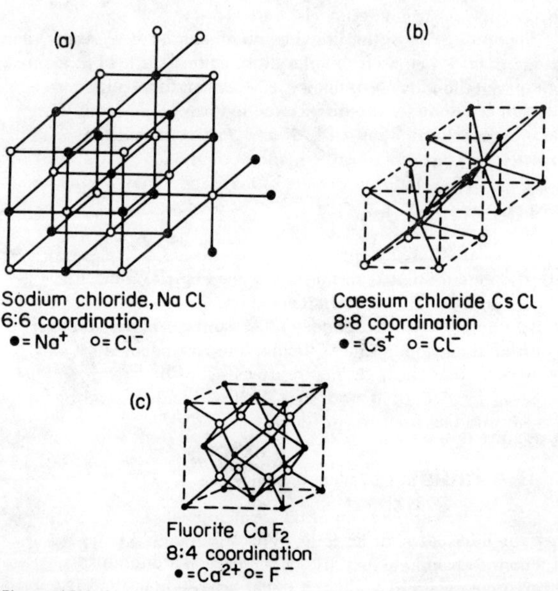

Figure 45.4

Table 45.1 The crystal structure of some ionic solids

Lattice type	*Examples*	
Sodium chloride or Rock-salt structure 6,6 co-ordination	Lif NaF KF LiCl NaCl KCl RbCl LiBr NaBr KBr RbBr LiI NaI KI RbI NH_4I	MgO CaO SrO BaO MnO NiO
Caesium chloride structure 8,8 co-ordination	CsCl RbF NH_4Cl CsBr NH_4Br CsI	
Fluorite structure 8,4 co-ordination	CaF_2 $SrCl_2$ SrF_2 CdF_2 BaF_2 PbF_2	

and *(b)* and examples of which are given in *Table 45.1*.

The other general formula for ionic solids, $\mathbf{AB_2}$ is represented by the **fluorite structure** of calcium fluoride shown in *Figure 45.4(c)*.

11 The **change in co-ordination numbers** in these ionic lattice structures and their different arrangements can be explained by considering that the arrangement of the ions depends on three main properties, firstly the **difference in charge**, secondly the **relative sizes of cations and anions** and thirdly the **ratio of cations to anions**.

The arrangement of ions

12 Since cations are positively charged and anions are negatively charged, any crystal lattice will experience **repulsive forces** between identical cations and identical anions of identical charge, whereas, it will experience **forces of attraction** between oppositely charged cations and anions. The arrangement of ions must be such that electrical neutrality exists in the structure. This is achieved by a cation surrounding itself with anions, and by an anion surrounding itself with cations.

The radius ratio

13 (i) A simple application of this principle would lead to the same lattice structure for all ionic solids but the ability of cations and anions to pack together also depends on the relative sizes of these particles. Cations are much smaller than anions and it is

easier to view the positioning of anions about a cation than vice versa. For example, by representing a simple arrangement of 4 anions around a central cation as shown in *Figure 45.5* the ratio of the size of the cation to the size of the anion can be found by considering the triangle ABC. It is constructed as a right angled triangle in which $C\hat{A}B = A\hat{B}C = 45°$.
Then by using trigonometry,

$$\frac{\mathbf{AC}}{\mathbf{AB}} = \cos 45°$$

Using the identity, $\cos 45° = \frac{1}{\sqrt{2}}$ gives

$$\frac{\mathbf{AC}}{\mathbf{AB}} = \frac{r_a + r_c}{2r_a} = \frac{1}{\sqrt{2}}$$

Rearranging this expression gives

$$\frac{r_a + r_c}{r_a} \cdot \frac{2}{\sqrt{2}} = \sqrt{2}$$

Simplifying gives

$$1 + \frac{r_c}{r_a} = \sqrt{3} \text{ or } \frac{r_c}{r_a} = \sqrt{2} - 1 = 0.414$$

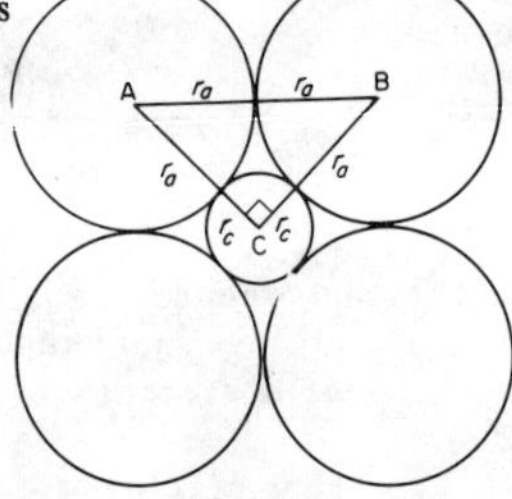

Figure 45.5

Hence for this site the **radius ratio** for touching contact is 0.414.

By placing 2 more anions directly above and below the cation produces an **octahedral site** in which the cation is in contact with **6 anions.**

(ii) For the sodium chloride cyrstal lattice (*Figure 45.4(a)*), each Na^+ ion has six Cl^- ions as near neighbours. This is because the ionic radius of the sodium ion is small enough to fit into such a site within the lattice. The geometrical calculations show that for the six Cl^- ions to touch the Na^+ ions, the radius ratio,

$$\frac{\textbf{cation radius}}{\textbf{anion radius}}, \frac{r_c}{r_a} \textbf{ is 0.414.}$$

Since the ionic radius of Na^+ is 0.98 A and that of Cl^- is 1.81 A then

$$\frac{r_c}{r_a} = 0.541.$$

This means that in sodium chloride the ions are not touching but are dispersed about the central Na^+ ion.

(iii) As the radius ratio increases the positioning of a cation in such a site causes the anions to be moved farther and farther apart. This results in a change of site for the cation when the radius ratio reaches 0.732. This crystal lattice structure is shown by caesium chloride (*Figure 45.4(b)*), where instead of six near neighbours **each Cs^+ ion** is surrounded by **eight Cl^- ions** and each Cl^- ion is surrounded by eight Cs^+ ions. The ionic radii of Cs^+ and Cl^- are 1.65 A and 1.81 A giving a value of

$$\frac{r_c}{r_a}=0.912.$$

The **size** of the **radius ratio** is responsible for a **change in co-ordination number from 6:6 to 8:8**. Hence the radius ratio for combinations of ions of equal charge can be used to predict the lattice structure of the compound. Some values of ionic radii together with **limiting** ratio values are given in *Table 45.2.*

14 The arrangement of ions of different charges is represented here by the fluorite structure of calcium fluoride CaF_2 in which each cation has a charge of $+2$ and each anion -1, and for electrical neutrality there must be **twice as many anions as cations**. The structure of this crystal lattice is such that each **Ca^{2+} ion is surrounded by eight F^-** ions and **each F^- ion is surrounded by four Ca^{2+} ions** and this results in a co-

Table 45.2

Cation	**Ionic** *radius* (nm)	*Cation*	**Ionic** *radius* (nm)	*Anion*	**Ionic** *radius* (nm)
Li^+	0.068	Mg^{+2}	0.065	F^-	0.133
Na^+	0.098	Ca^{+2}	0.094	Cl^-	0.181
K^+	0.133	Sr^{+2}	0.110	Br^-	0.196
Rb^+	0.149	Ba^{+2}	0.134	I^-	0.229
Cs^+	0.165	Mn^{+2}	0.080	O^{-2}	0.146
NH_4^+	0.148	Ni^{+2}	0.072		
		Cd^{+2}	0.097		
		Pb^{+2}	0.084		

$\frac{r_c}{r_a}$ AB type	sodium chloride	0.414 to 0.732
	caesium chloride	0.722 to 1.
$\frac{r_c+2}{r_a-1}$ AB_2 type	fluorite	0.65 to 1.

ordination number of **8 :4**. Other examples are given in *Tables 45.1* and *45.2*.

15 When two different elements combine to form a compound which is **not ionic**, then the bonding between them is called **covalent bonding**. The number of covalent bonds formed by a non-metal is discussed in section 44, para 5 for the elements. When different elements combine, the ratio is such that **each element** attains a stable electronic structure. For example, in methane, CH_4, the carbon atom requires a **share in four other** electrons to attain stability, and hydrogen requires a **share in one other** electron. Some examples of compounds held together by covalent bonding are shown in *Figure 45.6*.

16 The **major difference** between covalent bonds in elements and those in compounds is that in the elements, the two electrons forming the covalent bond are **equally shared** between the two

Elements	Ratio	Dot and cross diagram	Structure
H and F	1 : 1	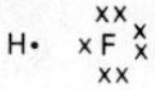 ×× H• ×F × ××	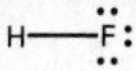 •• H——F: ••
H and O	2 : 1	 ×× H• ×O× •H ××	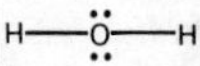 •• H——O——H ••
H and N	3 : 1	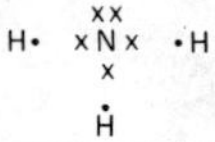 ×× H• ×N× •H × • H	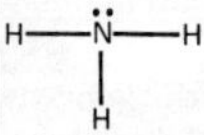 •• H——N——H | H
H and C	4 : 1	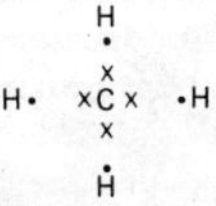 H • × H• ×C× •H × • H	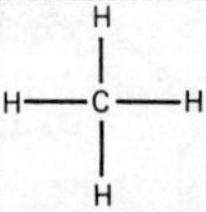 H | H——C——H | H
Cl and C	4 : 1	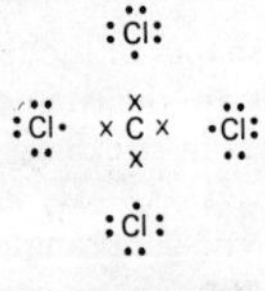 :Cl: • × :Cl• ×C× •Cl: × • :Cl:	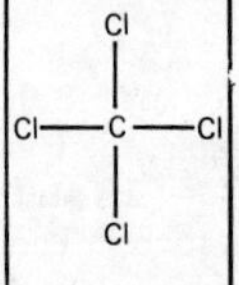 Cl | Cl——C——Cl | Cl

Figure 45.6

Table 45.3 The electronegatives of some elements

Element	F	O	N	Cl	Br	S	C	I	P	H
Electro-negativity	4.0	3.5	3.0	3.0	2.8	2.5	2.5	2.5	2.1	2.1

contributing atoms, but in compounds the **sharing of electrons is unequal**. This is because each element has a specific **attracting power** for other electrons in covalent bonds. This power is called **electronegativity** and the arbitrary values assigned to the principal non-metals are shown in *Table 45.3.*

When two different elements form a covalent bond there will be a difference in the electronegativity values of the two elements. The atom which has the **higher electronegativity** will **attract** the 2 electrons from one element to another and this is accompanied by the formation of **partial electrical** charges, positive at the lesser electronegative atom and negative at the greater electronegative atom. For example, hydrogen chloride, HCl, can be represeted as $\mathbf{H}\overset{\times}{\underset{\bullet}{—}}\mathbf{Cl}$ to show the two shared electrons one from each atom. Since chlorine is more electronegative than hydrogen a better representation is $\mathbf{H}\overset{\times}{\underset{\bullet}{—}}\mathbf{Cl}$.

In order to show the partial charges which are formed in the molecule the symbols $\delta+$ and $\delta-$ are used, i.e. $^{\delta+}\mathbf{H}—\mathbf{Cl}^{\delta-}$. This unequal sharing of electrons and the formation of **polarised bonds** has resulted in such molecules being called **polar molecules**. The factor which decides how polar a molecule becomes is the difference in the values of the electronegativities of the constituent elements.

17 Compounds which are covalently bonded can be gases, liquids or solids. The molecules which constitute these states have **specific shapes** which are dependent on the **electronic repulsion forces** within the molecule. Since each bond contains a pair of electrons and because **like charges repel** each other, the bonds will be distributed to **minimise** the repulsion forces in the molecule. The **non-bonding pairs** of electrons also give rise to repulsive forces. The shapes of some covalent compounds are given in *Figure 45.7.*

In addition to the shapes of the molecules the **bond angles** made between the atoms are dependent upon electronic effects. For example, the bonds angles in **methane (H—C—H), ammonia (H—N—H) and water (H—O—H)** are 109.5°, 107° and 104.5° respectively. These different bond angles can be explained by considering the dot and cross diagrams in *Figure 45.6.* This shows clearly that for methane there are four covalent bonds surrounding

Compound	Number of bonds	Number of non-bonding pairs	Shape	Description
$BeCl_2$	2	0	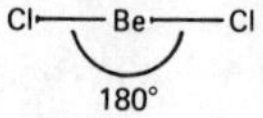 Cl—Be—Cl 180°	Linear
BCl_3	3	0	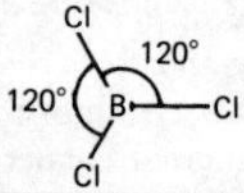 Cl, Cl, B—Cl 120°, 120°	Planar
CH_4	4	0	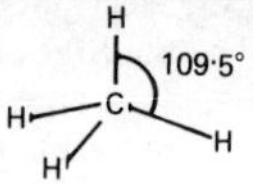 H, H—C—H, H 109·5°	Tetrahedral
NH_3	3	1	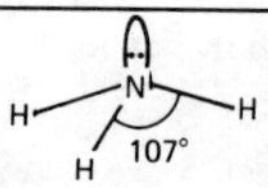 H—N—H, H 107°	Pyramidal
H_2O	2	2	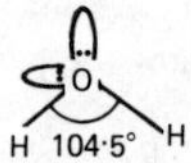 O, H 104·5° H	Planar, angular
HCl	1	3	 Cl—H 180°	Linear

Figure 45.7 The shapes of covalently bonded molecules

the carbon, each of which is composed of a bonding pair of electrons. The identical polarity of these bonding pairs leads to a repulsion between them.

In order to minimise these repulsive forces, bonds are distributed to the four corners of a **regular tetrahedron**. This results in each of the H–C–H bond angles being equal and 109.5° in magnitude, as shown in *Figure 45.6.* Similarly, dot and cross diagrams are also shown for ammonia and water in *Figure 45.6.* The difference between these compounds and methane are that (a), in ammonia, a non-bonding (**lone pair**) of electrons replaces one of the covalent bonds, and (b), in water two non-bonding pairs are present. The three covalent bonds around nitrogen, together with the non-bonding electron pair are distributed to the corners of an **irregular tetrahedron** to minimise repulsion forces.

However, since each of the three bond angles formed as H—N—H bonds are smaller than in methane, the repulsive force between the non-bonding pair and the bonding pairs must be slightly larger than the repulsive forces between the bonding pairs alone. This results in the smaller bond angle of 107°. Since the H—O—H bond angle is 104.5°, this is further evidence that the repulsive forces between non-bonding pairs are greater than those

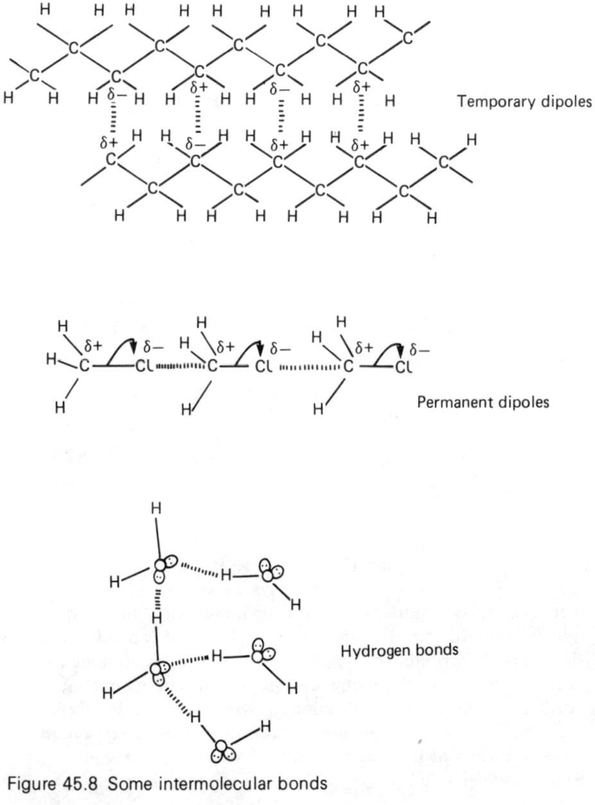

Temporary dipoles

Permanent dipoles

Hydrogen bonds

Figure 45.8 Some intermolecular bonds

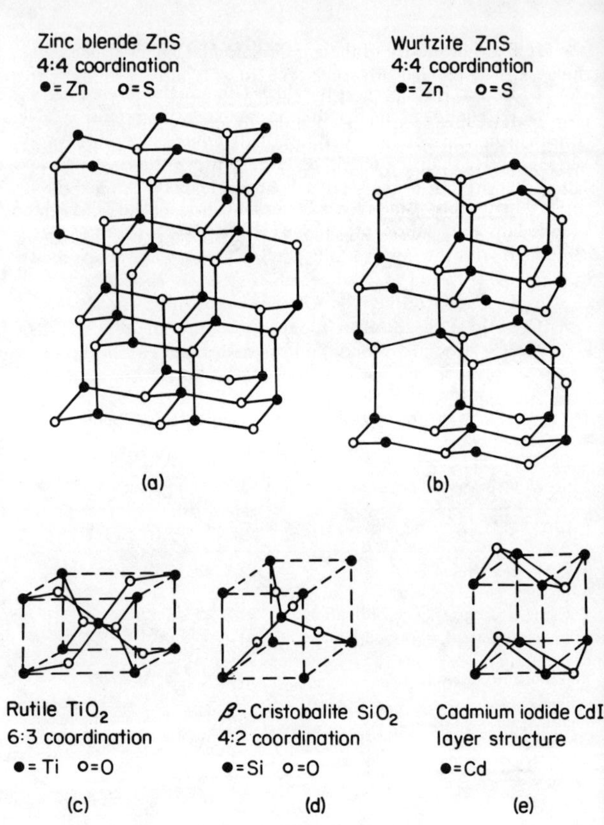

Figure 45.9

between a non-bonding pair and a bonding pair. This in turn is greater than the repulsion between two bonding pairs of electrons.

18 The majority of covalently bonded molecules are much **more volatile** than ionic compounds. This is because the forces of attraction between the molecules are **weak intermolecular forces**. These can be **temporary dipole forces**, **permanent dipole forces** or **hydrogen bonds**. Examples of these forces are shown in *Figure 45.8*.

It is the **intensity** of these intermolecular bonds which causes the covalently bonded substances to exist as liquids or solids.

19 Some metals combine with non-metals by covalent bonding to form a **giant lattice** structure, similar to ionic compounds, but with the elements present as **atoms not ions**. They can be subdivided into the types **AB** and $\mathbf{AB_2}$.

(i) Covalent solids of the type **AB** can be considered to be the **zinc blende** lattice structure and the **wurzite** lattice structure of the compound with the same formula ZnS. These structures are shown in *Figures 45.9(a)* and *(b)* and examples given in *Table 45.9.*
(ii) Covalent solids of the type $\mathbf{AB_2}$ can be classified into three types, the **rutile** or **titanium dioxide** structure, the **β-cristobalite** structure of SiO_2 and the **cadmium iodide layer** structure. These are shown in *Figures 45.9(c)* to *(e)* with examples given in *Table 45.4.*

Table 45.4 The crystal structure of some covalent compounds

Structure	*Compound*
Zinc blended structure 4,4 co-ordination	ZnS AgI CuBr HgS
Wurtzite struccture 4,4 co-ordination	ZnS NH_4F ZnO
Rutile structure 6,3 co-ordination	TiO_2 ZnF_2 MnF_2 CoF_2 SnO_2 MnO_2
β-Cristobalite structure 4,2 co-ordination	SiO_2 Cu_2O Ag_2O
Cadmium iodide structure layer structure	CdI_2

20 Covalent solids which exist as crystal lattice structures are equally as rigid as ionic crystals and have high melting and high boiling points. However, covalent solids are **not conductors** in the molten state and are not usually soluble in water.

21 The concept of co-ordination number is used to express the number of different nearest neighbours associated with an atom or ion. The number refers to the total number of near neighbours both in the same plane and above and below the particle under consideration.

The co-ordination numbers associated with the different lattice crystal structures are shown in *Figure 45.9* and also in *Table*

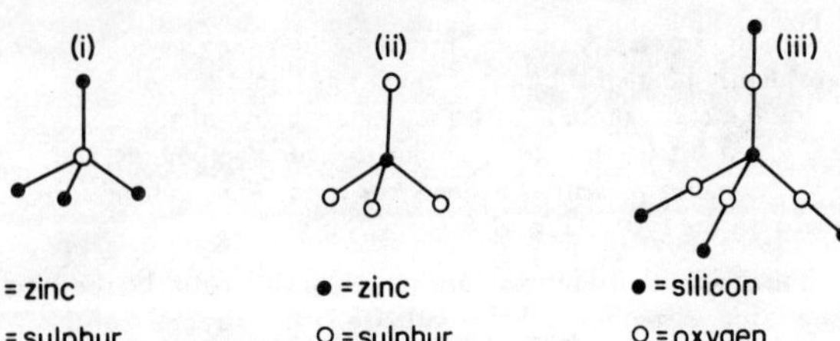

Figure 45.10

Donor	Acceptor	Structure	Name
NH_3 Ammonia	Cu^{2+} Copper (II)	$\left[NH_3 \rightarrow \overset{2+}{Cu} \leftarrow NH_3;\ NH_3 \rightarrow Cu \leftarrow NH_3 \right]^{2+}$	Tetraammine copper (II) complexion
CO Carbon monoxide	Ni Nickel	$\left[CO \rightarrow Ni \leftarrow CO;\ CO \rightarrow Ni \leftarrow CO \right]$	Nickel tetracarbonyl
CN^- Cyanide	Fe^{2+} Iron (II)	$\left[6\ CN^- \rightarrow \overset{2+}{Fe} \right]^{4-}$	Hexacyano ferrate (II) complexion
$AlCl_3$ Aluminium trichloride	$AlCl_3$ Aluminium trichloride	$Cl_2Al(\mu\text{-}Cl)_2AlCl_2$ (Cl → Al dative bonds)	Aluminium chloride dimer
NH_3 Ammonia	BF_3 Boron trifluoride	$H_3N \rightarrow BF_3$	Ammonia–boron trifluoride complex

Figure 45.11 Formation of dative covalent bonds

45.4. For example, in the zinc blende structure shown in *Figure 45.10*, the zinc atoms are represented by ● and the sulphur atoms by ○. By considering a single zinc atom and the sulphur atoms nearest to it, it can be seen that four sulphur atoms are distributed **tetrahedrally** about the zinc atoms as shown diagrammatically in *Figure 45.10(i)*. Similarly, four zinc atoms are distributed tetrahedrally about the sulphur atom also shown in *Figure 45.10(ii)*.

This means that both zinc and sulphur have four nearest neighbouring atoms and the **co-ordination number of 4:4**. Also in *Figure 45.9(d)* the silicon atoms in β-cristobalite are represented by ● and the oxygen atoms by ○.

By considering a single silicon atom and the oxygen atoms surrounding it shown in *Figure 45.10(iii)*, it can be seen that four oxygen atoms are distributed **tetrahedrally** about the silicon atom. Also included in *Figure 45.10(iii)* are the oxygen atoms which are shared **linearly** between two silicon atoms. Hence because each silicon atom has four atoms surrounding it and each oxygen atom is shared by two silicon atoms, **the co-ordination number is 4:2**.

22 Another type of covalent bonding that occurs between metals and non-metals and also between certain molecules is **dative covalent** bonding. Whereas in covalent bonds the two electrons which are shared come from different atoms, in dative bonding **both electrons are given from one atom to another**. The atom or group which gives both electrons is called the **donor** and the atom or molecule which accepts the electrons is called the **acceptor** molecule. Some examples of formation of dative bonds are given in *Figure 45.11*.

Table 45.5 The peoperties of ionic and covalent compounds

Property	*Ionic*	*Covalent*
State	Solid only	Solid, liquid gas
Melting point	High	Low
Boiling point	High	Low
Solubility in H_2O	Good	Poor
Solubility in organic solvents	Poor	Good
Conductivity of molten solids	Good	Poor
Conductivity of aqueous solutions	Good	Poor
Reaction with water	Hydration	Hydrolysis
	$NaCl(s)+H_2O=Na^+(aq)+Cl^-(aq)$	$SiCl_4+4H_2O=Si(OH)+4HCl$

The two requirements necessary for dative bonding are that the donor molecule must have a **non-bonding pair** of electrons (**lone pair**), which it can donate and that the acceptor molecule must have a **vacant ortital** (be **electron deficient**), into which the electron pair can be accepted. Although these bonds are strong attractive bonds, they are not as strong as covalent bonds.

23 The properties of ionic and covalently bonded substances are summarised in *Table 45.5*.

46 The kinetic theory of matter

1 The **kinetic theory of gases** was devised by Bernouilli to explain the gas laws, and was developed by Clausius, Joule and Maxwell to explain many other properties of gases.

2 (i) A gas is considered to be composed of **particles** which may be atoms (for example, He, Ne) or molecules (for example, O_2, SO_2).

(ii) In the gas state the particles are separated by **distances which are large compared to the particles themselves**; the distance is decreased in liquids and is even less in solids.

(iii) In the gas state, the particles are in **continuous motion** moving in straight lines until they **collide** with either each other or the walls of the containing vessel. In the liquid state the **velocity** of the particles is considered to be **less** than in the gas state. In the solid state particles have even lower velocities and can be regarded as simply **oscillating about a point** in space.

(iv) Thc particles are assumed to be **perfectly elastic** so that the collisions they undergo do not affect the **total kinetic energy** of the gas.

(v) The **average kinetic energy** for each particle is proportional to the **absolute temperature** of the state of the matter. This implies that a substance has **zero energy at absolute zero**.

3 The concept of the **pressure** of a gas in terms of the kinetic theory of matter is that as particles with a discrete amount of energy collide with the walls of a container they cause a **force** to be produced against the wall of the container. Since **pressure is the force per unit area**, the number of particles and their velocities combine to produce the **total pressure** on the surface.

The relationship between pressure, volume and the velocity of particles

(i) Consider N molecules of a gas, each of mass m, contained in a cube of side length l, moving with a velocity u.

(ii) It is assumed that the motion of the particles takes place parallel to the sides of the container, such that the particles are travelling back and forth between **two walls** of the container, and that because there are three pairs of walls, the particles are divided such that **one third** travels between each pair of walls. This is shown diagrammatically in *Figure 46.1.*

(iii) The time taken for a particle to travel from one face to another

$$=\frac{\textbf{distance}}{\textbf{velocity}}=\frac{l}{u},$$

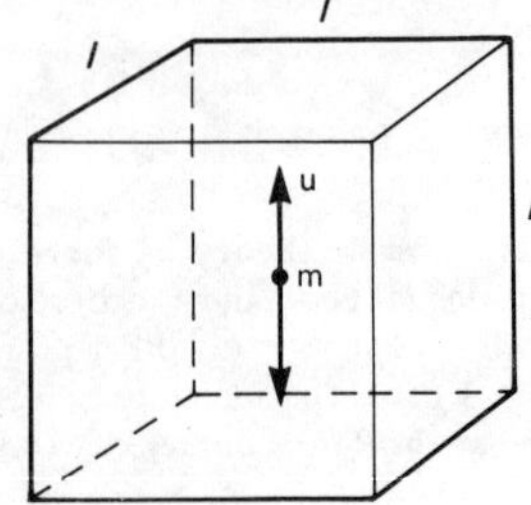

Figure 46.1 A molecule colliding with the faces of the cube

Thus at which each particle strikes the faces of the cube in a unit of time

$$=\frac{\textbf{velocity}}{\textbf{distance}}=\frac{u}{l},$$

(iv) The momentum of each molecule = **mass** × **velocity** = mu. For each collision the momentum changes from mu in one direction to mu in the opposite direction, a **change in momentum of 2**mu.

(v) The **rate of change** of momentum $=2mu\times\frac{u}{l}=\frac{2mu^2}{l}$.

This change in momentum is the **force** produced by one particle moving in this way, and hence is the force the face of the container produces on the particle. Since the total number of molecules is N the number of molecules moving in one direction is $N/3$.

The total force exerted by two faces of the container

$$=\frac{2mu^2}{l}\times\frac{N}{3}$$

where u^2 is the average value of u^2 for all of the N particles and is called the **mean square velocity** and is found by using the expression

$$\bar{u}=\sqrt{\left(\frac{u_1^2+u_2^2+u_3^2+u_4^2\ \ldots\ u_n^2}{n}\right)}\text{ for } n \text{ particles}$$

(vi) The total force exerted by the faces caused by one third of the particles is shared all over the area of the faces, i.e. $2(l\times l)=2l^2$.

Hence the pressures on the two faces is $\frac{\textbf{Force}}{\textbf{Area}}$.

$$\text{or } P=\frac{2m\bar{u}^2}{l}\times\frac{N}{3}\times\frac{l}{2l^2}=m\bar{u}^2\times\frac{N}{3}\times\frac{l}{l^3}$$

Substituting $l^3=V$ gives $P=\dfrac{1Nm\bar{u}^2}{3V}$ or $PV=\dfrac{1}{3}Nm\bar{u}^2$.

This is the **fundamental gas law**, that the pressure of a gas is inversely proportional to the volume occupied by the gas (since $\frac{1}{3}Nm\bar{u}^2$ is a constant value).

Charles' law

5 The fundamental equation $PV=\frac{1}{3}Nm\bar{u}^2$ can be rearranged such that

$$\text{volume } V=\frac{N}{3P}\times m\bar{u}^2 \text{ or } V=\frac{2N}{3P}\times\frac{1}{2}\bar{m}^2$$

Hence for a given mass of gas (N=constant) at a **constant pressure** (P=constant) then $V\propto\frac{1}{2}m\bar{u}^2$ and since the average kinetic energy is proportional to absolute temperature $\frac{1}{2}m\bar{u}^2\propto T$ which by substitution gives $V\propto T$, which is **Charles' law**.

Boyles' law

6 At a **constant temperature**, the average kinetic energy is a constant,

$$\text{or } \frac{1}{2}mu^2=\text{constant}$$

From the fundamental equation

$$PV=\frac{2N}{3}\times\frac{1}{2}m\bar{u}^2$$

it follows that PV=a constant for a constant mass of gas, or $P\propto 1/V$ which is **Boyles' law**.

Dalton's law of partitial pressures

7 If two different non-reacting gases A and B are mixed inside a particular volume at a constant temperature, then gas A produces a pressure

$$P_A=\frac{\frac{1}{2}N_A m_A\bar{u}^2}{V} \text{ and gas } B \text{ a pressure } P_B=\frac{\frac{1}{2}N_B m_B\bar{u}^2}{V}$$

The **total pressure** inside the volume must be the **sum** of these components

$$P=\frac{1}{3}\frac{N_A m_A \bar{u}^2}{V}+\frac{1}{3}\frac{N_B m_B \bar{u}^2}{V} \text{ or } P=P_A+P_B$$

Graham's law of diffusion

The density of a gas ρ is given by

$$\rho=\frac{\textbf{number of molecules}\times\textbf{mass of one molecule}}{\textbf{volume}}$$

$$\text{or } \rho=\frac{Nm}{V}$$

Using the fundamental equation

$$PV=\tfrac{1}{3}Nm\bar{u}^2$$

and substituting for $\frac{Nm}{V}$ gives

$$P=\frac{\bar{u}^2}{\rho} \text{ or } \frac{P}{\rho}=\frac{\bar{u}^2}{3}$$

At a constant pressure

$$\frac{1}{\rho}\propto\bar{u}^2$$

If the mean square velocity is assumed to be proportional to the **rate**, r, at which the gas is **diffusing** through a small hole, then $\bar{u}^2 \propto r^2$. By substituting this value in the derived equation

$$\frac{1}{\rho}\propto r^2 \text{ or } r\propto\sqrt{\frac{1}{\rho}}$$

This is Graham's diffusion law.

Avagadro's hypothesis

9 For two different gases A and B the fundamental equations are

$$P_A V_A=\tfrac{1}{3}N_A m_A \bar{u}_A^2 \text{ and } P_B V_B=\tfrac{1}{3}N_B m_B \bar{u}^2{}_B$$

At a **constant pressure**

$$P_A=P_B,$$

at **constant temperature**

$$\tfrac{1}{2}m_A\bar{u}_A^2=\tfrac{1}{2}m_B\bar{u}_B^2$$

and at **constant volume**

$$V_A=V_B.$$

By substituting these values into the equation for gas B and using

$$P_AV_A=P_BV_B$$

gives

$$\tfrac{2}{3}N_A\times\tfrac{1}{2}m_A\bar{u}_A^2=\tfrac{2}{3}N_B\times\tfrac{1}{2}m_A\bar{u}_A^2$$

and simplyfing this equation gives

$$N_A=N_B$$

which means that the **number of molecules in each gas must be equal**. This is **Avagadro's hypothesis** that at constant pressure and temperature the number of molecules in a volume of gas is a constant and that **1 mole of any gas occupies 22 400 cm^3 at 273 K and 101.3 kPa**.

The ideal gas law

10 In the fundamental equation

$$PV=\tfrac{1}{3}Nm\bar{u}^2$$

which can be written as

$$PV=\tfrac{2}{3}N\times\tfrac{1}{2}m\bar{u}^2$$

the average kinetic energy is proportional to the absolute temperature, thus the equation can be written as

$$PV\propto\tfrac{2}{3}NT \text{ or } \frac{PV}{T}\propto N$$

which means that the value PV/T only depends upon the **number of molecules of gas** under consideration. Under two different sets of conditions

$$\frac{P_1V_1}{T_1}\propto N \text{ and } \frac{P_2V_2}{T_2}\propto N$$

$$\frac{P_1V_1}{T_1}=\frac{P_2V_2}{T_2}$$

Alternatively for a **fixed mass** of gas n moles,

$$\frac{PV}{T}=Rn$$

where R is called the **gas constant** and is the same for all gases (i.e. $R=8.314\ \mathrm{JK^{-1}\ mol^{-1}}$).

For **1 mole** of any gas $\frac{PV}{T}=R$ or $PV=RT$.

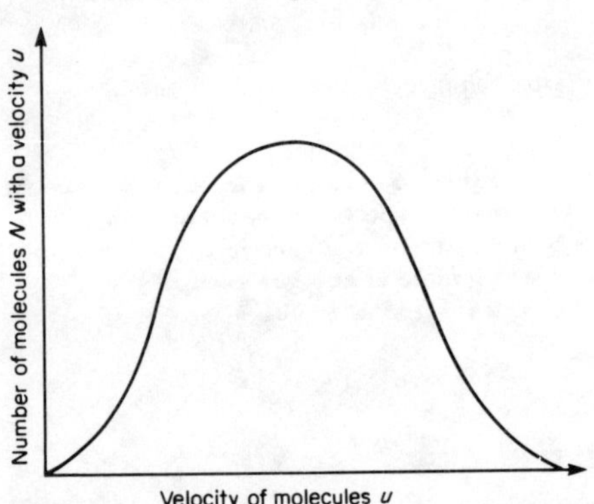

Figure 46.2 The distribution of the velocities of N_0 gas molecules

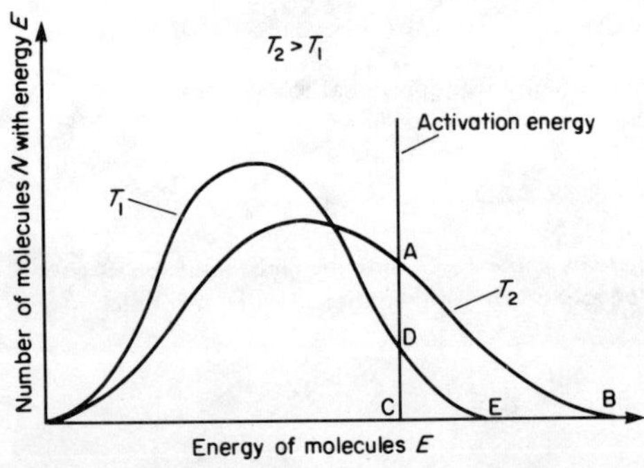

Figure 46.3 The normal distribution curves for the energies of the molecules of a gas at different temperatures

The distribution of molecular velocities

11 The **velocities** of the various individual molecules caused by the **numerous collisions** which take place within a gas, cover a **wide range** of values. Maxwell and Boltzmann calculated the **distribution of velocities of gas molecules** using the laws of probability and found the distribution to approximate to a normal distribution as shown in *Figure 46.2*.

The distribution of kinetic energy follows a similar distribution because $k.e. = \frac{1}{2}m\bar{u}^2$. When the **temperature changes** the distribution of both molecular velocities and kinetic energy also changes as shown in *Figure 46.3* for a change from T_1 to T_2 where $T_2 > T_1$.

47 Chemical reactions

1 A **chemical reaction** takes place when one or more substances undergo a **rearrangement** of their constituent atoms to form new substances.

2 This rearrangement of atoms taking part in a chemical reaction is represented by chemical shorthand in which the elements and compounds undergoing reaction are assigned by their **chemical symbols** of formulae. The reactions are usually written in the form of an **equation** in which the **number of atoms** of each element on **each side** of the equality sign **are the same**. A recent addition to these chemical equations has been the introduction of suffices to show the phase of the substances, (**s**) to represent **a solid**, (**1**) a **liquid**, (**aq**) an **aqueous solution** and (**g**) a **gas**. For example, the equation

$2HI(g)$	$=$	$H_2(g)$	$+$	$I_2(g)$
hydrogen iodide		hydrogen		iodine

represents 2 molecules of hydrogen iodide reacting in the gas phase to produce 1 molecule of hydrogen gas and 1 molecule of iodine gas. Similarly the equation

$CaCO_3(s)$	$=$	$CaO(s)$	$+$	$CO_2(g)$
calcium carbonate		calcium oxide		carbon dioxide

represents 1 molecule of calcium carbonate reacting in the solid state to give 1 molecule of solid calcium oxide and 1 molecule of gaseous carbon dioxide. As another example, the equation

$Na_2CO_3(aq)$	$+$	$H_2SO_4(aq)$		
sodium carbonate		sulphuric acid		
$= Na_2SO_4(aq)$	$+$	$CO_2(g)$	$+$	$H_2O(l)$
sodium sulphate		carbon dioxide		water

represents the reaction between aqueous solutions of sodium carbonate and sulphuric acid, to produce an aqueous solution of sodium sulphate, carbon dioxide gas and water as a liquid.

3 Most chemical reactions can be broadly classified as (i)

acid : base reactions, or (ii) **double decomposition** reactions, or (iii) **redox reactions**.

Acid: base reactions

4 (i) An **acid** can be defined simply as a substance which can dissolve in water to produce a solution containing **hydrogen ions**, for example,

$HCl(g)$	$+\ H_2O(l) =$	$HCl(aq)$	$+\ H_2O(l)$
hydrogen chloride	water	hydrochloric acid	water
	$=$	$H_3^+O(aq)$	$+\ Cl^-(aq)$
		hydrogen ions	chloride ions

It is **not necessary for hydrogen to be contained** in a substance which can act as an acid; for example, when phosphorus pentoxide P_4O_{10} dissolves in water, the solution **is acidic** because the phosphoric (v) acid solution

$P_4O_{10}(s)$	$+\ 6H_2O(l) =$	$4H_3PO_4(aq)$
phosphorus pentoxide	water	phosphoric (v) acid

contains **aqueous hydrogen ions** according to the equation

$H_3PO_4(aq)$	$=\ 3H_3^+O(aq)$	$+\ PO_4^{3-}(aq)$
	hydrogen ions	phosphate (v) ions

(iii) A **base** can be defined simply as a substance which can **neutralise an acid**. Using this definition many of the **metal oxides**, **hydrogencarbonates**, **carbonates** and **hydroxides** are bases. However, there are a special set of bases called **alkalis** which dissolve in water to produce **aqueous hydroxide ions** in solution. These bases are those formed when the **group I metals and their oxides** dissolve in water to form the **alkaline hydroxides**. For example

$Na_2O(s)$	$+\ H_2O(l) =$	$2NaOH(aq) =$	$2Na^+(aq)$	$+\ 2OH^-(aq)$
disodium oxide	water	sodium hydroxide	sodium ions	hydroxide ions

(iv) When an acid and an alkali react together so that no acid or alkali is left at the end of the reaction, the reaction is called a **neutralisation** and always results in the formation of **a salt** and water. Some examples of these reactions are:

$NaOH(aq)$	+ $HCl(aq)$	= $NaCl(aq)$	+ $H_2O(l)$
sodium hydroxide	hydrochloric acid	sodium chloride	water

and

$Ca(OH)_2(aq)$	+ $H_2SO_4(aq)$	= $CaSO_4(aq)$	+ $2H_2O(l)$
calcium hydroxide	sulphuric acid	calcium sulphate	water

(v) When an acid and a basic oxide react together a similar reaction takes place. For example

$MgO(s)$	+ $2HNO_3(aq)$	= $Mg(NO_3)_2(aq)$	+ $H_2O(l)$
magnesium oxide	nitric acid	magnesium nitrate	water

(vi) When an acid reacts with a hydrogencarbonate or a carbonate the acid is neutralised and a **salt**, water and **carbon dioxide** gas are formed. For example,

$KHCO_3(s)$	+ $CH_3COOH(aq)$	
potassium hydrogen carbonate	ethanoic acid	
= $CH_3COOK(aq)$	+ $H_2O(l)$ +	$CO_2(g)$
potassium ethanoate	water	carbon dioxide

and

$SrCO_3(s)$	+ $2HCl(aq)$	
strontium carbonate	hydrochloric acid	
= $SrCl_2(aq)$	+ $H_2O(l)$ +	$CO_2(g)$
strontium chloride	water	carbon dioxide

(vii) In all of these reactions the name of the salt which is formed is derived from the **metals name first** followed by **a name derived from the acid**. Some of the **common salts** and their so called **parent acids** are listed in *Table 47.1*.

Double decomposition of salt solutions

5 In these reactions **two soluble salts** are mixed together to form new salts, **one** of which is **insoluble in water**. Some salts are only sparingly soluble in water and when the **constituent ions** of such compounds are in the same solution, the **solid** is

Table 47.1

Acid		Salts formed	Ions
Hydrochloric	$HCl(aq)$	Chlorides	Cl^-
Hydrobromic	$HBr(aq)$	Bromides	Br^-
Hydroiodic	$HI(aq)$	Iodides	I^-
Hydrofluoric	$HF(aq)$	Fluorides	F^-
Sulphuric	$H_2SO_4(aq)$	Hydrogen sulphates and sulphates	HSO_4^- SO_4^{2-}
Nitric	$HNO_3(aq)$	Nitrates	NO_3^-
Phosphoric	$H_3PO_4(aq)$	Dihydrogen phosphates hydrogen phosphates and phosphates	$H_2PO_4^-$ HPO_4^{2-} PO_4^{3-}
Ethanoic*	$CH_3COOH(aq)$	Ethanoates	CH_3COO^-

* Ethanoic acid is the second member of a large number of organic carboxylic acids of the general formula, $CnH_{2n}O_2$.

formed when the **solubility product** of the solide is exceeded. The **solubility produce** Ksp is defined as

$$Ksp = [A^+]\ B^-] \text{ for a sparingly soluble ionic solid } AB$$

where $[A^+]$ and $[B^-]$ are the concentrations of this ions A^+ and B^- in **moles per dm^3**. Some solubility products are given in *Table 47.2*. Some examples of these reactions are

(i)

$Na_2CO_3(aq)$ aqueous sodium carbonate	+	$CaSO_4(aq)$ aqueous calcium sulphate	=	$CaCO_3(s)$ solid calcium carbonate
			+	$Na_2SO_4(aq)$ aqueous sodium sulphate

Table 47.2

Compound	$BaSO_4$	$CaSO_4$	$AgCl$	$AgBr$	AgI
Solubility product	10^{-10}	2×10^{-5}	2×10^{-10}	5×10^{-13}	8×10^{-17}
Compound	BaC_2O_4	CaC_2O_4	$PbCl_2$	$PbBr_2$	PbI_2
Solubility product	1.7×10^{-7}	2.3×10^{-9}	2×10^{-5}	4×10^{-5}	7×10^{-9}
Compound	$Cu(OH)_2$	$Fe(OH)_2$	$Zn(OH)_2$	$Mg(OH)_2$	$Ni(OH)_2$
Solubility product	2×10^{-19}	2×10^{-17}	2×10^{-17}	2×10^{-11}	6.3×10^{-18}

This reaction is important in the **removal of permanent hardness** from water using sodium carbonate, sometimes called **washing soda crystals.**

(ii)

$AgNO_3(aq)$ +	$NaBr(aq)$ =	$AgBr(s)$ +	$NaNO_3(aq)$
aqueous silver nitrate	aqueous sodium bromide	solid silver bromide	aqueous sodium nitrate

This reaction precipitates silver bromide which is light sensitive and in light decomposes slowly to give colloidal silver,

$$2AgBr(s) = 2Ag(s) + Br_2(l),$$

a reaction used in photography.

(iii)

$CuSO_4(aq)$ +	$2NH_3/H_2O$ =	$Cu(OH)_2(s)$ +	$(NH_4)_2SO_4(aq)$
aqueous copper sulphate	aqueous ammonia	solid copper hydroxide	aqueous ammonium sulphate

The **precipitation of hydroxides of metals** is used as a method of **separation and identification** of metals.

Reduction:oxidation reactions or Redox reactions

6 These reactions are those which take place between an **oxidising agent** and **a reducing agent**.

(i) An **oxidising agent** is a substance which can **accept electrons** and hence change its **oxidation number**;
(ii) A **reducing agent** is a substance which can **donate electrons** and change its **oxidation number**;
(iii) The concept of **oxidation number** was devised to **differentiate** between elements in **different environments**. Certain elements have **fixed** oxidation numbers whilst others **vary** depending on the particular compound the element is contained in. For example, when chlorine combines with any element except fluorine it has an oxidation number of -1, but with fluorine it is $+1$, since it forms a compound FCl in which fluorine is -1 and chlorine is $+1$. Some oxidation numbers are given in *Table 47.3.*

The oxidation number of an element in a compound is found by assigning the oxidation numbers of all the known elements and then giving the unknown element a value which will ensure the **sum of the oxidation numbers is zero**. For example, in $KMnO_4$ K is $+1$, O is -2,

Table 47.3

Element	*Oxidation numbers*
Li, Na, K, Rb, Cs	+1
Be, Ca, Mg, Sr, Ba	+2
Al	+3
Fe	+2, +3
Cu	+1, +2
Mn	+7, +6, +4, +3, +2
F	−1
H	−1, +1
O	−2, −1
Cl	−1, +1
S	−2, +1, +2, +4, +6
Br	−1, +1, +3
I	−1, +1, +3, +5, +7

Note: All elements in their natural states have an oxidation number of zero.

Hence $(+1)+(X)+(+2\times 4)=0$, or $X-7=0$, $X=+7$,

i.e. the **oxidation** number of **manganese in $KMnO_4$ is +7**.
However in K_2MnO_4, K = +1, O = −2,

Hence $(+1\times 2)+(X)+(+2\times 4)=0$, or $X-6=0$,

i.e. the **oxidation number of manganese in K_2MnO_4 is +6**.

Direct combination reactions

(iv) When two elements combine together by **direct reaction** there is a change in oxidation number for each of the elements involved; for example

$C(s)$	+ $O_2(g)$	= $CO_2(g)$	
carbon	oxygen	carbon dioxide	**oxidation**
C°	O°	C^{+4} O^{-2}	**numbers**

Hence carbon changes from 0 to +4, an **oxidation reaction**, and oxygen from 0 to −2, a **reduction reaction**. Other examples together with oxidation number changes are

$2Na(s)$	+ $Cl_2(g)$	= $2NaCl(s)$
sodium	chlorine	sodium chloride

Sodium changes from 0 to +1 and chlorine from 0 to −1, an **oxidation** and a **reduction** respectively, and

$S(g)$	+ $3F_2(g)$	= $SF_6(l)$
sulphur	fluorine	sulphur hexafluoride

S changes from 0 to +6 an **oxidation**, and F from 0 to −1 a **reduction**.

Displacement of metals from solutions of their salts

(v) When a metal *A*, usually in the powdered form, is added to a solution of the salt of a different metal *B*, where *B* is lower than *A* in the **redox potential series** (see Section 29), a reaction takes place in which metal *B* is **displaced** by metal *A*; for example, when powdered zinc is added to a solution of copper sulphate the reaction which occurs is

$$Zn(s) + CuSO_4(aq) = ZnSO_4(aq) + Cu(s)$$

Oxidation changes	Zn	0 to +2	an **oxidation**
	Cu	+2 to 0	a **reduction**

The change from Zn° to Zn^{+2} involves a **loss of two electrons** whereas the change from Cu^{+2} to Cu° involves the **gain of two electrons**, thus **the reaction ratio of Zn° to Cu^{+2} is 1:1** as can be seen from the equation of the reaction, and $CuSO_4$ is the **oxidising agent** and zinc metal is the **reducing agent**.

Transition metal compound reactions

(vi) There are many reactions between compounds which either contain **transition metal elements** as their **cations** or in their **anions**. Potassium manganate (VII) is a chemical which is used extensively as an **oxidising agent** in chemical analysis. It oxidies iron(II) to iron(III) according to the following equation

$$10FeSO_4(aq) + 2KMnO_4(aq) + 8H_2SO_4(aq) =$$
$$5Fe_2(SO_4)_3(aq) + K_2SO_4(aq) + 8H_2O(l)$$

Although this looks quite a complicated reaction there are only two oxidation number changes; they are

an **oxidation**, Fe from +2 to +3 involving **1 electron** and
a **reduction**, Mn from +7 to +2 involving **5 electrons**

This reduction takes place such that **equal numbers** of electrons are lost and gained, the **1 mole of Mn^{+7} oxidises 5 moles of Fe^{2+}** and the **reaction ratio is 1:5** as shown above. When written in the ionic form the equation is much more simple.

Disproportionation

(vii) This is a special type of redox reaction in which a substance undergoes **oxidation** and **reduction** in the **same reaction**. For example, when copper(I) oxide reacts with sulphuric acid the following changes occur

$$Cu_2O(s) + H_2SO_4(aq) = CuSO_4(aq) + Cu(s) + H_2O(l)$$

Oxidation changes	Cu^{+1} to Cu^{+2}	an **oxidation**
	Cu^{+1} to Cu^{0}	a **reduction**

In this reaction copper(I) oxide reacts to give copper(II) sulphate an **oxidation** and also to metallic copper a **reduction** reaction. Another reaction of this type is

$$2NaOH(aq) + Cl_2(g) = NaCl(aq) + NaOCl(aq) + H_2O(l)$$

Oxidation changes	Cl^{0} to Cl^{-1}	a **reduction**
and	Cl^{0} to Cl^{+1}	an **oxidation**

In this reaction chlorine gas is converted into chloride ions and hypochlorite ions. The first reaction is a **reduction** and the second an **oxidation.**

48 Rates of chemical reaction

1 The **rate of a chemical reaction** can be obtained by either measuring the amount of products formed or by measuring the amount of reagents used up in a given time.
2 Chemical reactions can be considered as a rearrangement of elements, or groups of elements, into new patterns. Each particular chemical reaction takes place at an **individual rate** which can be very slow or very rapid. Because of this large variation in reaction time, it is apparent that under a set of conditions one reaction might not take place whilst another does take place with ease.
3 Since reactions depend upon the **breaking** and **forming** of chemical bonds, **energy** must be of prime importance in considering rates of reaction. However, since reactions can take place in many different ways no single theory of reactions has been established.
4 Chemical reactions do not all take place in the same way. Some chemical reactions take place by a **one step** mechanism whilst others take place as a result of a **number of steps** each equivalent to the formation of an intermediate product before further reaction occurs. When a multi-stage reaction occurs, the rate of the reaction is taken to be that of the slowest step in the reaction mechanism.
5 The rate of a reaction has been found **experimentally** to depend upon certain factors. These factors are
(i) **temperature** (all reactions);
(ii) **concentration** (non-gaseous systems);
(iii) **pressure** (gaseous systems);
(iv) **catalysts** (all reactions);
(v) **particle size** (solids).
6 Methods used to compare rates of chemical reactions include the measurement of volumes of gas evolved, the titration of acids and bases or redox systems and colorimetric measurements.

(i) To investigate the effect of temperature, different thermostated heating baths must be used, all other factors being kept constant.
(ii) To investigate the effect of concentration, different

concentrations of one of the components is used, everything else being kept constant.
(iii) The effect of pressure is more difficult to monitor in a simple experiment; the variation of pressure must be the only variable in this investigation.
(iv) To investigate the effect of a catalyst, the rate of reaction can be compared with and without a catalyst.
(v) The effect of particle size can be investigated by using a single large lump of substance and comparing its rate to that of the same mass of substance crushed into smaller pieces.

7 The fact that chemical reactions take place over a wide range of rates (including no reaction to spontaneous explosive reactions), has been investigated by the application of the **kinetic theory of matter**.

The collision theory of chemical reactions

8 Consider **bimolecular reactions** of the type, **A + B → Products** for a reaction to occur, existing chemical bonds must be broken, and new bonds must be formed. In order to explain qualitatively why reaction rates can be increased by changing various factors, the **collision theory** has been formulated. It has been suggested that the reacting molecules must collide together, which is a statement of the kinetic theory of matter. Thus, the **rate of a reaction is dependent on collisions occurring between molecules**.

In the gas phase many reactions **do not** occur until the **temperature** of the reactants is **high enough**. The collision theory suggests that since the **increase** in temperature **increases** the motion of the molecules, **more collisions** will occur between molecules A and B, and hence the **reaction rate increases**. If a reaction is at a high enough temperature to occur, an increase in the pressure of the gas mixture also increases the rate of the reaction. This can be achieved either by **reducing** the volume available to the mixture or by **increasing** the number of molecules. Whichever process is used, the number of collisions must be increased.

For reactions in the liquid phase or in solution, reaction rates can be increased by increasing the temperature or the concentration of the reactants, both of these factors increasing the number of collisions which take place in the reaction.

For a reaction in which a solid is involved, the rate of reaction can be increased by increasing the state of division of the solid. This increases the **surface area** at which collisions can occur and hence the collision theory qualitatively explains this by an increased number of collisions.

9 A further point of interest is that it can be shown that for a particular number of molecules in a given volume and at a constant temperature, the number of collisions which occur, does not reflect the rate at which reaction takes place.

10 It was suggested by **Arrhenius** that reaction only takes place between **activated molecules** which have an **activation energy** in excess of a particular value. This theory implied that, unless the interacting molecules collide with suffcient energy, reaction will not take place. Alternatively, if they collide with an energy much greater than the energy required an extremely vigorous reaction would take place. This prevention of reactions occurring led to the idea of an **energy barrier** which has to be overcome before reactions can occur. This allows **reaction profile diagrams** to be drawn for reactions as shown in *Figure 48.1*.

11 The above theory was placed on a quantitative basis by a consideration of the number of molecules N of a total number of molecules N_o having the requisite amount of energy E (**the Activation energy**) for a reaction to occur. Maxwell and Boltzmann showed that the relationship between these values approximated to a normal distribution which can be expressed in a simplified equation as:

$$N = N_o e^{-E/RT}$$

where R is the gas constant and T is the absolute temperature.

12 It has already been stated that the **rate of a reaction** is dependent upon the **number of effective collisions** which occur in the reaction. Hence, the rate of a reaction is proportional to the ratio N/N_o but the ratio N/N_o is itself proportional to $e^{-E/RT}$. Thus, the **rate of reaction**, k, can be related to the **Activation energy** E by the equation

$k = Ae^{-E/RT}$ where A is called the **Arrhenius constant**.

Taking logarithms of this equation gives:

$$lgk = lg(Ae^{-E/RT}) = lgA + lge^{-E/RT}$$

$$\text{i.e. } lgk = lgA - \frac{E}{RT}\, lge = (\text{a constant} - E/RT(0.4343)$$

$$\text{or } lgk = \text{a constant} \times \frac{-E}{2.303RT}$$

For two different temperatures T_1 and T_2 with corresponding **rate constants** k_1 and k_2 the relationship between these values can be

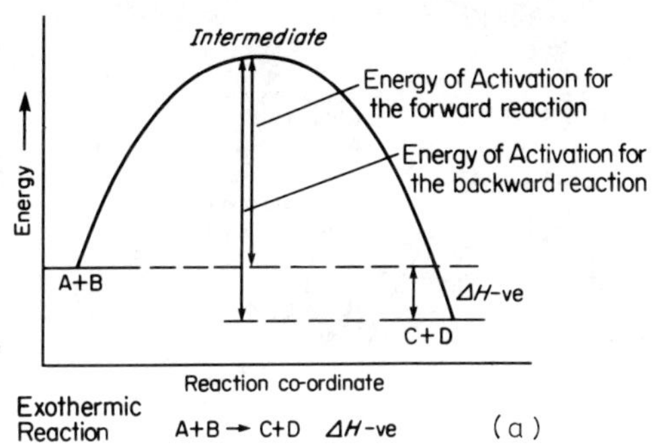

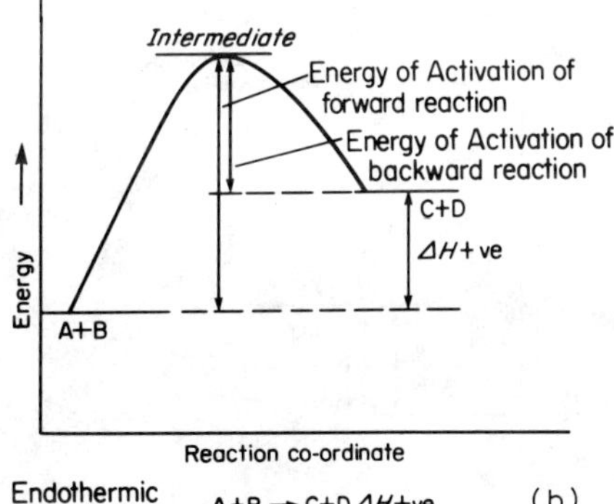

Figure 48.1 (a) Reaction propiles of exothermic and endothermic reactions and (b) a catalyst which decreases the rate of an endothermic reaction

expressed in the form

$$lgk_2 - lgk_1 = \frac{E}{2.303R}\left(\frac{T_2 - T_1}{T_2 T_1}\right) \tag{1}$$

This means that if T_1, E, R and k_1 are known, the rate of the

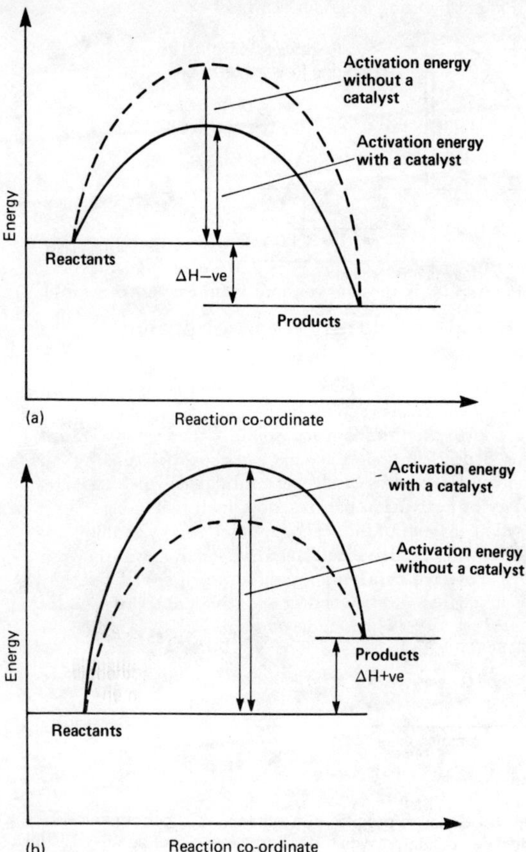

Figure 48.2

reaction at any other temperature T_2 can be found. For example, at 730 K, the rate constant k_1 for the reaction

$$2HI(g) \rightarrow H_2(g) + I_2(g)$$

is 5×10^{-1} mol^{-1} s^{-1} the energy of activation E is 105 kJ mol^{-1} and gas constant $R = 0.00832$ kJ mol^{-1} K^{-1}. The rate constant k_2 at

780 K is found by using equation (1), i.e.

$$\lg k_2 - \lg 0.5 = \frac{105}{2.303 \times 0.00832}\left(\frac{780-730}{780\times 730}\right)$$

$$\lg k_2 - \lg 0.5 = 0.4812$$

$$\lg k_2 = 0.4812 + \lg 0.5 = 0.4812 - 0.3010 = 0.1802$$

$$k_2 = \text{antilog}(0.1802)$$

$$k_2 = 1.514$$

Thus the **rate constant at 780 K is 1.514 mol^{-1} s^{-1}**. Hence for an **increase of 50 K** the rate constant is **increased threefold**.

The role of catalysts in chemical reactions

13 A **catalyst** can be defined as a substance which will **alter the rate** of a chemical reaction, but remaining chemically unchanged at the end of the reaction. Since the rate of a chemical reaction is dependent upon the energy of activation an alternative definition could be that a catalyst is a substance which **changes the energy of activation** of a reaction, itself remaining unchanged at the end of the reaction. Some energy profile diagrams showing **positive** and **negative** catalysts are shown in *Figure 48.2*. **Positive catalysts** are those which **speed up** reactions to a suitable rate whereas **negative catalysts** are used to **slow down** reactions which under normal conditions would be explosive or uncontrollable.

49 Energy of chemical reactions

1 It has been previously stated in Section 17, para 4 that energy is the capacity of a system to do work. The energy of atoms and molecules in a chemical system is made up of **kinetic energy** and **potential energy**. The kinetic energy is due to:

(a) **transitional** energy, i.e. the molecules in motion; for example, molecules in the liquid and gas phase,
(b) **rotational** energy, i.e. the rotation of the molecules and
(c) **vibrational** energy, i.e. the vibration of atoms in a molecule about a specific position. The potential energy is due to the **repulsive** and **attractive** forces between the particles in a molecule.

2 The **total energy** of a chemical system is called the **internal energy** and is given the symbol U. The total amount of internal energy associated with a system is difficult to measure absolutely, but changes in energy can readily be determined. The change in internal energy is signified as ΔU.
3 The unit of heat energy is the **joule J**, and the **molar heat capacity** of a system is measured in **joules per mole Kelvin** or **J mol^{-1} K^{-1}**.

The internal energy of a monatomic gas

4 A consideration of the noble gas which exists in the monatomic state shows that there is no internal energy due to potential energy. The only energy is kinetic energy. The **kinetic theory of matter** (Chapter 46) derives the fundamental equation:

$$PV=\tfrac{1}{3}Nm\bar{u}^2$$

for 1 mole of gas, and the ideal gas law for one mole of gas is:

$$PV=RT$$

By combining these equations:

$$RT=\tfrac{1}{3}Nm\bar{u}^2 \text{ or } Nm\bar{u}^2=3RT$$

The total internal energy is U, and since the average kinetic energy of 1 mole of gas is $\frac{1}{2}Nm\bar{u}^2$ then

$$U=\tfrac{1}{2}Nm\bar{u}^2 \text{ or } 2U=Nm\bar{u}^2$$

Since $3RT=Nm\bar{u}^2$ then

$$2U=3RT \text{ or } U=\tfrac{3}{2}RT.$$

Differentiating with respect to T gives:

$$\frac{dU}{dT}=\frac{3R}{2}$$

The expression dU/dT is the **rate of change** of the internal energy with temperature, which is the energy change associated with 1 degree Kelvin. This is defined as the **molar heat capacity of the gas at constant volume** and symbolised Cv. Hence

$$Cv=\tfrac{3}{2}R=\tfrac{3}{2}\times 8.314=12.465 \text{ J mol}^{-1}\text{ K}^{-1}$$

5 Let 1 mole of a monatomic gas be contained in a cylinder fitted with a weightless friction-free piston as shown in *Figure 49.1*, and let the pressure due to the gas above and below the piston be P and the volume of the gas be V. If the cylinder is heated, the kinetic energy of the gas **increases** and causes an **increase** in temperature. This increases the force acting on the piston and the piston will move to accommodate the increased volume of the gas. The heat energy given to the cylinder increases the internal energy of the gas and in addition **does work** in moving the piston.

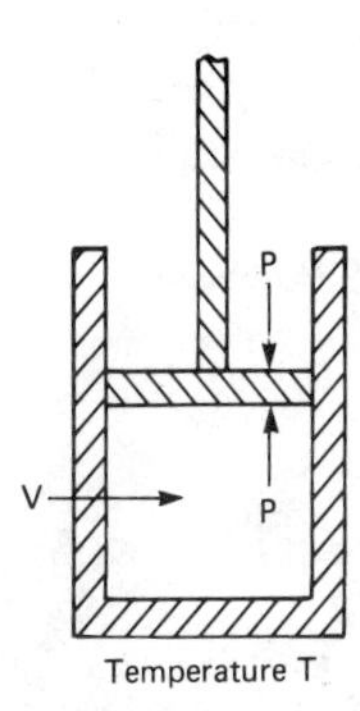

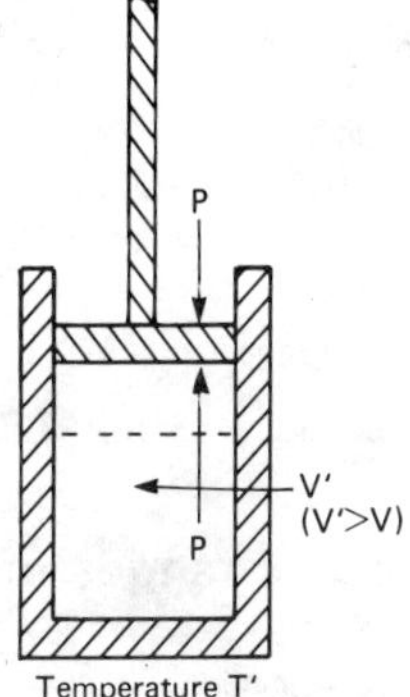

Figure 49.1

This can be represented by the equation

$$q = \Delta U + w,$$

the first law of thermodynamics, where q is the heat energy supplied, ΔU is the change in internal energy and w is the **work done** by the system. This means that the value of Cp, the **molar heat capacity at constant pressure**, is greater than Cv according to the equation:

$$\boldsymbol{Cp = Cv + \text{work done},}$$

where **the work done = pressure × increase in volume**.

For a pressure P the work done in changing the volume from V to V' is

$$P(V' - V) \text{ or } PV' - PV.$$

For 1 mole of a monatomic gas

$$PV = RT \text{ and } PV' = RT'.$$

The work done can be expressed as

$$RT' - RT = R(T' - T)$$

The temperature change for the **molar heat capacity** is 1 K hence

$$T' - T = 1.$$

This means that for a monatomic gas

$$\boldsymbol{Cp = Cv + \text{work done} = Cv + R}$$

Cv has been shown to be 12.465 $\text{J mol}^{-1}\text{ K}^{-1}$ and since $R = 8.314$ $\text{J mol}^{-1}\text{ K}^{-1}$ then $Cp = 20.779$ $\text{J mol}^{-1}\text{ K}^{-1}$. For a diatomic gas it can be shown that $Cv = 20.8$ and $Cp = 29.1$ $\text{J mol}^{-1}\text{ K}^{-1}$ and for a triatomic gas $Cv = 25.0$ and $Cp = 33.3$ $\text{J mol}^{-1}\text{ K}^{-1}$.

Hence at constant pressure, the heat energy absorbed is a combination of the internal energy of the system and the work done by the system. This is represented by the equation:

$$\Delta H = \Delta U + P\Delta V,$$

where ΔH is the **enthalpy change** of the system.

The energy changes in chemical reactions

6 A consideration of the internal energy of a chemical system involving several polyatomic molecules is difficult. In order to

obtain information about energy changes that occur, the measurements are made at constant pressure and hence are called the enthalpy changes, ΔH.

Exothermic and endothermic reactions

7 When a chemical reaction **releases heat energy** to its surroundings it is called an **exothermic reaction** and is represented by the equation:

Reactans = Products ΔH is negative

For example

$$C(s) + O_2(g) = CO_2(g) \ \Delta H = -393.4 \text{ kJmol}^{-1}$$

When a reaction absorbs heat from its surroundings it is called an endothermic reaction and is represented by the equation:

Reactants = Products ΔH is positive

For example,

$$NH_4Cl(s) + \text{water} = NH_4Cl(aq) \ \Delta H = +10.4 \text{ kJmol}^{-1}$$

8 There are many different types of enthalpy changes found in chemistry and they are usually stated under the **standard** conditions of 101.3 kPa and 298 K, and given the symbol $\Delta H^\ominus$.

Some definitions of standard ethalpy changes

9 **The standard enthalpy of combustion** $\Delta H_c^\ominus$ is the enthalpy change when one mole of a substance (the relative molecular mass, in grams) is **completely burned** in oxygen, under standard conditions. For example:

$$C(s) + O_2(g) = CO_2(g), \ \Delta H_c^\ominus = -393.4 \text{ kJ mol}^{-1}$$

The standard enthalpy of formation, $\Delta H_f^\ominus$ is the enthalpy change when one mole of a substance is **prepared from its elements**, under standard conditions. For example

$$H_2(g) + S(s) = H_2S(g), \ \Delta H_f^\ominus = -20.6 \text{ kJ mol}^{-1}$$

The standard enthalpy of solution, $\Delta H_s^\ominus$ is the enthalpy change when one mole of a substance is **dissolved in a solvent** to give a solution of defined concentration, under standard conditions. For example,

$$HCl(g) + aq = HCl(aq), \ \Delta H_s^\ominus = -72.38 \text{ kJ mol}^{-1}$$

Since the most common solvent used in chemistry is water, **the standard enthalpy of hydration** $\Delta H_h^\ominus$ is of importance.

The standard enthalpy of neutralisation is the enthalpy change when one mole of aqueous hydrogen ions are **neutralised** by a base in dilute solution. For strong acids and strong bases, this value is -57.33 kJ. This is because, independent of the strong acid or base, the neutralisation reaction is:

$$H_3^+O(aq) + OH^-(aq) = 2H_2O(l)$$

(For weak acids and bases, the enthalpy of neutralisation is lower because some of the heat energy produced is required by the reaction to **complete the ionisation** of the acids and bases).

The standard ehtalpy of atomisation is the enthalpy change which occurs when one mole of an element in its standard state at 298 K and 101.325 kPa is **converted into free atoms**.

The standard enthalpy of ionisation is the enthalpy change which occurs when one mole of gaseous atoms are converted into one mole of **gaseous ions** accompanied by a **loss** of one mole of electrons, measured under standard conditions. It should be noted that each element has a different number of ionisation energies dependent upon the number of electrons in the element. Usually the formation of stable metallic ions are associated with a loss of 1, 2, 3 or 4 electrons.

The standard enthalpy of lattice energy is the enthalpy change which occurs when one mole of an ionic solid is formed from **its constituent gaseous ions** measured under standard conditions.

The standard enthalpy of electron affinity is the enthalpy change which occurs when one mole of gaseous atoms are converted into one mole of **gaseous negative ions** by **accepting** one mole of electrons, measured under standard conditions. It should be noted that some non-metals can have more than one electron added: for example, oxygen has two **electron** affinity enthalpy values.

For **covalent molecules** the energy required to **separate** one mole of covalent bonds between two atoms in a diatomic molecule is called the **bond dissociation enthalpy**, or **bond dissociation energy**. Some typical examples of values are given in *Table 49.1*, where $B.D._{x-y}$ symbolises those terms.

In covalent molecules containing more than two atoms the bond dissociation energies are found for all of the bonds and

Table 49.1 Band dissociation energies

Molecule X–Y	*Enthalpy* $B.D._{x-y}$ kJ mol^{-1}
H–H	436
F–F	269.9
Cl–Cl	242.7
Br–Br	192.9
I–I	151
H–F	621.3
H–Cl	431.8
H–Br	366.1
H–I	298.7

an average value is derived. For example, in methane, CH_4 there are four carbon to hydrogen bonds. The enthalpy required to break the four bonds is 1666 kJmol^{-1}, and the average value is thus 416.5 kJmol^{-1}.

The law of energy changes

10 The law of Hess states that '**the total change in enthalpy in a chemical reaction is independent of the number of stages used to complete the reaction**'. Considerable use is made of this law in the calculation of enthalpy values that are difficult to measure directly.

The Born-Haber cycle for ionic solids

11 The **Born-Haber cycle** is an application of **Hess's law** used to relate together all of the enthalpy changes involved in the formation of an ionic solid. An example of such a Born-Haber cycle is shown in *Figure 49.2* for calcium sulphide. In this diagram the combined 1st and 2nd electron affinities have been omitted. To find the value, Hess's law is applied to the system and stated as:

$$\Delta H_1 + \Delta H_2 + \Delta H_3 + \Delta H_4 + \Delta H_5 = \Delta H_6$$

Substituting the known values of enthalpy changes into the equation gives:

$$+176.6 + 590 + 1100 + 238.1 + \Delta H_4(X) - 3084 = -482.4,$$

which gives a value for the enthalpy of electron affinity of

$$\Delta H_4 = +496.9 \text{ kJmol}^{-1}$$

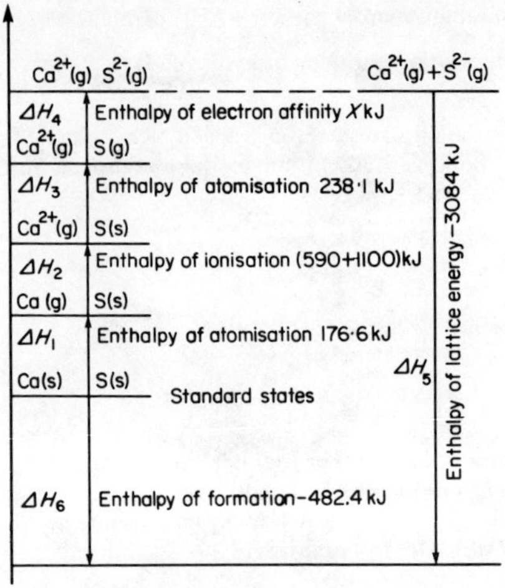

Ca2+(g) S2-(g)
Ca2+(g)+S2-(g)
ΔH4
Enthalpy of electron affinity X kJ
Ca2+(g) S(g)
ΔH3
Enthalpy of atomisation 238·1 kJ
Ca2+(g) S(s)
ΔH2
Enthalpy of ionisation (590+1100)kJ
Ca (g) S(s)
ΔH1
Enthalpy of atomisation 176·6 kJ
Ca(s) S(s)
Standard states
ΔH5
Enthalpy of lattice energy −3084 kJ
ΔH6
Enthalpy of formation−482.4 kJ

Figure 49.2

12 For covalently bonded molecules similar enthalpy cycles as above, can be constructed to obtain information about covalent reactions. The enthalpy cycle shown in *Figure 49.3* shows the interrelationships between methane and its elements. For example, the equation for the formation of methane can be written:

$$C(s) + 2H_2(g) \xrightarrow{\Delta H_1} CH_4(g)$$

where by definition, ΔH_1 is the enthalpy of formation of methane. Before covalent bonds are formed by atoms of carbon and hydrogen the carbon must be atomised and the hydrogen molecules dissociated into atoms. The enthalpies involved are ΔH_2 and ΔH_3 both of which are endothermic. Hence

$$C(s) \xrightarrow{\Delta H_2} C(g) \text{ and } 2H_2(g) \xrightarrow{\Delta H_3} 4H(g)$$

C (g) + 4 H (g)
4 x EC–H
CH4 (g)
ΔH°a (C)
2 x B.D.H–H
ΔH°f (CH4)
C (s)
2H2(g)
(Standard states)

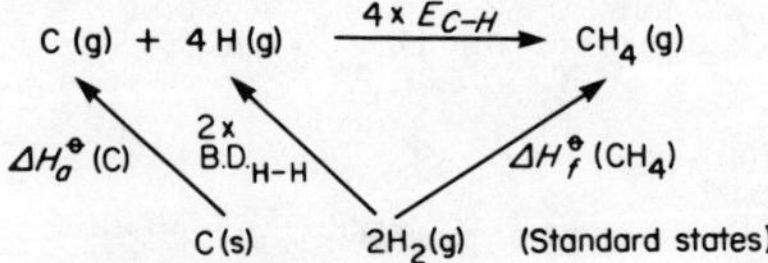

Figure 49.3

ΔH_2 = the enthalpy of atomisation of carbon = $\Delta H_a^{\ominus}$ carbon

$\Delta H_3 = 2 \times$ the bond dissociation enthalpy of hydrogen
$= 2 \times B.D._{H-H}$

The next step involves the formation of four carbon to hydrogen single covalent bonds. Bond formation is an exothermic process and ΔH_4 in the equation:

$$C(g) + 4H(g) \xrightarrow{\Delta H_4} CH_4(g)$$

is $4 \times$ the average bond dissociation energy of a C—H bond,

i.e. $\Delta H_4 = 4 \times E_{C\text{-}H}$

By applying Hess's law of constant heat summation:

$$\Delta H_1 = \Delta H_2 + \Delta H_3 + \Delta H_4$$

or $\Delta H_f = \Delta H_a$ carbon $+ 2 \times B.D._{H\text{-}H} + 4 \times E_{C\text{-}H}$

Hence by knowing the required enthalpy values, the **enthalpy of formation** can be calculated.

13 Another useful application of Hess's law is the determination of the **enthalpy of covalent reactions**. When any reaction takes place chemical bonds must be **first broken** and then **new bonds formed**. The bond breaking is an **endothermic** process and bond forming an **exothermic** process. By considering the energy associated with covalent bonds in all of the molecules, the difference between reactants and products will be the enthalpy of the reaction. For example, the **hydrogenation of ethene** takes place according to the equation:

$$H_2C{=}CH_2(g) + H{-}H(g) = H{-}CH_2{-}CH_2{-}H(g)$$

The **mean bond energies** of the bond to be broken and formed can be presented in tabular form as follows:

Bonds broken	**Total energy**
$4 \times C-H$	$4 \times 413 = +1652$
$1 \times C=C$	$+598$
$1 \times H-H$	$+436$
	$+2686$

Endothermic

Bonds formed	**Total energy**
$6 \times C-H$	$6 \times 413 = -2478$
$1 \times C-C$	-346
	-2824

Exothermic

The difference in enthalpy is $2686 - 2824 = -138$ kJ, and hence the enthalpy of the reaction ΔH_r is -138 kJmol^{-1}.

Measurement of enthalpy changes

14 The enthalpy changes which are measured are determined using **calorimetry**. A calorimeter is a container, which is insulated against heat loss, and contains a liquid of known specific heat capacity, c (usually water) in which the chemical reaction takes place. The rise in temperature of a known mass of liquid, m kg, from t_1° C to t_2° C, is associated with an energy change ΔH given by the expression:

$$\Delta H = mc(t_2 - t_1) \text{ joules}$$

(See Section 19, para 4.)

A simple calorimeter

15 The simple calorimeter shown in *Figure 49.4*, shows the precautions which are taken against heat loss. In addition to heating the water through the measured temperature range, the whole calorimeter experiences the same temperature rise using heat energy from the chemical reaction. The total energy from the reaction is a combination of these two quantities. The heat energy required to raise the temperature of the calorimeter through 1°C is called the **heat capacity of the calorimeter**. The heat capacity is calculated by placing a known mass of water m kg, into a calorimeter containing all of the apparatus required for a calorimetry experiment (see *Figure 49.4*).

An **electrical heating element** is then placed into the water and a steady current, I amperes, flows for a time t seconds from a supply voltage, V volts. The rise in temperature, $\Delta\theta$ K of the water is noted. By assuming that all of the electrical energy is converted into heat energy, and ignoring any heat lost from the outer

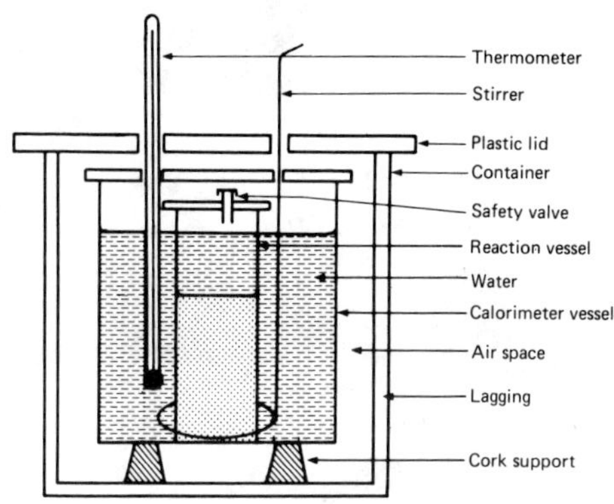

Figure 49.4 A simple calorimeter

container the electrical energy supplied is given by:

$$\textbf{Electrical energy} = \frac{I \times V \times t}{1000}\ \textbf{kJ} \qquad (1)$$

This energy heats the water and the calorimeter. The heat energy used in producing a temperature rise of $\Delta\theta$ K on m kg of water is given by:

$$\textbf{Heat energy} = m \times c \times \Delta\theta\ \textbf{kJ} \qquad (2)$$

where $c = 4.18$ kJ kg^{-1} K^{-1}, the specific heat capacity of water. The difference in the two values obtained from equations (1) and (2) is the heat energy required to raise the temperature of the calorimeter through $\Delta\theta$ K. The **heat capacity of the calorimeter** for a charge in temperature of $\Delta\theta$ K is given by the equation (1) minus equation (2), that is

$$\left(\frac{I \times V \times t}{1000}\right) - (m \times c \times \Delta\theta)\ \textbf{kJ}$$

The heat capacity for a rise in temperature $\Delta\theta$ is given by:

Heat capacity of the calorimeter

$$= \frac{\left(\dfrac{I \times V \times t}{1000}\right) - (m \times c \times \Delta\theta)\ \text{kJ K}^{-1}}{\Delta\theta}$$

For example, when 0.6 kg of water in a calorimeter of heat capacity 0.15 kJK^{-1} increases in temperature from 291 K to 311 K the enthalpy change for the water, ΔH_1, is given by the expression

$$\Delta H_1 = m.c.(t_2 - t_1)$$

where c for water is 4.18 kJ kg^{-1} K^{-1}.

Substituting in the values for m, c, t_2 and t_1 gives:

$$\Delta H_1 = 0.6 \times 4.18 \times (311 - 291) = 50.16 \text{ kJ}$$

The enthalpy change for the calorimeter ΔH_2 is given by

$$\Delta H_2 = \text{heat capacity of the calorimeter} \times (t_2 - t_1),$$

that is

$$\Delta H_2 = 0.15 \times 20 = 3 \text{ kJ}.$$

The total enthalpy change ΔH is given by:

$$\Delta H = \Delta H_1 + \Delta H_2 = 50.16 \text{ kJ} + 3 \text{ kJ},$$

that is,

$$\Delta H = 53.16 \text{ kJ}.$$

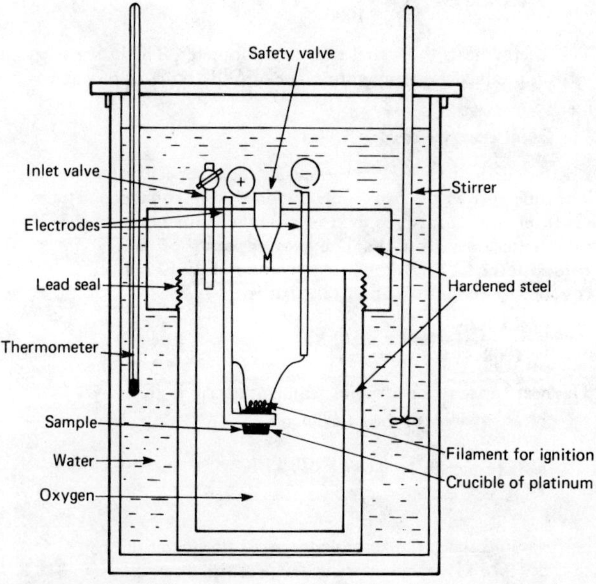

Figure 49.5 A bomb calorimeter

By finding the enthalpy change for a known number of moles of reactants, the molar enthalpy change for the reaction can be calculated.

The bomb calorimeter

16 When the reaction taking place produces a large change in volume the construction of the calorimeter must be robust. The **bomb calorimeter** shown in *Figure 49.5* has been constructed for measuring the enthalpy of combustion of compounds in oxygen, which are reactions causing a large change in volume of this type. The measurements are made in the same way as in the simple calorimeter.

17 A knowledge of the enthalpy changes of chemical reactions is very useful in giving an indication of how much heat energy is available from a chemical reaction.

50 Chemical equilibrium in liquids and solutions

1 When a chemical reaction takes place such that all of the reactants are converted into products the reaction is said to have **reacted to completion** and can be represented by the equation

$$A+B=C+D$$

However, many chemical reactions do not reach completion and these are classified as **reversible reactions**, and can be represented by the equation

$$A+B\rightleftharpoons C+D$$

assuming the reaction takes place in a **closed vessel**. The symbol $\rightleftharpoons$ shows that two reactions are taking place **simultaneously**, namely

(i) $A+B\rightarrow C+D$ and (ii) $C+D\rightarrow A+B$.

Reaction (i) is referred to as the **forward reaction** and reaction (ii) as the **backward reaction**. For example, when ethanol, CH_3CH_2OH reacts with ethanoic acid, CH_3COOH, in the presence of an acid catalyst, ethyl ethanoate $CH_3COOCH_2CH_3$ and water are formed according to the equation

$$CH_3COOH(l)+CH_3CH_2OH(l)\xrightarrow{\text{acid catalyst}} CH_3COOCH_2CH_3(l)+H_2O(l)$$

Similarly, when ethyl ethanoate reacts with water in the presence of a base, ethanol and ethanoic acid are formed according to the equation

$$CH_3COOCH_2CH_3(l)+H_2O(l)\xrightarrow{\text{base catalyst}} CH_3COOH(l)+CH_3CH_2OH(l)$$

Both of these reactions lead to the production of a **mixture of reactants and products**, such that ethanol, ethanoic acid, ethyl ethanoate and water are **all present**.

2 When calcium carbonate, $CaCO_3$, is heated in a stream of air it is converted into calcium oxide, CaO, together with the volution of carbon dioxide, CO_2, according to the reaction:

$CaCO_3(s)$ heat $CaO(s) + CO_2(g)$

If this reaction is performed in a **closed vessel** the reaction does not reach completion. Instead, a reversible reaction takes place, i.e.

$$CaCO_3(s) \xrightarrow{\text{heat}} CaO(s) + CO_2(g)$$

Thus, although some reactions reach completion in an **open vessel** they become reversible reactions in **closed vessels**.

3 When a reversible reaction takes place at a constant temperature, the reaction reaches a stage when, **independent of the time** allowed for the reaction to take place, the **composition** of the reaction mixture becomes **constant**. At this stage the **rate** of the forward reaction has become **equal** to the **rate** of the backward reaction, and the reaction exists in a stage of what is called **dynamic equilibrium**.

4 The rate of a chemical reaction is proportional to the **active masses** of the reactants. These **active masses** approximate to the **concentrations** of the reactants measured in **moles per litre** which in S.I. units is **mol dm**$^{-3}$. Considering the general equation

$$A + B \rightleftharpoons C + D$$

the two simultaneous reactions which are occurring are

(i) the forward reaction, $A + B \rightarrow$ **Products**, and
(ii) the backward reaction, $C + D \rightarrow$ **Products**.

The rate of the forward reaction is therefore proportional to $[A][B]$, where the square brackets [ ], signify the concentration of a chemical measured in mol dm^{-3}. Represented mathematically, this becomes

$$\textbf{Rate of reaction (i)} \propto [A][B]$$

and the introduction of a constant into the equation gives:

$$\textbf{Rate of reaction (i)} = k_1[A][B],$$

where k_1 is the velocity constant of reaction (i). Similarly

$$\textbf{Rate of reaction (ii)} = k_2[C][D],$$

where k_2 is the velocity constant of reaction (ii). At this stage of dynamic equilibrium the rates of reaction are equal, therefore

$$\textbf{Rate of reaction (i)} = \textbf{Rate of reaction (ii)}$$

$$\text{or } k_1[A][B] = k_2[C][D]$$

By (a) rearranging the equation, and (b) combining the constants k_1 and k_2 into a single constant K_c the above equation becomes,

$$\text{(a)}\ \frac{k_1}{k_2}=\frac{[C][D]}{[A][B]},\ \text{or (b)}\ K_c=\frac{[C][D]}{[A][B]}$$

This latter expression is called the **equilibrium law**, and it relates together the concentrations of the reactants and products, and K_c is called the **equilibrium constant**. For example, the equilibrium law for the reaction

$$CH_3COOH(l) + CH_3CH_2OH(l) \rightleftharpoons CH_3COOCH_2CH_3(l) + H_2O(l)$$

$$\text{is } K_c=\frac{[CH_3COOCH_2CH_3]\times[H_2O]}{[CH_3COOH]\times[CH_3CH_2OH]}$$

and the equilibrium law for the reaction

$$PCl_5(g) \rightleftharpoons PCl_3(g) + Cl_2(g)$$

$$\text{is } K_c=\frac{[PCl_3]\times[Cl_2]}{[PCl_5]}$$

5 For an equation of the type

$$wA + xB\ .yC + zD$$

the equilibrium law becomes modified such that

$$K_c=\frac{[C]^y[D]^z}{[A]^w[B]^x}$$

For example, given

$$N_2(g) + 3H_2(g) \rightleftharpoons 2NH_3(g)$$

then

$$K_c=\frac{[NH_3]^2}{[N_2]\times[H_2]^3}$$

and given

$$3Fe(s) + 4H_2O(g) \rightleftharpoons Fe_3O_4(s) + 4H_2(g)$$

then

$$K_c=\frac{[Fe_3O_4][H_2]^4}{[Fe]^3[H_2O]^4}$$

6 The value of the equilibrium constant K_c of a reversible reaction is **dependent upon the concentrations** of the reactants and products at equilibrium. The general equation of a reversible reaction can be expressed as $A + B \rightleftharpoons C + D$ for which the

equilibrium constant is given by

$$K_c=\frac{[C][D]}{[A][B]}$$

If the concentrations at equilibrium are $[A]=1$ mol dm^{-3}, $[B]=1$ mol dm^{-3}, $[C]=10$ mol dm^{-3} and $[D]=10$ mol dm^{-3}, substituting these values into the equilibrium law expression gives

$$K_c=\frac{10\times 10}{1\times 1}=100$$

This shows that if the **concentrations** of the products in the equilibrium reaction **are high by comparison** with the concentrations of the reactants, the value of the **constant K_c is high**. The **larger** the value of K_c for an equilibrium reaction the **nearer to completion** the reaction becomes.

Similarly, $[A]=10$ mol dm^{-3}, $[B]=10$ mol dm^{-3}, $[C]=1$ mol dm^{-3} and $[D]=1$ mol dm^{-3}, then the value of K_c becomes

$$K_c=\frac{[C][D]}{[A][B]}=\frac{1\times 1}{10\times 10}=\frac{1}{100}$$

If the value of K_c for an equilibrium reaction is **low**, the **concentration of products** in the reaction mixture will be **low**.

7 When a reversible reaction reaches equilibrium, the ratio of products and reactants must obey the equilibrium law regardless of the amounts of starting materials taken initially. For example, if 1 mole of ethanol and 1 mole of ethanoic acid are allowed to react to equilibrium, the amount of ethanol and ethanoic acid present in the equilibrium mixture is 0.33 moles of each. The amount of ethyl ethanoate and water is 0.66 moles of each, hence the value of K_c is found from the equation

$$CH_3CH_2OH(l)+CH_3COOH(l)\rightleftharpoons CH_3COOCH_2CH_3(l)+H_2O(l),$$

from which

$$K_c=\frac{[CH_3COOCH_2CH_3]\times[H_2O]}{[CH_3CH_2OH]\times[CH_3COOH]}=\frac{0.66\times 0.66}{0.33\times 0.33}=4$$

The effect of changing the concentration of one of the reactants on the composition of the equilibrium mixture is summarised in *Table 50.1* as the results of three experiments.

Experiment (ii) shows the effect of **increasing** the concentration of ethanoic acid, and experiment (iii) shows the effect of **decreasing** the concentration of ethanoic acid with

Table 50.1 The composition of the equilibrium mixture of the ethanol, ethanoic acid reaction, from different initial concentrations

Compound	*Experiment (i)* *Initial Concn*	*Equilibrium Concn*	*Experiment (ii)* *Initial Concn*	*Equilibrium Concn*	*Experiment (iii)* *Initial Concn*	*Equilibrium Concn*
CH_3COOH	1	0.33	2	1.155	0.5	0.08
CH_3CH_2OH	1	0.33	1	0.155	1	0.58
$CH_3COOCH_2CH_3$	0	0.66	0	0.845	0	0.42
H_2O	0	0.66	0	0.845	0	0.42

Note. For each set of values K_c is constant, i.e. 4.

respect to the concentration used in experiment (i). In general, if the concentration of **a reactant is increased**, the concentration of **the other reactant** (if present), **is decreased**, whereas the concentrations of the **products are increased** (c.f. experiments (i) and (ii)). Conversely, if the concentration of a **reactant is decreased**, the concentration of the **other reactant** (if present) **is increased**, while the concentration of the **products is decreased** (c.f. experiments (i) and (iii)).

8 When the temperature of any chemical reaction is increased the rate of the reaction increases. Considering the general equation for an equilibrium reaction established at temperature T_1°C:

$$A+B \rightleftharpoons C+D$$

there are two reactions taking place for which the rate of reaction equations are shown below.

(i) The forward reaction, $\boldsymbol{A+B\rightarrow}$**Products, Rate (i)**$\boldsymbol{=k_1[A][B]}$
(ii) The backward reaction, $\boldsymbol{C+D\rightarrow}$**Products, Rate (ii)** $\boldsymbol{=k_2[C][D]}$

An increase in temperature T_2°C increases the rate of both reactions. This results in the **velocity constants**, k_1 and k_2 increasing to k_1' and k_2' respectively. Since, for the equilibrium reaction at

$$T_1°\text{C},\ K_c=\frac{k_1}{k_2} \text{ and at temperature } T_2°\text{C},\ K_c=\frac{Kk_1'}{k_2'}$$

then, unless the velocity constants **change by equal amounts** (which is very unlikely), K_c^1 at T_1°C is different to K_c^2 at T_2°C. Hence the equilibrium constant K_c of a reversible reaction is **temperature dependent**.

9 The effect of temperature on an equilibrium reaction is given by the equation:

$$\log\frac{k_c^1}{k_c^2}=\frac{\Delta H^o}{2.303R}\left(\frac{I}{T_1}-\frac{1}{T_2}\right)$$

where k_c^1 is the equilibrium constant at T_1, k_c^2 is the constant at T_2. ΔH^o is the **standard enthalpy of the reaction** and R is the gas constant. The change in K_c is also dependent upon the **exothermic** $(-\Delta H^{\ominus})$ or **endothermic** $(+\Delta H^{\ominus})$ nature of the reaction. These effects are shown in *Table 50.2* and endothermic reaction.

Table 50.2 The effect of temperature on the value of the equilibrium constants for (a) an exothermic reaction and (b) on endothermic reaction

(a) Exothermic Reaction	$H_2(g)+I_2(g) \rightleftharpoons 2HI(g) \Delta H=-9.6kJ\ K_c^1$				
Temperature K	298	500	700	1000	1300
Equilibrium Constant K_c^1	794	160	54	31	12
(b) Endothermic Reaction	$H_2(g)+CO_2(g) \rightleftharpoons H_2O(g)+CO(g) \Delta H=+41kJ\ K_c^2$				
Temperature K	298	500	700	1000	1300
Equilibrium Constant K_c^2	1×10^{-6}	2.52×10^{-7}	2.82×10^{-3}	3.72	2.04×10^{2}

10 The effect of a **catalyst** on an equilibrium reaction is to **change** the rate of the forward and backward reactions by **equal amounts**. For example, in the general equation

$$A+B \rightleftharpoons C+D$$

the rate of the forward reaction can be written as $k_1[A][B]$. If the **catalyst doubles the rate** of reaction, then the expression for the rate of reaction can be written as $\mathbf{2k_1[A][B]}$. Since the rate of the backward reaction is also **doubled**, the rate of reaction can also be expressed as $\mathbf{2k_2[C][D]}$. The equilibrium constant for the unctalysed reaction will be

$$K_c=\frac{k_1}{k_2}=\frac{[C][D]}{[A][B]}$$

and the equilibrium constant for the catalysted reaction will be

$$K_c=\frac{2k_1}{2k_2}=\frac{[C][D]}{[A][B]}$$

and by division

$$K_c=\frac{2k_1}{2k_2}, \text{ i.e. } K_c=\frac{k_1}{k_2}$$

This shows that the equilibrium constant K_c is **independent of the effect of a reaction catalyst**. However, the stage of dynamic equilibrium will be **achieved more rapidly** by the introduction of a catalyst.

51 Chemical equilibrium in gaseous reactions

1 For **homogeneous reactions** in the **gas phase** it is more convenient to measure the **pressures** of the mixture of gases than to measure their concentrations in moles per cubic decimetre. For mixtures of gases in equilibrium the concentrations of the gases can be expressed in terms of their pressures.

2 The **ideal gas law** relates the **pressure**, **temperature**, **volume** and **number of moles** of an ideal gas by the equation $PV=nRT$ where P=pressure, V=volume, n=number of moles of gas, R=the gas constant and T=the absolute temperature. At a constant volume and temperature the pressure of the gas is proportional to the number of moles of that gas. By rearranging the ideal gas law, $PV=nRT$ can be written as,

$$P=n\frac{RT}{V} \text{ or } P=\textbf{a constant}\times \boldsymbol{n}$$

that is, $P \propto n$.

3 When a mixture of gases occupy a given volume at a constant temperature the pressure of the individual gases are called the **partial pressures** and the total pressure of the mixture is the sum of the partial pressures. The partial pressure of a gas is defined as **the mole fraction of the gas multiplied by the pressure**.

4 When a gaseous reaction forms an equilibrium mixture, the total pressure of the mixture of gases is equal to the sum of the partial pressures. Since the partial pressure of a gas is proportional to the number of moles of that gas, the **relative numbers of moles** of the gases are **proportional** to their **partial pressures**. Thus, the composition of an equilibrium mixture can be expressed in terms of the partial pressures of the gases.

5 For the general equilibrium equation

$$wA(g)+xB(g)\rightleftharpoons yC(g)+zD(g)$$

the **equilibrium constant** in terms of partial pressures is

$$K_p=\frac{(pC)^y\times(pD)^z}{(pA)^w\times(pB)^x}$$

Table 51.1

Equilibrium	K_p
$2HI(g) \rightleftharpoons H_2(g) + I_2(g)$	$\frac{(pH_2) \times (pI_2)}{(pHI)^2}$
$2CO(g) + O_2(g) \rightleftharpoons 2CO_2(g)$	$\frac{(pCO_2)^2}{(pCO)^2(pO_2)}$
$N_2(g) + 3H_2(g) \rightleftharpoons 2NH_3(g)$	$\frac{(pNH_3)^2}{(pN_2)(pH_2)^3}$
$N_2O_4(g) \rightleftharpoons 2NO_2(g)$	$\frac{(pNO_2)^2}{(pN_2O_4)}$
$2SO_3(g) \rightleftharpoons 2SO_2(g) + O_2(g)$	$\frac{(pSO_2)^2(pO_2)}{(pSO_3)^2}$

where pA, pB, pC and pD are the partial pressures of the gases A, B, C and D respectively. Some typical examples are given in *Table 51.1*.

6 The effect of **increasing the pressure** on a gaseous equilibrium is equivalent to **reducing the volume** to the gaseous mixture. If possible, the composition of the equilibrium will change to allow the mixture of gases to **exist in the smallest possible volume**. For example, in the equilibrium

$$N_2O_4(g) \rightleftharpoons 2NO_2(g)$$

the effect of reducing the pressure should **favour** the formation of dinitrogen tetroxide, N_2O_4. This can be confirmed by considering that for the equilibrium

$$N_2O_4(g) \rightleftharpoons 2NO_2(g)$$

at 350 K and a pressure of 200 kPa, if 1 mole of $N_2O_4(g)$ is taken then at equilibrium the mixture contains 0.14 moles of nitrogen dioxide NO_2. At the same temperature if 1 mole of N_2O_4 is brought to equilibrium at 100 kPa, the composition of the equilibrium changes. The composition of the equilibrium can be found by the following method.

First it is necessary to calculate K_p for the reaction

$$N_2O_4(g) \rightleftharpoons 2NO_2(g) \text{ using } K_p = \frac{(pNO_2)^2}{(pN_2O_4)}$$

Since 1 mole of N_2O_4 produces 2 moles of NO_2 on complete reaction, the production of 0.14 moles of NO_2 requires 0.07 moles of N_2O_4. The total number of moles at equilibrium is $(1-0.07)+0.14=1.07$ moles. The mole fractions and partial pressures can be expressed as

Gas	Mole fraction	Partial pressure
N_2O_4	$\frac{0.93}{1.07}$	$\frac{0.93}{1.07}\times 200$ kPa
NO_2	$\frac{0.14}{1.07}$	$\frac{0.14}{1.07}\times 200$ kPa

Substituting into the equilibrium expression

$$K_p=\frac{\left(\frac{0.14}{1.07}\times 200\right)^2}{\left(\frac{0.93}{1.07}\times 200\right)}=\frac{684.78}{173.83}=3.94$$

Hence $K_p=3.94$ kPa. At 100 kNm^{-2} pressure, let x moles of N_2O_4 be used up at equilibrium. The constitution of the equilibrium mixture can be expressed as

$$N_2O_4(g) \rightleftharpoons 2NO_2(g)$$

	N_2O_4	NO_2	
Initially	1	0	moles
At equilibrium	$\mathbf{1-x}$	$\mathbf{2x}$	moles

The total number of moles at equilibrium is $\mathbf{(1-x)+2x=(1+x)}$ moles. The mole fractions and partial pressures can be expressed as

Gas	**Mole fraction**	**Partial pressure**
N_2O_4	$\frac{(1-x)}{(1+x)}$	$\frac{(1-x)}{(1+x)}\times 100$ kPa
NO_2	$\frac{2x}{(1+x)}$	$\frac{2x}{(1+x)}\times 100$ kPa

Substituting these values into the expression for K_p and using $K_p=3.94$ gives

$$3.94=\frac{\left(\frac{2x}{(1+x)}\right)^2\times 100^2}{\frac{(1-x)}{(1+x)}\times 100}=\frac{400x^2}{(1+x)(1-x)}=\frac{400x^2}{(1-x^2)}$$

Substituting for x gives

$$3.94(1-x^2)=400x^2.$$

$$3.94=403.94x^2.$$

$$x=\sqrt{\frac{3.94}{403.94}}=0.099$$

Thus 0.099 moles of N_2O_4 are used up at equilibrium to produce $2 \times 0.099 = 0.198$ moles of NO_2.

By **halving** the pressure the number of moles of nitrogen dioxide **increases** from 0.14 moles to 0.198 moles. The **composition of any mixture of gases can be investigated in this way**.

52 Ionic equilibrium reactions

Weak acids and weak bases

1 All **weak acids** and **weak bases** can be represented by **equilibrium** reactions in their reactions with water.

(i) The **Brönsted-Lowry** theory defines an **acid** as **a molecule which can donate a proton** to another molecule. For example, the general acid **HA** dissociates or ionises according to the equation

$HA \rightleftharpoons$	H^+	$+ A^-$
Acid	**Proton**	**Base**

(ii) The Bronsted-Lowry theory defines a **base** as **a molecule which can accept a proton** from another molecule. Using the general equation given above

$A^- +$	H^+	$\rightleftharpoons HA$
Base	**Proton**	**Acid**

(iii) For any acid to behave as an acid it must have a base to which a proton can be **donate**, and for any base to behave as a base it must have an acid from which it can accept a proton.

2 When an acid and a base react together they produce **another acid and base**. For example,

$HNO_3 +$	$H_2O \rightleftharpoons$	$H_3^+O +$	NO_3^-
Acid 1	**Base 2**	**Acid 2**	**Base 1**

The **base** produced by the loss of a proton is called the **conjugate base of the acid**, and the **acid** and **base** are called a **conjugate acid base pair**. For the example shown, the loss of a proton from nitric acid, HNO_3, gives the nitrate ion, NO_3^-, thus nitric acid and the nitrate ion are a conjugate acid base pair. Similarly, when the water molecule accepts a proton, it becomes the hydrated hydrogen ion H_3^+O, and hence water is the base and the hydrogen ion is the acid of a conjugate acid base pair. Some examples of conjugate acid base pairs are given in *Table 52 1*

Table 52.1

Equilibrium	$CH_3COOH + H_2O \rightleftharpoons CH_3COO^- + H_3^+O$
Conjugate acid base pairs	$(CH_3COOH{:}CH_3COO^-)(H_3^+O{:}H_2O)$
Equilibrium	$HCl + H_2O \rightleftharpoons H_3^+O + Cl^-$
Conjugate acid base pairs	$(HCl{:}Cl^-)(H_3^+O{:}H_2O)$
Equilibrium	$NH_3 + H_2O \rightleftharpoons NH_4^+ + OH^-$
Conjugate acid base pairs	$(NH_4^+{:}NH_3)(H_2O{:}OH^-)$
Equilibrium	$H_2O + H_2O \rightleftharpoons H_3^+O + OH^-$
Conjugate base pairs	$(H_2O{:}{}^-OH)(H_3^+O{:}H_2O)$
Equilibrium	$NH_3 + NH_3 \rightleftharpoons NH_4^+ + NH_2^-$
Conjugate base pairs	$(NH_3{:}NH_2^-)(NH_4^+{:}NH_3)$

3 (i) For the general reaction of an acid HA with water, the equation is

$$\mathbf{HA + H_2O \rightleftharpoons H_3^+O + A^-}$$

The equilibrium constant for this reaction is given by

$$K_c = \frac{[H_3^+O][A^-]}{[HA][H_2O]}$$

In this equilibrium, the **concentration of water** is combined tc be **constant** and the equilibrium constant is called the **dissociation or ionisation constant**. K_a, of the acid, i.e.

$$K_a = \frac{[H_3^+O][A^-]}{[HA]}$$

(ii) For a base in solution, for example,

$$\mathbf{BOH + H_2O \rightleftharpoons B^+ + OH^-}$$

the equilibrium constant K_b is called the **dissociation constant (ionisation constant) of the base**, i.e.

$$K_b = \frac{[B^+][OH^-]}{[BOH]}$$

(iii) The numerical value of both K_a and K_b is a measure of the **degree of dissociation (or ionisation) of** the acid or base.
(iv) For large values of K_a and K_b the acids and bases are considered to be **completely** dissociated and represented by a suitable equation.

For example, for an acid $HCl + H_2O = H_3^+O + Cl^-$
or for a base, $NaOH + H_2O = Na^+ + OH^-$

These acids and bases are called **strong acids and bases**.

(v) For small numerical values of K_a and K_b the acids and bases are only **partially dissociated** and are called **weak acids** and **weak bases**, and are represented by equilibrium equations.

4 (i) The **degree of dissociation (or ionisation)** is given the symbol α and expressed either as a fraction of unity or as a percentage. *Table 52.2* shows the relationship between the values of α, K_a, and volume.

Table 52.2 Ethanoic acid at 298 K has a K_a value of 1.8×10^{-5} mol dm^{-3}

Molarity	*Volume containing 1 mole dm³*	*Degree of dissociation*
0.1	10	1.3×10^{-2}
0.01	100	4.1×10^{-2}
0.001	1000	1.3×10^{-1}
0.0001	10 000	4.1×10^{-1}

(ii) The degree of dissociation, α, is used to calculate the **concentration of hydrogen ions** for weak acids according to the expression

$$\underset{\text{(Equil)}}{[H_3^+O]} = \underset{\text{(Initial)}}{[HA]} \times \alpha$$

(iii) Similarly, the **concentration of hydroxide ions** for weak bases is given by the expression

$$\underset{\text{(Equil)}}{[OH^-]} = \underset{\text{(Initial)}}{[BOH]} \times \alpha$$

These definitions allow the hydrogen ion concentrations of weak acids or the hydroxide ion concentration of weak bases to be found. For example, at 298 K ethanoic acid is 1.8% ionised in a 0.1 M solution. This means that for the equilibrium

$$CH_3COOH(aq) \rightleftharpoons CH_3COO^-(aq) + H_3^+O(aq)$$

instead of the 0.1 moles being completely ionised to give 0.1 moles of aqueous hydrogen ions **only 1.8% are ionised**. Thus the true hydrogen ion concentration is given by

$$[H_3^+O]_{Equil} = [H_3^+O]_{soln} \times \text{degree of ionisation, } \alpha$$

$$[H_3^+O]_{Equil} = 0.1 \times \frac{1.8}{100} = \frac{0.18}{100} = 1.8 \times 10^{-3} \text{ mol dm}^{-3}$$

Thus, instead of a $[H_3^+O]$ value of 10^{-1} for a strong acid which is 0.1 M, for this weak acid the value of $[H_3^+O]$ is 1.8×10^{-3} mol d m^{-3}.

5 The dissociation constant, K_a, the degree of dissociation α, and the concentration of the acid, expressed as the **number of litres, V containing 1 mole** of the weak acid, are related by the expression

$$K_a = \frac{\alpha^2}{V(1-\alpha)}$$

which is called the **Ostwald Dilution Law**. (*Table 52.2* shows the variation in α for changing concentrations of ethanoic acid.)

6 The above equation can be used to find the value of K_a, the acid dissociation constant, if V and α are known. For example, for a 0.2 M solution of ethanoic acid which is 0.94% ionised, the acid dissociation constant K_a can be written as:

$$K_a = \frac{[CH_3COO^-][H_3^+O]}{[CH_3COOH]}$$

where K_a is the ionisation (or dissociation) constant of the acid.

Since the original concentration of the ethanoic acid is 0.2 M and $\alpha = 0.0094$, using the relationship $[H_3^+O] = \alpha[CH_3COOH]$ gives $[H_3^+O] = 0.0094 \times 0.2 = 0.00188$ or 1.88×10^{-3} mol dm^{-3}.

Since $[H_3^+O] = 1.88 \times 10^{-3}$ mol dm^{-3}, from the equation of the reaction $[CH_3COO^-] = 1.88 \times 10^{-3}$ mol dm^{-3}.

Since 1.88×10^{-3} mol dm^{-3} of the acid has been dissociated the concentration of the acid at equilibrium is given by

$[CH_3COOH]$ = Initial concentration − Equilibrium concentration,

that is, $0.2 - (1.88 \times 10^{-3})$ mol dm^{-3} = 0.198 mol dm^{-3}.

Substituting the concentration values in the equation gives

$$K_a = \frac{(1.88 \times 10^{-3}) \text{ mol dm}^{-3} \times (1.88 \times 10^{-3}) \text{ ,ol dm}^{-3}}{0.198 \text{ mol dm}^{-3}}$$

$$= \frac{3.534 \times 10^{-6}}{0.198} = 1.78 \times 10^{-5} \text{ mol dm}^{-3}.$$

Hence the ionisation (or dissociation) constant of ethanoic acid is 1.78×10^{-5} mol dm^{-3}.

The ionic product of water, K_w

7 (i) Although water is a **covalently bonded** molecule, in the pure state it is **self-ionising**, forming the equilibrium mixture

given by the equation

$$H_2O(l) + H_2O(l) \rightleftharpoons H_3O^+(aq) + OH^-(aq)$$

Applying the equilibrium law to this equation gives

$$K_c = \frac{[H_3O^+][OH^-]}{[H_2O][H_2O]}$$

By assuming that the concentration of water is **constant**, then the equation is simplified to become

$$\boldsymbol{K_w = [H_3O^+][OH^-]}$$

where the equilibrium constant K_w is called the **ionic product of water**, and at the temperature of 298 K the value of K_w **is 10^{-14} mol^2 dm^{-6}**.

(ii) At a temperature of 298 K the concentration of hydrogen ions, $H_3^+O(aq)$, and hydroxide ions, $OH^-(aq)$ are equal, i.e. $[H_3O^+] = [OH^-]$. Since

$$K_w = [H_3O^+][OH^-] = 10^{-14} \text{ mol}^2 \text{ dm}^{-6}$$

then

$$[H_3O^+] = [OH^-] = \sqrt[2]{10^{-14} \text{ mol}^2 \text{ dm}^{-6}} = 10^{-7} \text{ mol dm}^{-3}.$$

8 When water is made acidic by the addition of a solution of hydrogen ions, the hydroxide ion concentration must decrease such that $[H_3O^+][OH^-]$ **remains constant** at 10^{-14} mol^2 dm^{-6}. For example, if the hydrogen ion concentration of a solution is 10^{-2} mol dm^{-3}, the hydroxide ion concentration becomes 10^{-12} mol d m^{-3}. The product of these ion conccentrations $[H_3O^+][OH^-] = 10^{-2}$ mol dm$^{-3} \times 10^{-12}$ mol dm$^{-3} = 10^{-14}$ mol^2 dm^{-6}.

9 When a solution is made alkaline by the addition of a solution of hydroxide ions, the hydrogen ion concentration must change in order that the value of K_w remains constant at 10^{-14} mol^2 dm^{-6}.

The *pH* scale

10 The hydrogen ion concentration of one molar hydrochloric acid, pure water and one molar sodium hydroxide are 1, 10^{-7} and 10^{-14} mol dm^{-3} respectively. It is convenient to construct a scale of simpler numbers to represent these values. This can be achieved by taking the **reciprocal of the logarithm to the base ten of the hydrogen ion concentration** of the solution. When the conversion is made the value is called the ***pH* value** of the solution as shown by the conversions below.

When $[H_3O^+]=1\ mol\ dm^{-3}$,

$$\textbf{\textit{pH}}\ \textbf{value}=\frac{1}{\log_{10}[H_3O^+]}=-\log_{10}[H_3O^+]=-\log_{10}1=0.$$

When $[H_3O^+]=10^{-7}\ mol\ dm^{-3}$,

$$\textbf{\textit{pH}}\ \textbf{value}=\frac{1}{\log_{10}[H_3O^+]}=-\log_{10}[H_3O^+]=-\log_{10}10^{-7}=7$$

When $[H_3O^+]=10^{-14}\ mol\ dm^{-3}$,

$$\textbf{\textit{pH}}\ \textbf{value}=\frac{1}{\log_{10}[H_3O^+]}=-\log_{10}[H_3O^+]=-\log_{10}10^{-14}=14$$

the *pH* scale has been selected with values between 0 and 14 corresponding to hydrogen ion concentration of 1 mol dm^{-3} and 10^{-14} mol dm^{-3}. A knowledge of the *pH* value of a solution gives a value for the hydrogen ion concentration of that solution. For example, if a solution of hydrochloric acid, HCl(aq), has a *pH* of 4, this means that for the solution, using the equation

$$\boldsymbol{pH=-\log_{10}[H_3^+O]\ \text{then}\ 4=-\log_{10}[H_3^+O]}$$

Rearranging this expression gives

$$\log_{10}[H_3^+O]=-4.$$

and taking antilogs gives

$$[H_3^+O]=10^{-4}.$$

Hence the **concentration of the acid must be 10^{-4} M**.

The calculation of the hydrogen ion concentrations in weak acids from *pH* values does not give the concentration of the weak acid **directly** but if the degree of ionisation is known then the concentration can be found. For example, the *pH* of a solution of chloroethanoic acid is 1.72 when $\alpha=0.064$. This means that the hydrogen ion concentration of the solution given by the equation

$$pH=-\mathbf{log}_{10}\ [\mathbf{H_3^+O}]$$

and in the form $[\mathbf{H_3^+O}]=$**antilog** $(-pH)$ is $[H_3^+O]=\text{antilog}(-1.72)$ or $[H_3^+O]=0.019\ mol\ dm^{-3}$. By using the relationship

$$\boldsymbol{[H_3^+O]_{equil}=[H_3^+O]_{solution}\times\alpha}$$

$$\text{then}\ 0.019=[H_3^+O]_{solution}\times0.064$$

$$\text{or}\ [H_3^+O]_{solution}=\frac{0.019}{0.064}=0.3$$

Hence the **concentration** of the chloroethanoic acid must be **0.3 mol dm^{-3}**.

11 An **equivalent scale** can be applied to the concentration of hydroxide ions using the definition

$$pOH=\frac{\mathbf{1}}{\mathbf{log_{10}\,[OH^-]}}=-\mathbf{log_{10}\,[OH^-]}$$

12 On these combined scales of *pH* and *pOH* it can be shown that because for water when $pH=pOH=7$ that $pH+pOH=\mathbf{14}$. This relationship is useful in the interconversion of values. For example, the *pOH* at a 0.01 M solution of sodium hydroxide is 2, the *pH* of the same solution must be $14-2=12$.

Buffer solutions

13 (i) A **buffer solution** is a solution having a **specific pH value** which remains **approximately unchanged** when small amounts of an acid or alkali are added to it.

(ii) A buffer solution with a *pH* **value less than 7** is called an **acidic buffer solution**, whereas if the **pH value is greater than 7** it is called an **alkaline buffer solution**. The composition of some buffer solutions together with their *pH* values are given in *Table 52.3*.

(iii) A buffer solution is usually composed of a **weak acid** or **base and a salt of the acid or base** as shown in *Table 52.3*. Since an acidic buffer contains a weak acid, the buffer system must be in equilibrium. For the acidic buffer in *Table 52.3*, the equations of the reactions are

$$CH_3COOH(l)+H_2O(l)\rightleftharpoons CH_3COO^-(aq)+H_3O^+(aq) \quad (1)$$

$$CH_3COONa(s)+H_2O(l)=CH_3COO^-(aq)+Na^+(aq) \quad (2)$$

A combination of these two solutions causes a **change in the position of equilibrium** of the weak acid leading to a reduction

Table 52.3

Buffer system	*Ethanoic acid + sodium ethanoate*				
Concentration of ethanoic acid	1 M	1 M	1 M	0.1 M	0.01 M
Concentration of sodium ethanoate	0.01 M	0.1 M	1 M	1 M	1 M
pH	2.8	3.8	4.8	5.8	6.8
Buffer system	*Aqueous ammonia + ammonium chloride*				
Concentration of aqueous ammonia	0.01 M	0.1 M	1 M	1 M	1 M
Concentration of ammonium chloride	1 M	1 M	1 M	0.1 M	0.01 M
pH	7.2	8.2	9.2	10.2	11.2

in the hydrogen ion concentration. The addition of a small concentration of hydrogen ions to the buffer causes some ethanoate ions to react according to the equation

$$CH_3COO^-(aq) + H_3O^+(aq) = CH_3COOH(l) + H_2O(l)$$

Neither of the products causes a change in *pH* value. The addition of a small concentration of hydroxide ions causes a reaction with the hydrogen ions present according to the equation

$$H_3O^+(aq) + OH^-(aq) = 2H_2O(l)$$

The buffer solution reacts to this by displacing the equilibrium so that the hydrogen ions are replaced and no appreciable change occurs in the *pH* value.

14 The *pH* **value of a buffer solution** is given by the equation

$$pH = pK_a + \log_{10}\frac{[\textbf{concentration of the salt}]}{[\textbf{concentration of the acid}]}$$

The *pOH* value of a buffer solution is given by the equation

$$pOH = pK_b + \log_{10}\frac{[\textbf{concentration of the salt}]}{[\textbf{concentration of the base}]}$$

These relationships allow the concentrations of salts and acids or bases to be found for solutions of a required *pH* value. For example, to obtain a solution of *pH* 5 using ethanoic acid ($K_a = 1.8 \times 10^{-5}$) and potassium ethanoate, the **ratio** of the salt and acid can be found by using the equation

$$pH = pK_a + \log\frac{[CH_3COO^-]}{[CH_3COOH]}$$

the *pH* of the solution = 5. Since $K_a = 1.8 \times 10^{-5}$, then pK_a which is $-\log_{10}(1.8 \times 10^{-5}) = 4.74$. Substituting these values into the equation gives

$$5 = 4.74 + \log\frac{[CH_3COO^-]}{[CH_3COOH]}$$

Simplifying and taking antilogs gives

$$\frac{[CH_3COO^-]}{[CH_3COOH]} = \text{antilog } 0.26 = 1.8$$

Hence if the ethanoic acid is 0.1 M then the potassium ethanoate must be 0.18 M in order that the **ratio shall be 1.8 :1**.

53 The structure of materials

1 **Materials** is the collective name given to substances that exist in nature in a stable state. They can be **pure elements**, e.g. chlorine gas, bromine liquid or metallic copper solid, **pure compounds of elements**, e.g. hydrogen chloride gas, ethanol liquid or sodium chloride (rock salt) solid, or **mixtures** of different substances, e.g. brass, an alloy of metals and tungsten carbide, an interstitial compound.

2 The **structure of materials** is an extension of the bonding ideas discussed in Sections 44 and 45. The combination of atoms into elements or compounds considers only the constituent particles. However, the structure of materials and their properties takes into account the **vast number** of particles that make up **observable masses** of material. For example, when sodium chloride is considered, only enough particles are used to define the **unit cell** (shown in Section 45) but a grain of salt used for catering purposes will contain approximately 10^{16} particles. It is the properties of this and larger numbers of particles that are considered here.

3 There are **three states of matter** which material can adopt, solid liquid and gas. Of these the solid state offers a wider variety of structures than the liquid or gas states.

Solids

4 The structures of solid materials are all based upon particles being held in a regular **three-dimensional** pattern. The properties of a solid depend on the **nature** and **intensity** of the **forces** holding the particles in place.

Ionic materials

5 When a solution of an ionic solid is allowed to evaporate very slowly, very small **well defined crystals** will be deposited and all of the crystals will be the **same shape**. Some examples are shown in *Figure 53.1*.

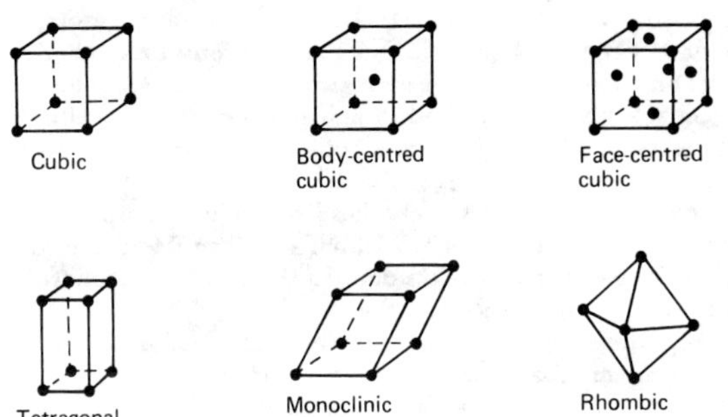

Figure 53.1 Some capital shapes

When the rate of evaporation increases, the well defined shapes disappear into **clusters** of crystals which grow over each other in various directions. Ionic crystals can be **cleaved** using a sharp cutting edge but not as easily in all directions. This is because the forces holding the solid together are not equal throughout the structure. The properties of ionic solids are summarised in *Table 53.1*.

Covalent solids

6 Unlike ionic solids the properties of **covalently bonded solids** are quite **diverse**. In the elemental state carbon can exist as diamond which is one of the hardest structures known, and graphite which is a very soft material sometimes used as a lubricating agent. The difference in properties is explained by their

Table 53.1

Ionic solids	*Covalent solids*
1. High melting point and high boiling point.	1. Can be high or low melting point and boiling point.
2. Soluble in water.	2. Usually insoluble in water.
3. Conduct electricity when molten or in aqueous solution.	3. Usually non-conductors except for graphite.
4. Can be cleaved along certain axes.	4. Extremes of hardness and softness, e.g. diamond and graphite.

respective structures (see Section 44). Diamond is a **network** of carbon atoms **all of which** are held in place by **four links to other carbon atoms** giving a very **rigid** structure. In graphite the carbon atoms are only **linked to three other atoms** resulting in the formation of **layers of atoms** only weakly attracted to each other; these layers can slip over each other and relieve the friction between two metal surfaces. Other solids like sulphur, phosphorus and iodine are not extensive networks, but **small molecules** held in place by **weak attractive van der Waal** forces. These solids are brittle, and easily crushed to a fine powder.

Compounds which are covalently bonded can also exist in **crystalline form**, quartz being an ideal example. In quartz, silicon and oxygen are extensively covalently bonded into a diamond type network. Crystals of this type are not as easily cleaved as ionic crystals. The properties of covalent solids are summarised in *Table 53.1*.

Metals

7 The **close packing** of metal atoms together is discussed in Section 44. On a much larger scale, the close-packing of atoms together as a molten metal solidifies is considered to take place over a series of steps shown in *Figure 53.2*. The **nuclei (a)**, form

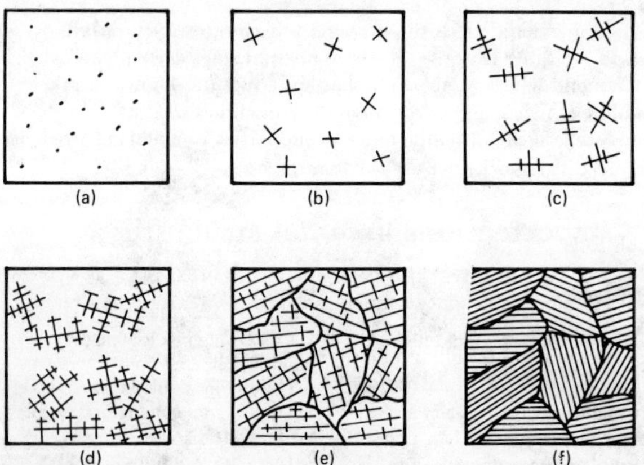

Figure 53.2 The shapes of crystallisation of a metal

into small crystals (**b**), which then form combinations of atoms called **dendrites** (**c**), which continue growing in **a certain orientation** (**d**) until they meet a set of dentrites growing in a different orientation when **a boundary is formed** (**e**) around the dentrites. This structure is called a **grain structure** (**f**). The grain structure of metals determines their properties. The size of the grains in some metals can be changed by **cold-working**, using such methods as **rolling**, **extruding** and **pressing**. In all cases the grains are **elongated** as shown in *Figure 53.3.* After such

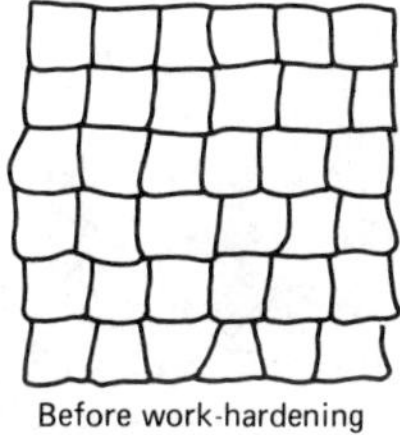

Figure 53.3

treatment the metals become **harder** and **less ductile**. Metals treated in this way can be **softened** by **annealing**, a reheating process. The internal stresses introduced by cold-working are relieved in some metals by the formation of **fine grains** in their new shape producing a softening of the metal. Some metals behave in this way at room temperature explaining why metals like tin and lead **cannot be work hardened**.

Imperfections in metal structures

8 (a) **Vacant sites**. If an atom is missing from the regular order of atoms, the space created is called a **vacant site**, shown diagrammatically in *Figure 53.4*. Metals' atoms can move from one position to another, filling one site and leaving another vacant. This process is called **diffusion**, and is increased by a rise in temperature.

(b) Dislocations When part of a plane of atoms is missing rather than a single atom, an **edge dislocation** is formed as shown in *Figure 53.5.* The movement of these dislocations within the metal is

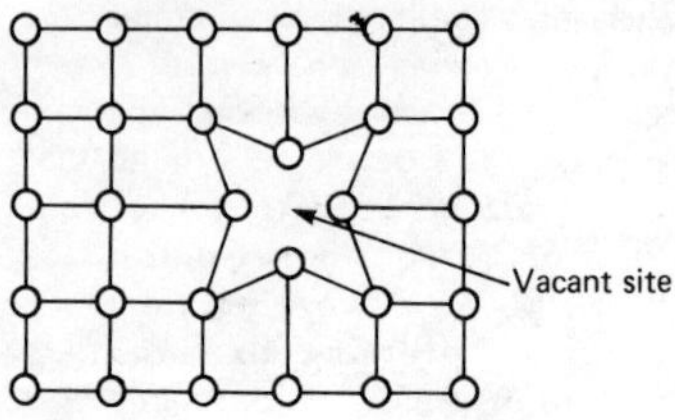

Figure 53.4

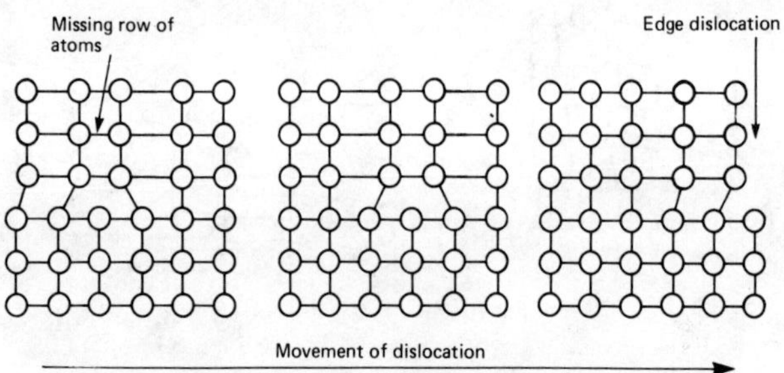

Figure 53.5 The movement of a dislocation of a metal structure

called **slip**. Under stress, the dislocations tend to accumulate resulting in a **deformation** of the metal structure.

9 When materials are subjected to repeated stresses they may break more easily than expected due to **metal fatigue**, caused by the accumulation of dislocations in one particular orientation.

Alloys

10 When two or more metals are melted, mixed together and allowed to cool, the atoms of the metals can become **dispersed** in each other and form **alloys**, for example, copper and zinc in brass. In other alloys the metals **crystallise out** separately, for example, tin and lead in plumbers solder. The properties of an alloy are different to the metals used to make them, for example, brass is much harder than copper or zinc. Some examples of alloys are given in *Table 53.2*.

Table 53.2 The composition of some alloys.

Alloy	*Composition*
Plain carbon steel	Fe, 0.45%C, small amounts of Si, Mn and S.
Cobalt steel	Fe, 2% Co, 4% Cr, 0.9% C
Bronze	Cu, 0–15% Sn
Brass	Cu, 0–36% Zn
High-tensile brass	57% Cu, 37% Zn, small amounts of Al, Mn, Fe
Monel, cupronickel	Ni, 30% Cu
Plumbers solder	70% Pb, 30% Sn
Gunmetal	Cu, 5% Sn, 5% Zn

Steel

11 **Steel** can be considered to be an alloy of iron and carbon, the carbon occupying **certain sites** in the iron metallic structure. Mild steel used in industry contains about 0.3% carbon. The carbon atoms **reduce the amount of movement** which can occur in the metallic structure, and increases the strength of the structure. High carbon steel contains up to 1% carbon and are very hard steels used for cutting tools. Other types of steel contain other elements, for example, **stainless steel** contains up to 18% chromium as well as traces of carbon. The heat treatment of steel has the same effect as in iron, rapid or gradual cooling giving different sizes of grains and hence different properties.

Interstitial compounds

12 The presence of carbon in steel, changes the properties considerably. Other small atoms like hydrogen and nitrogen can also be enclosed within the structure of the metal, and cause similar effects. For example, nitrogen and carbon in metals leads to the formation of **hydrides**, **nitrides** and **carbides**. The nitrides and carbides are **very hard** materials; for example, tungsten carbide is used to drill holes in masonry and rocks. The presence of these elements in metals **prevents the movement** of dislocation through a structure causing more rigid structures.

Polymer structures

13 **Polymers** are very large molecules formed by the joining together of smaller molecules called **monomers**, by the process of **polymerisation**. There is a large variety of polymers showing

many different properties. They can be broadly classified in terms of their physical properties as, **thermoplastics, thermosetting plastics, fibre polymers** and **rubbers**. The structure of a polymer depends upon its **method of preparation** and the **monomer units** it contains. For example, the long chain molecules which constitute thermoplastics are held together in the solid state by van der Waal forces, whereas in thermosetting plastics extensive cross-linking occurs within the polymer structure.

Thermoplastics

14 These are usually made from monomer units which contain a carbon to carbon bond. Some examples are given in *Table 53.3.* An example of this type of reaction is the formation of polyvinyl chloride P.V.C. from chloroethane,

$$\mathrm{nCHCl{=}CH_2} \xrightarrow[\text{high temperature catalyst}]{\text{high pressure}} \begin{array}{ccccccccccccccc} & \mathrm{Cl} & & \mathrm{H} & & \mathrm{Cl} & & \mathrm{H} & & \mathrm{Cl} & & \mathrm{H} & & \mathrm{Cl} & \\ & | & & | & & | & & | & & | & & | & & | & \\ - & \mathrm{C} & - & \mathrm{C} & - & \mathrm{C} & - & \mathrm{C} & - & \mathrm{C} & - & \mathrm{C} & - & \mathrm{C} & - \text{ etc.} \\ & | & & | & & | & & | & & | & & | & & | & \\ & \mathrm{H} & & \mathrm{H} & & \mathrm{H} & & \mathrm{H} & & \mathrm{H} & & \mathrm{H} & & \mathrm{H} & \end{array}$$

These polymers can be changed back to the monomer by distillation though it is not a useful process at this time. The long chain molecules are packed together in a somewhat ordered fashion, giving a certain degree of tensile strength. The materials can be **hot moulded** into shape by melting the plastics and blowing them into the mouldings. When cooled, this shape will be maintained unless the plastic is heated in which case it becomes **deformed**. The forces holding the chains of molecules together are not **rigid** and can be **broken** fairly easily.

Thermosetting plastics

15 These are all **co-polymers** and form large extensively **cross-linked** molecules as shown in *Table 53.4.* These plastics as the name suggests form a **rigid solid** as the reaction occurs. This means the plastics must either be moulded during reaction, or by machining after reaction. The plastics are brittle and do not melt but decompose. They are used for making **heat resistant materials** like ashtrays and electrical fittings.

Fibre polymers

16 These polymers can be co-polymers or made from a single monomer, some examples of which are shown in *Table 53.5.* They are similar to thermoplastics in that they can be **extruded** under

Table 53.3 Some thermoplastic polymers

Monomer		*Polymer*	
Name	*Structure*	*Name*	*Structure*
ethene	$\mathrm{(H)(H)C{=}C(H)(H)}$	polyethene (polythene)	$\mathrm{-\!\left(C(H)(H)-C(H)(H)\right)_n\!-}$
propene	$\mathrm{(CH_3)(H)C{=}C(H)(H)}$	polypropene (polypropylene)	$\mathrm{-\!\left(C(CH_3)(H)-C(H)(H)\right)_n\!-}$
styrene	$\mathrm{(C_6H_5)(H)C{=}C(H)(H)}$	polystyrene	$\mathrm{-\!\left(C(C_6H_5)(H)-C(H)(H)\right)_n\!-}$
chloroethene	$\mathrm{(Cl)(H)C{=}C(H)(H)}$	polychloroethene (polyvinyl chloride)	$\mathrm{-\!\left(C(Cl)(H)-C(H)(H)\right)_n\!-}$
methyl, 2-methyl propenoate	$\mathrm{(CH_3)(H)C{=}C(H)(COOCH_3)}$	polymethyl-2, methyl propenoate (perspex)	$\mathrm{-\!\left(C(CH_3)(H)-C(H)(COOCH_3)\right)_n\!-}$
tetra fluoro-ethene	$\mathrm{(F)(F)C{=}C(F)(F)}$	polytetrafluoroethene (P.T.F.E.)	$\mathrm{-\!\left(C(F)(F)-C(F)(F)\right)_n\!-}$

Note. Popular names in brackets

heating, but are much more stronger than thermoplastics. After extrusion as **fine fibres** the polymers are twisted into much thicker fibres and then formed into a variety of materials ranging from **monofilament fishing line** to **mooring ropes** used in the shipping industry.

Rubbers

17 Natural rubber is obtained from the sap of the rubber tree. After suitable treatment it is obtained as a clear flexible solid. In

Table 53.4 Some thermosetting polymers

Monomers		Polymer
$H_2N-\underset{\underset{O}{\parallel}}{C}-NH_2$ Urea	$H-\underset{\underset{O}{\parallel}}{C}-H$ Methanal	$-CH_2$, $N-CO-N$, $-CH_2$; $CH_2-N-CO-N-$; $CH_2-N-CO-N-$; CH_2 CH_2
NH_2, N, N, NH_2, N, NH_2 Melamine	$H-\underset{\underset{O}{\parallel}}{C}-H$	CH_2; $-CH_2-N$; N; N; N; $-CH_2-N$; CH_2; CH_2; CH_2-N; N; N; CH_2-N-CH_2; N; N; N; N; N; $N-$; $N-$; CH_2
OH Phenol	$H-\underset{\underset{O}{\parallel}}{C}-H$	OH, OH, CH_2, CH_2, CH_2, CH_2, CH_2, CH_2, OH, OH

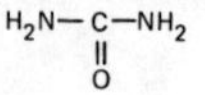

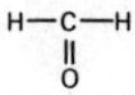

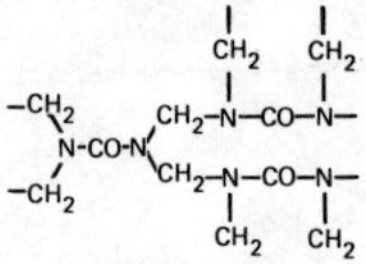

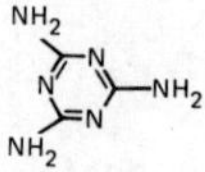

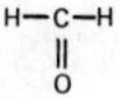

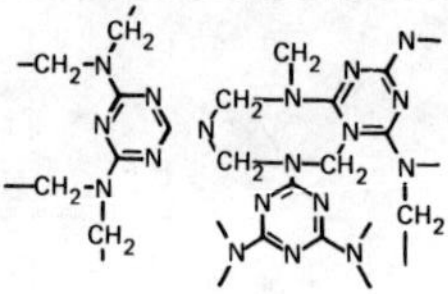

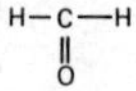

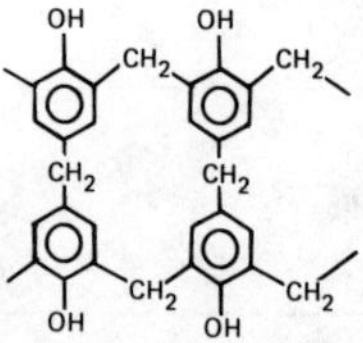

this, and in the synthetic rubbers like **styrene-butadiene rubber** as shown in *Table 53.6*, the structure of the material is such that under tension the solid **elongates**, but after the tension is released the **original shape** is reformed. This is due to the **retangling** of the polymer chains that are **partially untangled** under tension as shown in *Figure 53.6*. However, if the elastic limit is exceeded the original shape will not be reformed, but remains **deformed**.

The process of **vulcanisation** which is the addition of sulphur to the rubber causes cross-linking to occur between the polymer chains. The greater the amount of vulcanisation the more **rigid** the rubber becomes.

18 The structure of plastics and polymers can be changed dramatically by the introduction of **additives**. The major additives are **fillers** which increase **bulk** and **rigidity**, **plasticiser** which increase the **flexibility**, **dyes** which control

Table 53.5 Some fibre polymers

Monomers		Polymer
$H_2N-(CH_2)_6-NH_2$ +	$HO_2C(CH_2)_4COOH$	$H[NH(CH_2)_6NH-CO(CH_2)_4CO]_nOH$
1, 6- diamino hexane	**1, 6- hexandioc acid**	**6, 6- Nylon**
$H_2N(CH_2)_6NH_2$ +	$ClOC(CH_2)_8COCl$	$H[NH(CH_2)_6-NH-CO(CH_2)_8CO]_nOCl$
	1, 10- decandioyl chloride	**6, 10- Nylon**
CH_2-CH_2, CH_2, $C=O$, CH_2, $N-H$, CH_2 (ring)		$[NH(CH_2)_5CO]_n$
ω - caprolactam		**6, Nylon**
$CH_3O_2C-C_6H_4-CO_2CH_3$ +	$HO(CH_2)_2OH$	$CH_3O[CO-C_6H_4-CO-CH_2-CH_2-O]_nH$
dimethyl terephthalic acid	**ethan-1, 2-diol**	**Terylene**
$CH_2=CH-CN$	$CH_2=CHCO_2CH_3$	$[CH_2-CH(CN)-CH_2-CH(CO_2CH_3)]_n$
propeno-l-nitrile	**methyl propenoate**	**Acrilan**

Without tension

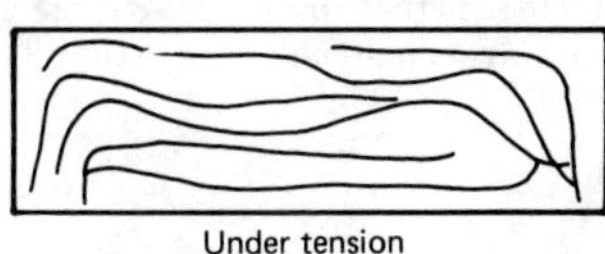
Under tension

Figure 53.6 The elongation of a rubber

Table 53.6 Some synthetic rubbers

Monomers	*Polymer*
$CH_2{=}CH{-}\underset{\displaystyle CH_3}{\underset{\vert}{C}}{=}CH_2$ 2-methylbuta-1,3-diene	$\left(CH_2{-}\underset{\displaystyle H}{\underset{\vert}{C}}{=}\underset{\displaystyle CH_3}{\underset{\vert}{C}}{-}CH_2\right)_n$ poly 2, methyl buta-1,3-cliene
$CH_2{=}CH{-}\underset{\displaystyle C}{\underset{\vert}{C}}{=}CH_2$ 2-chlorobuta-1,3-diene	$\left(CH_2{-}\underset{\displaystyle H}{\underset{\vert}{C}}{=}\underset{\displaystyle Cl}{\underset{\vert}{C}}{-}CH_2\right)_n$ poly 2, chlorobuta-1,3-diene
$CH_2{=}CH{-}CH{=}CH_2$ buta-1,3-diene $CH_2{=}CH{-}C_6H_5$ phenylethene (styrene)	$\left[\left(CH_2{-}CH{=}CH{-}CH_2\right)_4CH_2{-}\overset{\displaystyle C_6H_5}{\overset{\vert}{\underset{\displaystyle H}{\underset{\vert}{C}}}}\right]_n$ styrene butadiene rubber S.B.R.
$CH_2{=}CH{-}CH{=}CH_2$ buta-1,3-diene $CH_2{=}\underset{\displaystyle CN}{\underset{\vert}{CH}}$ propenonitrile	$\left[\left(CH_2{-}CH{=}CH{-}CH_2\right)_2CH_2{-}\overset{\displaystyle CN}{\overset{\vert}{\underset{\displaystyle H}{\underset{\vert}{C}}}}\right]_n$ nitrile rubber

colour and **stabilisers** which increase the **useful life** in the atmosphere or in water. The use of different combinations of additives can change the structure to the extent that polyvinychloride can be used as a flexible clothing material or the rigid pipes used for transportation of materials below the earth's surface.

Liquids and gases

19 When solids undergo phase changes to become liquids and gases, their structural properties change completely and behave in the same way as materials that exist as liquids and gases at room temperature.

54 Organic chemistry

1 **Organic chemistry** is regarded as the chemistry of carbon excluding the oxides, metallic carbonates, metallic hydrogen carbonates and metallic carbonyls.
2 The bonding of organic chemicals is **covalent** and the covalencies of carbon and the other atoms is discussed in Chapter 45.
3 Organic chemicals are classified by the content and arrangement of atoms within the molecules. There is a vast family of organic chemicals and only a limited selection of these are considered in this text.
4 **Hydrocarbons** are organic chemicals which contain **only carbon** and **hydrogen**. A detailed study of these compounds provides a basis for a study of other organic chemicals containing atoms in addition to carbon and hydrogen, details of which are given in tabular format later in this chapter.
5 There are four classes of hydrocarbons, the **alkanes**, the **alkenes**, the **alkynes** and the **aromatic** hydrocarbons, of which, the alkanes, alkenes and aromatic hydrocarbons are discussed in this chapter. The main source of the compounds is from crude oil via the petroleum industry.

The alkanes

6 The class of chemicals called the alkanes is recognised from the other hydrocarbons by the following features:

(a) Each carbon atom has four covalent bonds directed to the corners of a regular tetrahedron (see *Figure 54.1*). This results in the **repulsive forces** between the four covalent bonds being minimised,
(b) all carbon to carbon bonds are single bonds,
(c) they form a **homologous series** (meaning of the same essential nature), having the formula

$$C_nH_{2n+2}, \text{ where } n=1, 2, 3, \text{ etc.}$$

7 Every class of organic chemicals forms a homologous series in

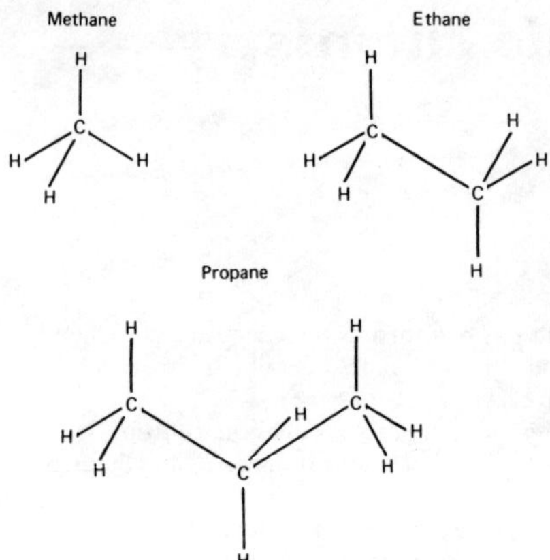

Figure 54.1 The structure of some alkanes

which each successive member is different from the former by **one carbon** and **two hydrogens** as shown in *Table 54.1*.

8 Every class of organic chemicals is named in accordance with the **International Union of Pure and Applied Chemists (IUPAC)** system. The names of organic chemicals refer to both the **number of carbon atoms** in the molecule and the **class** of the chemical. Thus C_5H_{12}, is called **pentane**, the **pent-** indicating **five carbon atoms** and the **-ane** indicating the **alkane class**. The names of some alkanes are shown in *Table 54.1*.

9 The alkanes themselves are subdivided into three categories,

(a) the **straight chain** alkanes in which the carbon atoms are joined together in a single continuous chain, as shown in *Table 54.1 (b,c,d,f)*.

(b) the **branched chain** alkanes in which extra carbon atoms are joined onto the straight chain of carbon atoms, as shown in *Table 54.1 (e,g,h)*, and

(c) the **cycloalkanes** in which the carbon atoms are joined together in a continuous ring as shown in *Table 54.1(i)*.

10 It can be seen from *Table 54.1 (d and e)*, that it is possible for a straight chain alkane, and a branched chain alkane, to have the same molecular formula, namely C_4H_{10}. **Two or more different compounds having the same formula are called isomers**, and a further example of **isomerism** is shown in *Table 54.1 (f, g and h)* by the molecular formula C_5H_{12}.

Table 54.1

Formula	Structure	Type	Name	Boiling point °C
(a) CH_4	H H—C—H H	Straight chain	Methane	-162
(b) C_2H_6	H H H—C—C—H H H	Straight chain	Ethane	-89
(c) C_3H_6	H H H H—C—C—C—H H H H	Straight chain	Propane	-44
(d) C_4H_{10}	H H H H H—C—C—C—C—H H H H H	Straight chain	Butane	-0.5
(e)	H H H H—C^3—C^2—C^1—H H H—C—H H H	Branched chain	2-Methyl propane	-12
(f) C_5H_{12}	H H H H H H—C—C—C—C—C—H H H H H H	Straight chain	Pentane	36
(g)	H H H H H—C^4—C^3—C^2—C^1—H H H H—C—H H H	Branched chain	2-Methyl butane	28
(h)	H H H—C—H H H—C^3—C^2—C^1—H H H—C—H H H	Branched chain	2, 2- Dimethyl propane	9.6
(i) C_6H_{12}	H H H C H H—C C—H H—C C—H H C H H H	Cyclic	Cyclohexane	81

14 In every homologous series, a gradual increase in physical properties, such as the temperature of the boiling point and density, occurs, with the increase in the number of carbon atoms in the chain. For example, the boiling points of the straight chain

alkanes shown in *Table 54.1* increases from $-162°C$ for CH_4 to $36°C$ for C_5H_{12}.

12 In the case of a straight chain alkane being one of several isomers, it is always the straight chain alkane which has the highest boiling point. Branched chain isomers have a lower boiling point depending on how many branches are present along the straight chain. For example, pentane is a straight chain and has a b.p. of 36°C, and 2,2-dimethylpropane has two branches and a b.p. of 9.6°C.

Reactions of the alkanes

13 (a) **Combustion** When any hydrocarbon burns in air the products are carbon dioxide and water, for example, $CH_4(g) + 2O_2(g) = CO_2(g) + H_2O(g)$ $\Delta H - ve$. The production of a considerable amount of heat during this reaction allows the alkanes to be used for fuels, for example, Calor Gas is propane C_3H_8. The general equation for the combustion of any hydrocarbon is

$$C_xH_y + \left(x + \frac{y}{4}\right)O_2 = xCO_2 + \frac{y}{2}\ H_2O$$

For example, for butane, C_4H_{10} shows that $x = 4$ and $y = 10$. The general equation becomes

$$C_4H_{10}(g) + 6\tfrac{1}{2}O_2(g) = 4CO_2(g) + 5H_2O(g)$$

(b) **Halogenation** Methane reacts with chlorine in the presence of bright sunlight (the catalyst to promote reaction) at room temperature and produces a mixture of products in a **four stage synthesis** (i.e. four stages of 'building up'). The stages are:

(i) $CH_4(g) + Cl_2(g) = CH_3Cl(l) + HCl(g)$
chloromethane

(ii) $CH_3Cl(l) + Cl_2(g) = CH_2Cl_2(l) + HCl(g)$
dichloromethane

(iii) $CH_2Cl_2(l) + Cl_2(g) = CHCl_3(l) + HCl(g)$
trichloromethane

(iv) $CHCl_3(l) + Cl_2(g) = CCl_4(l) + HCl(g)$
tetrachloromethane

Ethane C_2H_6 and propane C_3H_8 react in a similar way with chlorine.

The alkenes

14 The hydrocarbons known as the alkenes are recognised by the following features:

(a) the presence in the molecule of a **carbon to carbon double bond** shown as a **C═C linkage** in *Figure 54.2*.

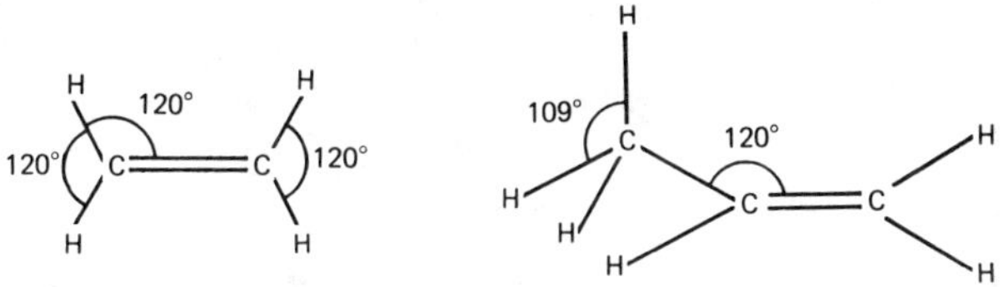

Figure 54.2 The structure of some alkenes

The covalent bonds of these two carbon atoms are directed to the three corners of an equilateral triangle as shown in *Figure 54.2*. This means that the molecule is **planar** (i.e. on a flat surface) in the vicinity of the carbon to carbon double bond. Alkenes with more than two carbon atoms in their structure, contain a planar part of the molecule (C═C), connected to the other carbon atoms dispersed in an identical way to that of the alkanes shown in *Figure 54.2*.

(b) The **open chain** alkenes form a homologous series having the general formula $\mathbf{C_nH_{2n}}$, **where** n=**2, 3, 4** etc.; note that the compound formed there when n=1, i.e. CH_2 is not an alkene. The alkenes can be subdivided into three categories as were the alkanes, namely, the straight chain alkenes, the branched chain alkenes and the cycloalkenes. Examples of the three categories are shown in *Table 54.2*.

15 The **nomenclature** (i.e. system of names or terminology used) of the alkenes is shown in *Table 54.2*. The presence of the carbon to carbon double bond in the molecule is signified by the letters **ene** at the end of the name. The first part of the name refers to the number of carbon atoms present. When there is **more than one possible position for the carbon to carbon double bond, it is identified by a number**. For example, the alkene with the molecular formula C_4H_8 can exist in two different structures. These may be written as (i) $\mathbf{C^1H_2{=}C^2HC^3H_2C^4H_3}$ which is named **but-l-ene** and (ii) as $\mathbf{C^1H_3C^2H{=}C^3HC^4H_3}$ which

Table 54.2

Formula	Name	Type	Structure	Boiling point °C
C_2H_4	Ethene	Straight chain	H H C=C H H	-103.6
C_3H_6	Propene	Straight chain	CH_3 H C=C H H	-57.5
C_4H_8	But-*1*-ene	Straight chain	H H $C^2=C^1$ CH_3CH_2 H	-6
	trans But-*2*-ene	Straight chain	*1* CH_3 H $C^2=C$ H CH_3	1
	cis But-*2*-ene	Straight chain	*1* CH_3 CH_3 $C^2=C$ H H	3.9
	2-methylpropene	Branched chain	CH_3 H $C^2=C^1$ CH_3 H	2.4
C_6H_{10}	Cyclohexene		H H H C H C C H H H C C H C H H H	83

is named **but-2-ene**. These are examples of **structural isomers**.

16 In addition to the straight chain/branched chain isomerism shown by the alkenes, another type of isomerism is shown by the alkenes called **geometrical isomerism**. Since the double bond

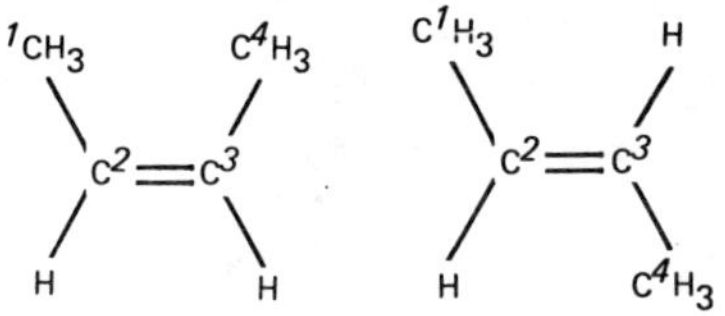

$^{1}CH_3$ \ $C^2{=}C^3$ / C^4H_3; H / \ H

C^1H_3 \ $C^2{=}C^3$ / H; H / \ C^4H_3

cis but-2-ene   *trans* but-2-ene

Figure 54.3 The geometrical isomers of butane

part of the molecule is planar and the carbon atoms are joined together twice, **this part of the molecule is rigid**. The compound with the structural formula $\mathbf{CH_3CH{=}CHCH_3}$ can exist in **two forms** which are shown in *Figure 54.3*. The suffix **cis** is used to show that the hydrogen atoms present on carbon atoms 2 and 3 are on the **same side of the molecule**, whereas the suffix **trans** is used to show that the hydrogen atoms are on **opposite sides of the molecule**.

Reactions of the alkenes

17 In contrast to the alkenes, the alkenes undergo many more reactions due to the presence of the carbon to carbon double bond in the molecules. Some examples of the reactions are given below.

(a) **Combustion**

$$C_2H_4(g) + 3O_2(g) = 2CO_2(g) + 2H_2O(g) \quad \Delta H\text{-}ve$$

ethene

(b) **Reaction with chlorine**

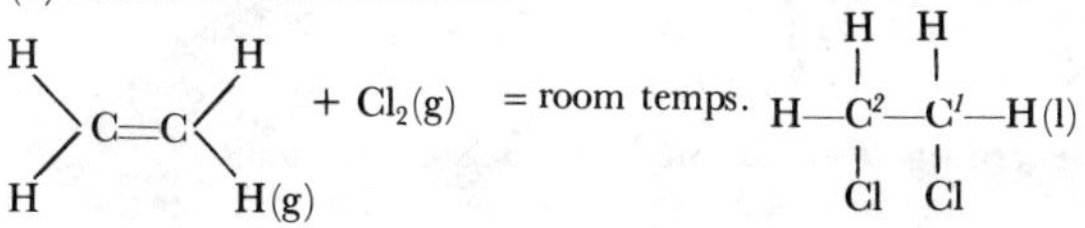

H, H $\rangle$C=C$\langle$ H, H(g) + $Cl_2(g)$ = room temps. H—C^2(H)(Cl)—C^1(H)(Cl)—H(l)

1,2-dichloroethane

(c) **Reaction with bromine**

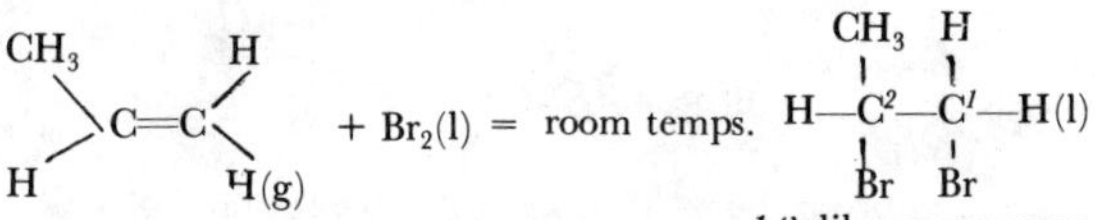

CH_3, H $\rangle$C=C$\langle$ H, H(g) + $Br_2(l)$ = room temps. H—C^2(CH_3)(Br)—C^1(H)(Br)—H(l)

1,2-dibromopropane

(d) **Reaction with hydrogen chloride**

$$\begin{array}{c}HH\\ \!\!>C{=}C<\!\!\\ HH(g)\end{array} + HCl(g) \;=\; R.T. \qquad \begin{array}{ccccccc} & & H & & H & & \\ & & | & & | & & \\ H & - & C^{2} & - & C^{1} & - & H(l)\\ & & | & & | & & \\ & & H & & Cl & & \end{array}$$

1,-chloroethane

(e) **Reaction with hydrogen bromide**

$$\begin{array}{c}CH_3H\\ \!\!>C{=}C<\!\!\\ HH\end{array}(g) + HBr(g) \;=\; R.T. \qquad \begin{array}{ccccccc} & & CH_3 & & H & & \\ & & | & & | & & \\ H & - & C^{2} & - & C^{1} & - & H(l)\\ & & | & & | & & \\ & & Br & & H & & \end{array}$$

2,-bromopropane

(f) **Reaction with hydrogen**

$$\begin{array}{c}HH\\ \!\!>C{=}C<\!\!\\ HH(g)\end{array} + H_2(g) \;\overset{\text{Ni catalyst}}{\underset{150°C}{=}}\; \begin{array}{ccccccc} & & H & & H & & \\ & & | & & | & & \\ H & - & C & - & C & - & H(g)\\ & & | & & | & & \\ & & Cl & & Cl & & \end{array}$$

ethane

The reactions (b) to (f) are called **addition reactions**. This is because two molecules, i.e. an alkene and another molecule are reacting together to form a **single addition product**.

18 **Unsaturation in the alkenes**. If a hydrocarbon contains carbon to carbon bonds which are not single bonds, but double or triple bonds, they are classified as **unsaturated hydrocarbons**. Thus, since all alkenes contain double covalent bonds, they are all unsaturated hydrocarbons. The presence of unsaturation in a hydrocarbon can be detected by classification tests. These tests are:

(a) **Reaction with bromine in tetrachloromethane**.
Bromine is present in tetrachloromethane, CCl_4, as Br_2 molecules. When an unsaturated compound is reacted with bromine the reaction which takes place is

$$\begin{array}{c}HH\\ \!\!>C{=}C<\!\!\\ HH\end{array}(g) + Br_2(l) \;=\; \begin{array}{ccccccc} & & H & & H & & \\ & & | & & | & & \\ H & - & C^{1} & - & C^{2} & - & H(l)\\ & & | & & | & & \\ & & Br & & Br & & \end{array}$$

1,2,-dibromoethane

The bromine in CCl_4 is an **orange colour** which, on reaction, **fades** until the resultant solution is **colourless**. This loss of colour is a test for unsaturation.

(b) **Reaction with dilute alkaline potassium permanganate.** When an unsaturated hydrocarbon is added to an alkaline solution of potassium permanganate, the **pale pink colour** of the solution **disappears**. The equation can be summarised as

$$\begin{matrix} H & & H \\ & C{=}C & \\ H & & H \end{matrix}(g) + H_2O(l) + O \underset{KMnO_4}{\overset{\text{alkaline}}{=}} \begin{matrix} & H & H & \\ & | & | & \\ H{-} & C^1 & {-}C^2 & {-}H \\ & | & | & \\ & OH & OH & \end{matrix}$$

ethan-*1,2*-diol

The loss of the pink colour is another test for unsaturation.

The aryl hydrocarbons

19 The simplest **aryl hydrocarbon (or arene)** is **benzene**, having the formula C_6H_6, and the second **homologue** in the series is **toluene**, C_7H_8, both names being trivial names rather than systematic I.U.P.A.C. names. The structure of benzene and the other aryl hydrocarbons was the object of considerable debate when they were discovered. A cyclic double bonded structure was assigned to benzene by a chemist called Kekulé as shown in *Figure 54.4*.

However, the reactions of the aryl hydrocarbons are not consistent with a structure containing carbon to carbon double bonds. This, together with a knowledge of the different types of carbon to carbon bond lengths, has led to **delocalised structures** being assigned to benzene and it's homologues as shown in *Figure 54.4*.

Reactions of the aryl hydrocarbons

20 The reactions of the aryl hydrocarbons (or arenes) can be broadly classified as **addition reactions** and **substitution reactions**.

21 **Addition reactions of benzene** Benzene undergoes only a few addition reactions, the two most useful examples being reactions with chlorine and hydrogen. When writing equations involving benzene, symbols are used for convenience. Both complete and symbolic equations are shown in *Figure 54.5*.

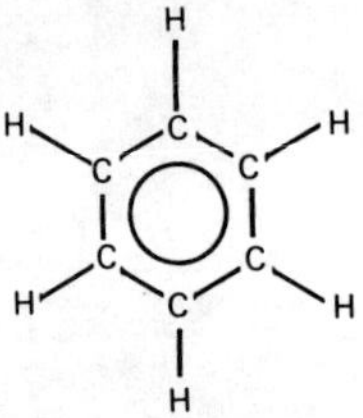

Kekulé structures

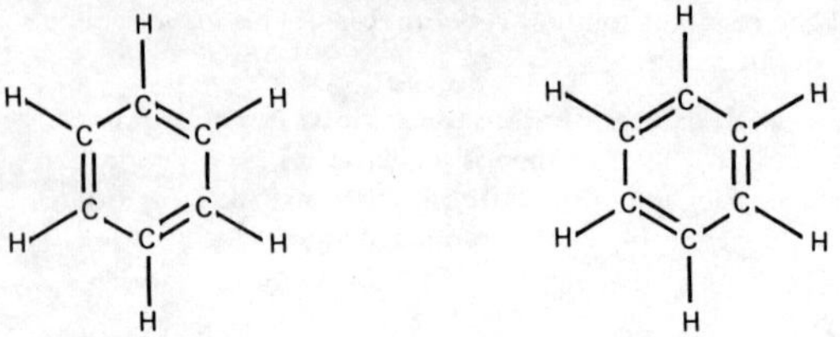

Modern structure Benzene

Kekulé structures

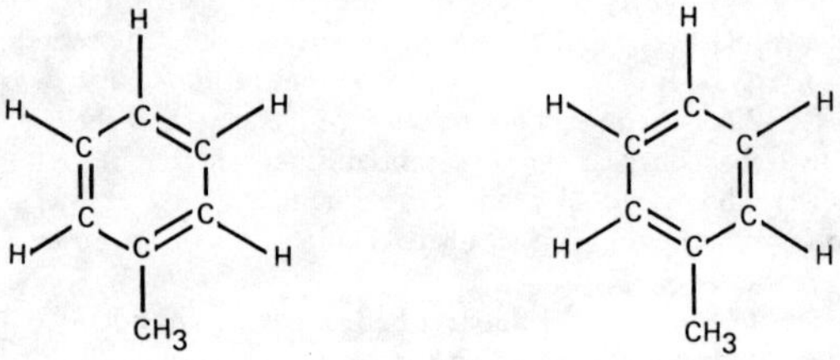

CH_3 CH_3

Modern structure Toluene

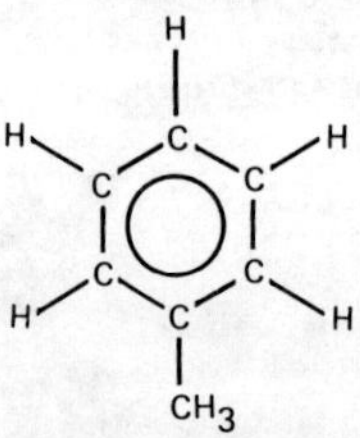

CH_3

Carbon to carbon bond lengths.

Alkanes 0.154 nm. Alkenes 0.135 nm. Benzene 0.139 nm.

Figure 54.4 The structure of benzene and toluene

Complete equation

$(l) + 3H_2(g)$ Ni. Catalyst = 105° (l)

Symbolic equation

Cyclohexane

$(l) + 3H_2(g)$ = (l)

Benzene

Cyclohexane

Complete equation

$(l) + 3Cl_2(g)$ U.V. = light

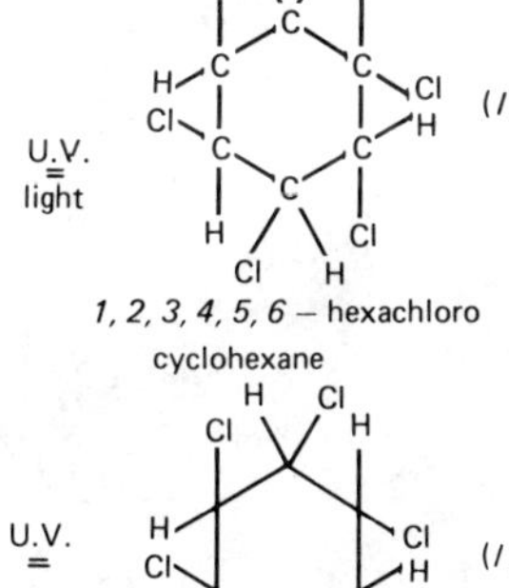

1, 2, 3, 4, 5, 6 – hexachloro cyclohexane

Symbolic equation

$(l) + 3Cl_2(g)$ U.V. = (l)

1, 2, 3, 4, 5, 6 - hexachloro cyclohexane

Figure 54.5 Some addition reactions of benzene

22 **Substitution reactions of benzene** When benzene undergoes substitution reactions, the groups which are substituted into the benzene ring are called **electrophiles** which means **electron seeking**. Benzene is a molecule which attracts such groups.

(a) **Reaction with nitric acid in the presence of sulphuric acid.**

$$C_6H_5H(l) + HNO_3(aq) \xrightarrow[20°C]{H_2SO_4} C_6H_5NO_2(l) + H_2O(l)$$

nitrobenzene

The reaction between concentrated nitric and sulphuric acids produces the NO_2^+ ion which is the electrophile in this reaction.

$$2H_2SO_4(aq) + HNO_3(aq)$$
$$\rightleftharpoons 2HSO_4^-(aq) + H_3^+O(aq) + NO_2^+(aq)$$

(b) **Reaction with chlorine**

$$C_6H_5H(l) + Cl_2(g) \xrightarrow{AlCl_3\ \text{catalyst}} C_6H_5Cl(l) + HCl(g)$$

chlorobenzene

The aluminium trichloride, $AlCl_3$, acts as a catalyst with chlorine to produce the Cl^+ ion which is the electrophile in this reaction; benzene also acts as the solvent.

$$AlCl_3 + Cl_2 \rightleftharpoons AlCl_4^- + Cl^+$$

(c) **Reaction with bromine**

$$C_6H_5H(l) + Br_2(l) \xrightarrow{FeBr_3\ \text{catalyst}} C_6H_5Br(l) + HBr(g)$$

bromobenzene

The iron (III) bromide, $FeBr_3$ acts as a catalyst with bromine to produce the Br^+ ion which is the reaction electrophile.

$$FeBr_3 + Br_2 \rightleftharpoons FeBr_4^- + Br^+$$

(d) **Reaction with fuming sulphuric acid**
Fuming sulphuric acid contains **sulphur trioxide**, which is a polarised molecule able to act as an electrophile according to the equation:

$$C_6H_5H(l) + SO_3(l) \xrightarrow{H_2SO_4} C_6H_5SO_3H(s)$$

benzene sulphonic acid

(e) **Reaction with halogenalkanes**

$$C_6H_5H(l) + CH_3Cl(l) \xrightarrow{AlCl_3 \text{ catalyst}} C_6H_5CH_3(l) + HCl(g)$$

chloromethane

The aluminium trichloride acts as a catalyst with chloromethane to form the carbonium ion CH_3^+.

$$AlCl_3 + CH_3Cl \rightleftharpoons AlCl_4^- + CH_3^+$$

(f) **Reaction with acid chlorides**

$$C_6H_5H(l) + CH_3COCl(l) \xrightarrow{AlCl_3} C_6H_5COCH_3(l) + HCl(g)$$

ethanoyl chloride — methyl phenyl ketone

In this reaction aluminium trichloride is acting as a catalyst with the ethanoyl chloride to produce the electrophile CH_3C^+O according to the equation

$$AlCl_3 + CH_3COCl \rightleftharpoons AlCl_4^- + CH_3C^+O$$

23 When atoms other than carbon and hydrogen are present in an organic compound they are classified by name depending upon the **particular functional group** present in the compound. A selection of C_3 compounds is given in *Figure 54.6* showing some of the bond angles. It should be noted that the **geometry of the molecules only changes near the functional group**. Elsewhere the geometry is identical to the **parent hydrocarbon**.

24 A selection of the reactions of functional groups is given in the form of tables and diagrams in the remainder of this chapter.

The alcohols and phenol

25 The **alcohols** and **phenol** all have an **–OH group** of atoms in their structures. The alkyl alcohols are classified as **primary, secondary and tertiary alcohols** depending upon the **number of hydrogen atoms on the carbon atom to which the OH group is attached**. The ability of a molecular formula to exist in different structural forms is another example of **isomerism**. Some examples of structure and nomenclature are given in *Table 54.3*.

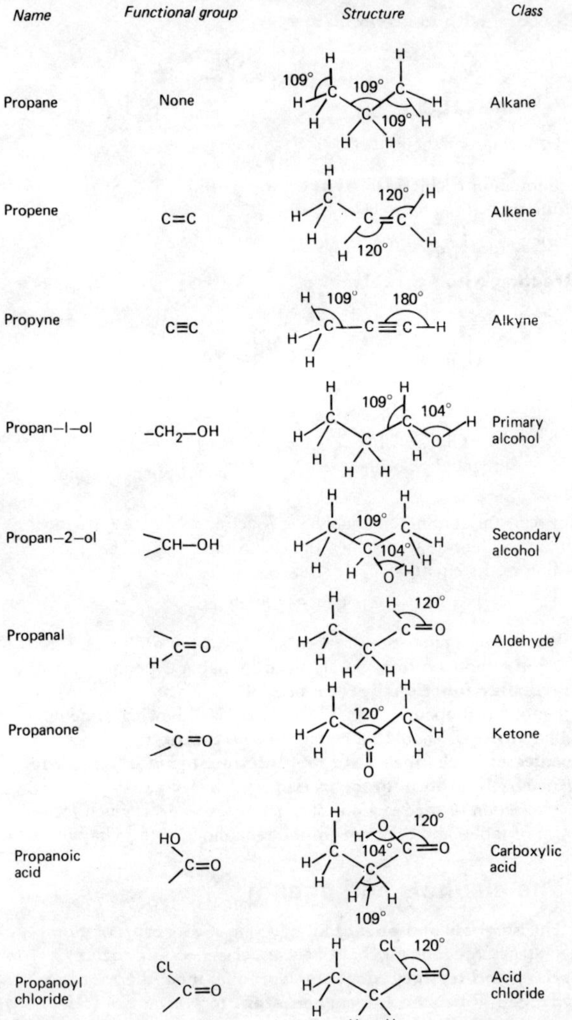

Figure 54.6 Some functional groups of organic compounds

Name	Functional group	Structure	Class
Propanamide	$-C(=O)NH_2$		Acid amide
Ethyl propanoate	$-C(=O)O-$		Ester
Propylamine	$\geq C-NH_2$		Amine
Bromopropane	$\geq C-Br$		Alkyl halide
Propanonitrile	$-C\equiv N$		Acid nitrile

Figure 54.6 (continued)

26 The **reactions of alcohols** depend to a certain extent upon the type of alcohol under consideration as shown in the reactions of primary and secondary alcohols shown in *Figure 54.7*.

Phenol also contains an OH group, but, **because it is attached to a benzene ring**, the reactions of phenol are quite different to the alcohols as shown in *Figure 54.8*. This is an effect which is shown by many of the compounds of benzene. The chemistry of compounds of benzene is generally more concerned with **reactions of the benzene ring than reaction involving the functional group attached to benzene**.

The aldehydes and ketones

27 The structure and nomenclature of some aldehydes and ketones are given in *Table 54.4*.
28 Both types of compound contain a carbonyl group, C═O,

Table 54.3 The structure, names and isomers of some alcohols

Formulae	Name	Type	Structure	B. Pt (°C)
CH_4O CH_3OH	Methanol	Primary	H H—C—O—H H	66
C_2H_6O CH_3CH_2OH	Ethanol	Primary	H H H—C—C—O—H H H	78
C_3H_8O $CH_3CH_2CH_2OH$	Propan-*1*-ol	Primary	H H H H—C—C—C*1*—O—H H H H	97
$CH_3CH(OH)CH_3$	Propan-*2*-ol	Secondary	H H H H—C—C*2*—C*1*—H H O H H	82
$C_4H_{10}O$ $CH_3CH_2CH_2CH_2OH$	Butan-*1*-ol	Primary	H H H H H—C—C—C—C—O—H H H H H	118
$CH_3CH_2CH(OH)CH_3$	Butan-*2*-ol	Secondary	H H H H H—C—C—C*2*—C*1*—H H H O H H	108
$(CH_3)_3$-COH	*2*-Methyl propan-*2*-ol	Tertiary	H H—C—H H H H—C—C*2*—C—H H O H H	83
C_7H_8O $C_6H_5CH_2OH$	Benzyl alcohol	Primary aryl	H H H H—⌬—C—O—H H H H	206

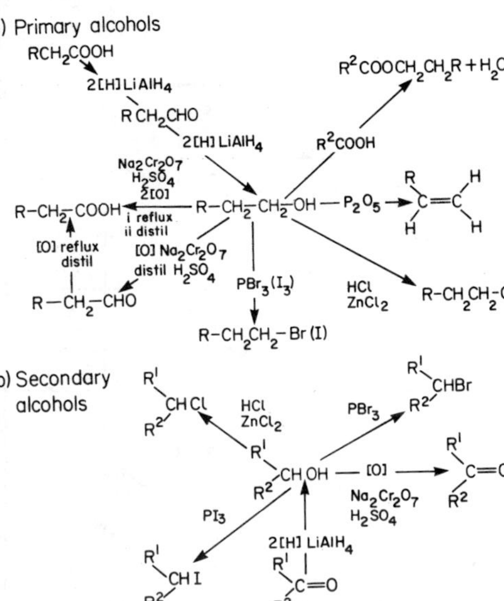

Figure 54.7 A summary of the reactions of the alcohols

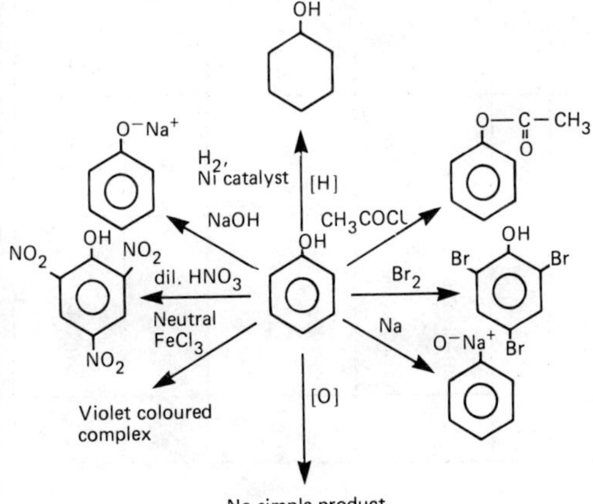

Figure 54.8 Some reactions of phenol

Table 54.4 Names, structure and formulae of the aldehydes and betones

Formulae		Name	Structure	b.pt. °C	Formula	Name	Structure	b.pt. °C
CH_2O	HCHO	Methanal	H\\ C=O H/	-21	None			
C_2H_4O	CH_4CHO	Ethanal	CH_3\\ C=O H/	21	None			
C_3H_6O	CH_3CH_2CHO	Propanal	CH_3CH_2\\ C=O H/	50	CH_3COCH_3	Propanone	CH_3\\ C=O CH_3/	56
C_4H_8O	$CH_3CH_2CH_2CHO$	Butanal	$CH_3CH_2CH_2$\\ C=O H/	74	$CH_3CH_2COCH_3$	Butanone	CH_3CH_2\\ C=O CH_3/	80
	$CH_3CH(CH_3)CHO$	2-Methyl propanal	CH_3CHCH_3\\ C=O H/	71				
$C_5H_{10}O$	$CH_3(CH_2)_3CHO$	Pentanal	$CH_3(CH_2)_2CH_2$\\ C=O H/	92	$CH_3CH_2COCH_2CH_3$	Pentan-3-one	CH_3CH_2\\ C=O CH_3CH_2/	101
	CH_3CH_3CH- -$(CHO)CH_3$	2-Methyl butanal	$CH_3CH_2CH\ CH_3$\\ C=O H/	89	$CH_3CH_2CH_2COCH_3$	Pentan-2-one	$CH_3CH_2CH_2$\\ C=O CH_3/	102
C_7H_6O	C_6H_5CHO	Benzaldehyde	C_6H_5\\ C=O H/	179	None			
C_8H_8O	$C_6H_5CH_2CHO$	Phenylethanal	C_6H_5–CH_2\\ C=O H/	194	$C_6H_5COCH_3$	Phenylmethyl ketone	C_6H_5\\ C=O CH_3/	202

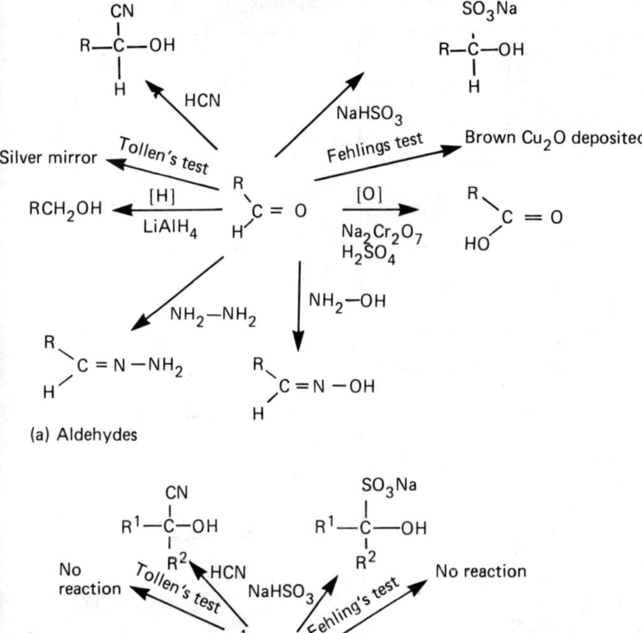

(a) Aldehydes

CN
R^1—C—OH
R^2
HCN
SO_3Na
R^1—C—OH
R^2
$NaHSO_3$
No reaction
Tollen's test
Fehling's test
No reaction
R^1
CH—OH
R^2
[H]
$LiAlH_4$
R^1
C=O
R^2
[O]
No simple acid formed
$NH_2—NH_2$
$NH_2—OH$
R^1
C = N — NH_2
R^2
R^1
C = N — OH
R^2

(b) Ketones

Figure 54.9 Some reactions of aldehydes and ketones

but aldehydes with a hydrogen atom attached, CHO, usually show the ability to act as reducing agents. In this respect aldehydes are unlike the ketones. Some reactions of the compounds are shown in *Figure 54.9*.

Table 54.5 The formulae of the carboxyllic acids

Molecular Formula	Name	Structure	b.pt°C
HCOOH	Methanoic acid	H–C(=O)–HO	101
CH_3COOH	Ethanoic acid	CH_3–C(=O)–HO	118
CH_3CH_2COOH	Propanoic acid	CH_3CH_2–C(=O)–HO	141
$CH_3CH_2CH_2$–COOH	Butanoic acid	$CH_3CH_2CH_2$–C(=O)–HO	163
			m.pt°C
C_6H_5 COOH	Benzoic acid	C_6H_5–C(=O)–HO	122
C_6H_5 CH_2– –COOH	*2*-phenylethanoic acid.	C_6H_5–C^2H_2–C(=O)–HO	146

The carboxylic acids

29 The structure and nomenclature of **some carboxylic acids** are given in *Tqble 54.5.*

30 The acidic properties of these compounds are due to the **hydrogen of the COOH group being easily lost to bases.** Some reactions of the acids are shown in *Figure 54.10.*

The amines

31 The **amines** are classified as shown in *Table 54.6*, together with some structures and nomenclatures.

32 The **nitrogen atom in the amines causes the compounds to behave as bases.** This and some other reactions

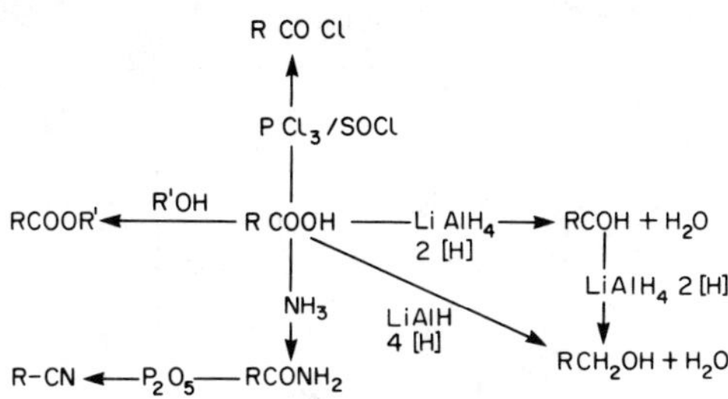

Figure 54.10 A summary of the reactions of the carboxylic acid

are shown in *Figure 54.11* together with some important reactions of phenylamine.

The alkyl halides or halogeno alkanes

33 There are four stable halogens, fluorine, chlorine, bromine and iodine. Organic compounds containing one of these elements is called an **alkyl halide** or **halogeno alkanes**. Some examples of structure and nomenclature are given in *Table 54.7*.
34 The compounds are very reactive, the order of reactivity for a particular alkyl group being R—I > R—Br > R—Cl > R—F. The reactivity of **bromoalkanes** is often ideal for the purposes of **controlled** organic synthetic reactions.
35 The compounds are versatile and can be converted into many other compounds some examples of which are given in *Figure 54.12*.
36 **Glossary of reactions** Organic reactions are given names to describe the type of reaction taking place.
37 **Oxidation** A variety of oxidising agents can be used to bring about an oxidation, two examples being:

(i) $C_3H_8(g) + 5O_2(g) = 3CO_2(g) + 4H_2O(l)$

(ii) $C_2H_5OH(l) + [O] \underset{H_2SO_4}{\overset{Na_2Cr_2O_7}{=}} CH_3CHO(l) + H_2O(l)$

Examples of these reactions are found in *Figure 54.7* and *54.9*.

Table 54.6 The formulae of the amines

Formula	Structural Formula	Name	Structure	Class	B.pt °C
CH_5N	CH_3NH_2	methylamine	$CH_3-\ddot{N}(H)-H$	Primary	6°
C_2H_7N	$CH_3CH_2NH_2$	ethylamine	$CH_3-CH_2-\ddot{N}(H)-H$	Primary	19°
	CH_3NHCH_3	dimethylamine	$CH_3-\ddot{N}(H)-CH_3$	Secondary	7°
C_3H_9N	$CH_3CH_2CH_2NH_2$	propylamine	$CH_3CH_2CH_2-\ddot{N}(H)-H$	Primary	49°
	$CH_3CH_2NHCH_3$	*N* methylethylamine	$CH_3CH_2-\ddot{N}(H)-CH_3$	Secondary	34°
	$(CH_3)_3N$	trimethylamine	$CH_3-\ddot{N}(CH_3)-CH_3$	Tertiary	3°
C_6H_7N	$C_6H_5NH_2$	phenylamine	$C_6H_5-\ddot{N}(H)-H$	Primary	184°

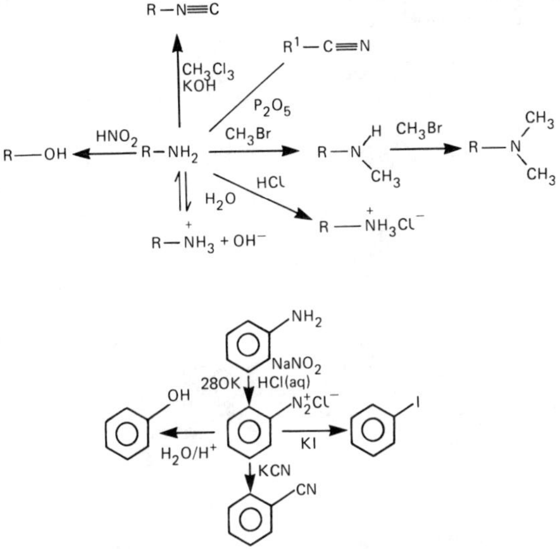

$$R{-}NH_2 \xrightarrow[KOH]{CH_3Cl_3} R{-}N{\equiv}C$$

$$R{-}NH_2 \overset{P_2O_5}{\text{———}} R^1{-}C{\equiv}N$$

$$R{-}OH \xleftarrow{HNO_2} R{-}NH_2$$

$$R{-}NH_2 \xrightarrow{CH_3Br} R{-}N(H)CH_3 \xrightarrow{CH_3Br} R{-}N(CH_3)_2$$

$$R{-}NH_2 \xrightarrow{HCl} R{-}\overset{+}{N}H_3Cl^-$$

$$R{-}NH_2 \underset{H_2O}{\rightleftharpoons} R{-}\overset{+}{N}H_3 + OH^-$$

$$C_6H_5NH_2 \xrightarrow[280K]{NaNO_2,\ HCl(aq)} C_6H_5N_2^+Cl^-$$

$$C_6H_5OH \xleftarrow{H_2O/H^+} C_6H_5N_2^+Cl^- \xrightarrow{KI} C_6H_5I$$

$$C_6H_5N_2^+Cl^- \xrightarrow{KCN} C_6H_5CN$$

Figure 54.11 Some reactions of amines

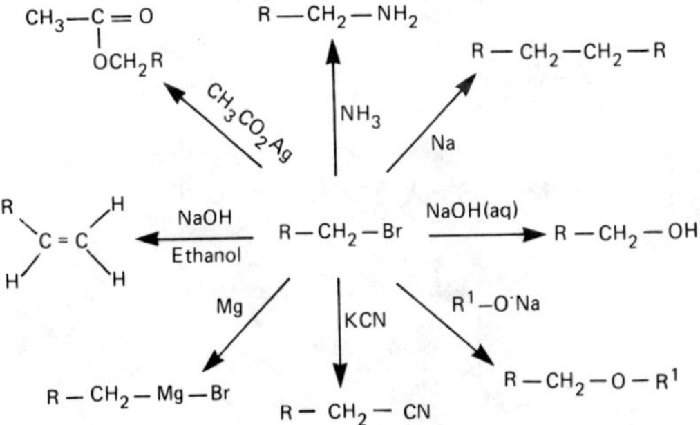

Figure 54.12 Some reactions alkyl halides

38 **Reduction** A variety of reducing agents can be used to cause reduction, two examples being:

(i)
$$\begin{matrix}CH_3 \\ \quad\diagdown \\ \quad C{=}O(l) + 2[H] \\ \quad\diagup \\ CH_3\end{matrix} \quad \underset{\text{dry ether}}{\overset{LiAlH_4}{=}} \quad \begin{matrix}CH_3 \\ \quad\diagdown \\ \quad CH{-}OH(l) \\ \quad\diagup \\ CH_3\end{matrix}$$

(ii)
$$C_6H_6(l) + 3H_2(g) \underset{\text{catalyst}}{\overset{Ni}{=}} C_6H_{12}(l)$$

Other examples are found in *Figures 54.7* to *54.10.*

39 **Substitution** This is the term to describe the **removal of an** atom or group of atoms from a compound and replacing it by a different atom or group. Substitution can be further subdivided into **electrophilic** and **nucleophilic** substitution depending upon the nature of the substituent. Two examples are

(i)
$$C_6H_5{-}H(l) + HNO_3(aq) \overset{H_2SO_4}{\longrightarrow} C_6H_5{-}NO_2(l) + H_2O(l)$$

Table 54.7 The structure and names of some alkyl halides

Formula	Name	Structure	Type
CH_3F	Fluoromethane	$H{-}CH_2{-}F$	Primary
C_2H_5Cl	Chloroethane	$H{-}CH_2{-}CH_2{-}Cl$	Primary
C_3H_7Br	l,Bromopropane	$H{-}CH_2{-}CH_2{-}CH_2{-}Br$	Primary
	2,Bromopropane	$H{-}CH_2{-}CHBr{-}CH_2{-}H$	Secondary
C_4H_9I	l,Iodobutane	$H{-}CH_2{-}CH_2{-}CH_2{-}CH_2{-}I$	Primary
	2,Iodobutane	$H{-}CH_2{-}CH_2{-}CHI{-}CH_2{-}H$	Secondary
	2,Iodo,2,methyl propane	$H{-}CH_2{-}CI(CH_3){-}CH_2{-}H$	Tertiary

an **electrophilic substitution** (for other examples see *Figures 54.8* and *54.11*) and,

(ii) $C_2H_5-Br(l) + NaOH(aq) = C_2H_5-OH(l) + NaBr(aq)$

a nucleophilic substitution (for other examples see *Figures 54.7* to *54.12*).

40 **Addition** Addition reactions can only take place in compounds which have **a multiple bond**. The reactions are easily recognised — only one reaction product is formed. Two examples are

(i) $CH_3CH{=}CH_2(g) + Cl_2(g) = CH_3CHCl-CH_2Cl(l)$, and

(ii) $CH_3CHO(l) + HCN(aq) = CH_3-\underset{\displaystyle H}{\overset{\displaystyle CN}{\overset{|}{\underset{|}{C}}}}-OH(l)$

Other examples can be found in *Figure 54.9*.

41 **Dehydration** When **water is removed** from an organic compound such that **H** and **OH** come from **adjacent atoms**, the reaction is called dehydration. For example

(i) $CH_3-CH_2-CH_2-OH(l) \overset{P_2O_5}{=} CH_3-CH{=}CH_2(g) + H_2O(l)$

(ii) $CH_3-\underset{\displaystyle O}{\overset{\displaystyle C}{\|}}-NH_2(s) \overset{P_2O_5}{=} CH_3-C{\equiv}N(l) + H_2O(l)$

For other examples, see *Figure 54.7*.

42 **Esterification** This is a specific type of substitution reaction which results in the formation of an ester. Two examples are

(i) $C_2H_5OH(l) + CH_3COOH(l) \overset{H^+}{\rightleftharpoons} CH_3COOC_2H_5(l) + H_2O(l)$

(ii) $C_6H_5-OH(s) + C_6H_5-COCl(l) =$

$C_6H_5-O-\underset{\displaystyle O}{\overset{\|}{C}}-C_6H_5(s) + HCl(g)$

Other examples are given in *Figures 54.7, 54.8* and *54.10*.

55 Inorganic chemistry

1 **Inorganic chemistry** is the name given to the study of the elements other than carbon and their reactions. The study of inorganic chemistry is a wide ranging topic which can be undertaken from three main approaches:

(a) the study of an **individual element** and of it's compounds;
(b) a comparative study of the elements within their **groups in the periodic table**; or
(c) a comparative study of the elements in **blocks of elements**, i.e. the ***s*-block, *p*-block and *d*-block elements**.

The method used in this text is a combination of approaches (b) and (c).
2 The assignment of the elements into groups, periods and blocks of elements is discussed in Chapter 45 and shown in *Table 45.1* of that chapter.
3 The comparative approach to inorganic chemistry in this text is based on the **physical properties of the elements** and the **chemical properties** of the **chlorides**, **oxides** and **hydrides** of the elements. This is based on the broad division of the elements into **metals and non-metals**. The physical properties of the metals are those of **conductivity, density, malleability, ductility** and a shiny **lustrous** appearance. Of these, the property of conductivity is the most distinguishing property. However, for comparison purposes, chemical metallic character is also considered. The metallic chemical character of an element can be considered from the following properties:

(a) the **bonding** in the **elements**;
(b) the **bonding** in the **chlorides** of the elements;
(c) the **basic nature** of the **oxides** of the elements; and
(d) the **bonding** in the **hydrides** of the elements.

(a) The bonding in metals is called **metallic bonding**, in which the atoms are held together in a close packed arrangement, but in which there are no bonds formed between the atoms. The structures of the metals are regarded as **giant** structures. The non-metals form a variety of structures, **giant, layer, molecular crystals** and **simple molecules**, in which the atoms are held in position by covalent bonds between the atoms. X-ray crystallographic data is most useful in assigning the structure of solid elements.

(b) If the bonding shown by the **chloride** of the element is **predominantly ionic**, the element is considered to be **metallic**. The degree of ionic bonding can be considered as a measure of how metallic an element might be. The non-metals form covalent chlorides with the property of being **hydrolysed** by water, a reaction not shown by ionic chlorides. The equations of the reactions are:

Ionic chloride $MCl_n(s) + H_2O = M^{n+}(aq) + nCl^-(aq)$

Covalent chloride $ACl_n + nH_2O = A(OH)_n(aq) + nHCl(aq)$

where M, is any metal and A is any non-metal.

(c) If the **oxide** of the element forms an **alkaline solution** when it dissolves in water, the element is considered to be **metallic**. Some elements form oxides which are only sparingly soluble in water, but which will **react with both acids and alkalis**; these are called **amphoteric oxides**, and show that the element can be considered to have only **partial** metallic character. The elements, whose oxides are soluble in water to give **acidic solutions**, are called **acidic oxides** and the elements are considered to be **non-metallic**. Examples of these oxides are:

(i) $Na_2O(s) + H_2O(l) = 2NaOH(aq)$ **basic**

(ii) (a) $Al_2O_3(s) + 3H_2SO_4(aq) = Al_2(SO_4)_3(aq) + 3H_2O(l)$

basic
amphoteric

(b) $Al_2O_3(s) + 2NaOH + 3H_2O = 2NaAl(OH)_4(aq)$

acidic

(iii) $SO_3(g) + H_2O(l) = H_2SO_4(aq)$ **acidic**

(d) The **hydrides** of the elements can themselves be classified as (i) **ionic** hydrides, (ii) **covalent** hydrides and (iii) **interstitial** hydrides. Ionic hydrides are formed by the elements of group I and II and the elements are classified as metallic. Covalent hydrides are formed by non-metals and interstitial hydrides by the transition metal elements. The interstitial hydrides are not **true compounds** but are best regarded as **inclusion** compounds. For some hydrides which are inclusion compounds the application of a vacuum to the compound will degrade it, leaving the pure metal after the removal of hydrogen. A consideration of these properties allows the metallic character of the elements to be identified.

The *s*-block elements

4 The **s-block elements** are the group I elements called the **alkaline metals** and the group II elements called the **alkaline-earth metals**. They occupy these positions as a result of their **electronic configurations** which are shown in *Table 45.2*, Chapter 45. The elements of group I have one electron in its final *s*-orbital, for example, Li is $\mathbf{1s^2 2s^1}$, Na is $\mathbf{1s^2 2s^2 2p^6 3s^1}$, and K is $\mathbf{1s^2 2s^2 2p^6 3s^2 3p^6 4s^1}$,
whereas the elements in group II have two electrons in its final *s*-orbital, for example, Be is $\mathbf{1s^2 2s^2}$, Mg is $\mathbf{1s^2 2s^2 2p^6 3s^2}$ andCa is $\mathbf{1s^2 2s^2 2p^6 3s^2 3p^6 4s^2}$.The elements in group I all have a valency of +1, and group II elements, a valency of +2. This can be explained by the values of the ionisation energies shown in *Table 55.1*. The group I elements have a relatively low 1st ionisation energy (IE) but a high 2nd IE. This results in the formation of the stable M^+ cation when forming compounds. The group II elements have relatively low 1st and 2nd IE values but a high 3rd IE resulting in the formation of the stable M^{2+} cation when forming compounds. One exception is beryllium, which has a small atomic radius and whose **compounds are more stable as covalent compounds**. The formation of ionic chlorides can be explained by the ease of **electron exchange** between *s*-block elements and chlorine, due to the 1st IE values of the metals and the low electron affinity of chlorine. The **lattice energies** of the group I chlorides are most stable for the $\mathbf{MCl}$ structure, and for the group II chlorides they are most stable for the $\mathbf{MCl_2}$ formula.

Table 55.1 Some physical properties of the *s*-block elements

Group		*Covalent radius*, nm	*Ionic radius*, nm	1st I.E. kJ	2nd I.E. kJ	3rd I.E.
	Li	0.123	0.068	520	7300	11800
	Na	0.157	0.098	500	4600	6900
I	K	0.203	0.133	420	3100	4400
	Rb	0.216	0.148	400	2400	3800
	Cs	0.235	0.167	380	2400	3300
	Be	0.106	0.030	900	1800	14800
	Mg	0.140	0.065	740	1500	7700
II	Ca	0.174	0.094	590	1100	4900
	Sr	0.191	0.110	550	1100	5500
	Ba	0.198	0.134	500	1000	5300

The **basic nature** of the oxides can be shown by their reaction with water to produce solutions containing hydroxide ions as shown by the equations:

$$M_2O + H_2O = 2MOH = 2M^+ + 2OH^- \quad \text{(group I)}$$

and $MO + H_2O = M(OH)_2 = M^{2+} + 2OH^-$ (group II)

The group I oxides are more basic than those of group II, and in each group, the basic strength increases on descending the group. The properties of the *s*-block elements are given in *Table 55.2*.

Table 55.2 The properties of the *s*-block elements

	Property	*Exceptions*
1	All *s*-block elements display metallic bonding in the solid state, usually as body centred cubic structures.	Be and Mg are hexagonal close packed Ca and Sr are cubic close packed
2	The halides of Group I are ionic in character displaying the sodium chloride or caesium chloride structure.	Lithium halides display some covalent character
	The halides of Group II are ionic in character, usually as fluorite or rutile.	$BeCl_2$ is a covalent polmeric solid
3	The oxides are basic, dissolving to differing extents in water to form hydroxides.	BeO is covalent and insoluble in water
4	The hydrides are ionic, and react with water to form hydrogen gas.	Be and Mg form metallic hydrides. LiH is insoluble
5	The relative thermal stabilities of the nitrates of Group I are greater than those of Group II, which all decompose to the oxide.	$4LiNO_3 \rightarrow 2Li_2O + 2N_2O_4 + O_2$
6	The relative thermal stabilities of the carbonates of Group I are greater than those of Group II which all decompose to the oxide.	$Li_2CO_3 \rightarrow Li_2O + CO_2$
7	The elements show a valency of +1 and +2 for Group I and Group II in all of their compounds.	None
8	The elements undergo many direct combination reactions due to their high relative reactivity.	

The *p*-block elements

5 Data for some of the *p*-**block elements** is given in *Table 55.3*. These elements of group III to O, by comparison with the *s*-block elements are **more diverse** in their properties. The major differences are the tendency of the elements to form **covalent**

Table 55.3 The physical properties of some *p*-block elements

Group		r_a (nm)	r_i (nm)	*Cations*	*Anions*	*Oxidation states*	*Melting point*, (°K)
	B	0.088	0.016	–	–	+3	2300
III	Al	0.126	0.045	Al^{3+}	–	+3	2720
	Ga	0.126	0.062	Ga^{3+}	–	+3, +1	303
	C	0.077	0.016	–	–	+4, –4	4000
IV	Si	0.177	0.038	–	–	+4, –4	2950
	Ge	0.122	0.093	Ge^{2+}	–	+2, +4	1210
	N	0.070	0.171	–	N^{3-}	+3, +4, +5, –3, –2, –1, +1, +2,	63
V	P	0.110	0.212	–	P^{3-}	–3, –2, –1 +1, +2, +3, +4, +5	317
	As	0.121	0.069	–	–	+3, +5	886†
	O	0.066	0.146	–	O^{2-}, O_2^{2-}	–2, –1	54
VI	S	0.104	0.190	–	S^{2-}, S_2^{2-}	–2, –1 +2, +3, +4, +5, +6	392
	Se	0.117	0.198	–	–	+2, +4, +6	490
	F	0.064	0.133	–	F^-	–1	53
VII	Cl	0.099	0.181	–	Cl^-	–1, +1, +3, +4, +5, +6, +7	172
	Br	0.111	0.196	–	Br^-	–1, +1, +4, +5, +6	266
	I	0.128	0.219	–	I^-	–1, +1, +3, +5, +7	397
	Ne	0.160	–	–	–	–	25
O	Ar	0.192	–	–	–	–	84
	Kr	0.197	–	–	–	+2, +4	116

†Sublimation

bonds in their compounds and the ability of the elements to show **variable oxidation states**. The formation of positive ions can only be achieved if sufficient electrons can be removed to produce a stable noble gas electronic configuration. When the **size of the atom** is small this becomes particularly **difficult** from an **energy point of view**. Consequently, beryllium (an *s*-block element) and boron form only covalent compounds. As any ***p*-block group is decended**, the size of the atom increases, and for groups III, and IV the lower elements are able to form **positive ions**, as shown in *Table 55.3*.

The elements in groups V, VI and VII are able to accept electrons and form **negative ions**. The ability of an element to form negative ions is greatest for the first element in each group and can be considered to reflect the attractive force the element shows toward electrons. The values of electronegativities of the elements is a good measure of this attraction, being highest for fluorine. (See Chapter 45, *Table 45.3.*) The ability of *p*-block elements to show variable oxidation states is usually due to the use of **available vacant orbitals** of the elements. However, the *p*-block element, nitrogen also shows variable oxidation states by the formation of a mixture of covalent and dative bonding.

The electronic configuration of the elements places the elements in their respective *p*-block groups. Those elements which have electrons in energy levels greater than 2*p* will have vacant orbitals available for the so called **expansion of the octet**, used to explain oxidation numbers larger than expected.

6 The groups at the **extremities of the *p*-block**, group III and group VII, show similar properties within the group, but in the middle of the *p*-block, for example in the group IV, the properties change considerably on descending the group.

The group VII elements

7 The **group VII elements are fluorine, chlorine, bromine and iodine** which exist as the diatomic molecules, F_2, Cl_2, Br_2 and I_2. The electronic configurations of the elements are F, **2.7**, Cl, **2.8.7**, Br, **2.8.18.7**, and I, **2.8.18.18.7**. The outermost shell of electrons in each case is one less than that of the nearest inert gas. The reactions between *s*-block metals and the **halogens** results in the formation of ionic compounds in which the metallic elements lose one or more electrons to the halogens. For example,

$$Ca(s) + Cl_2(g) = CaCl_2(s)$$

When the halogens react with *p*-block non-metals, the outer shell of electrons of the halogens is completed by the sharing of electrons. For example,

$$H_2(g) + Br_2(g) = 2HBr(g)$$

When the halogens form covalent bonds, the bonds are usually **highly polarised** because of the high electronegativity values of the halogens. For example, hydrogen bromide dissociates in water to give a strongly acidic solution of hydrobromic acid.

$$HBr(g) + H_2O(l) = \underbrace{H_3O^+(aq) + Br^-(aq)}_{\text{hydrobromic acid}}$$

With the *d*-block metals the halogens form a variety of compounds which can be covalent or ionic. In general, the **higher the oxidation state of the metal the greater the covalent character of the compound**. For example, $FeCl_3$ is covalent but $FeCl_2$ is ionic. Thus the group VII elements can form both ionic and covalent compounds. The relative reactivity of the elements is easily shown by the reactions of chlorine gas with aqueous solutions of bromide and iodine. (Fluorine gas is not used because it is too reactive.) When chlorine gas is bubbled into a solution of potassium bromide the following reaction takes place.

$$Cl_2(g) + 2KBr(aq) = 2KCl(aq) + Br_2(l)$$

In this reaction chlorine has **oxidised** the potassium bromide in solution. The equation can be simplified to

$$Cl_2(g) + 2Br^-(aq) = 2Cl^-(aq) + Br_2(l)$$

The chlorine atoms in Cl_2 have been converted into ions by removing the electrons from the two bromide ions. Similarly, when chlorine is bubbled into a solution of potassium iodide the following reaction takes place

$$Cl_2(g) + 2KI(aq) = 2KCl(aq) + I_2(s)$$

The chlorine has **oxidised** the iodine ions. When bromine liquid is added to potassium chloride no reaction occurs, but when added to potassium iodide solution the following reaction takes place

$$Br_2(l) + 2KI(aq) = 2KBr(aq) + I_2(s)$$

These reactions show that chlorine is a **stronger oxidising agent** than bromine, which in turn is a **stronger oxidising agent** than iodine. Thus, all of the elements in this group are similar in their structure and reactions.

The group IV elements

7 The change in properties from **non-metal at the top to metallic at the bottom** of the group are shown in *Table 55.4*. The elements of group IV are **carbon, silicon, germanium tin and lead**. The characteristics of the group can be determined by considering the structure of the element, the bonding of the chloride, the bonding of the hydride and the basic or acidic nature of the oxide. The structures of the elements are given in *Table 55.4* showing that **only the last two elements are true metals** and germanium is a **metalloid**. The existence of **allotropes** for tin shows that it has certain elements of non-metallic character.

The chlorides of the elements which have the formulae CCl_4, $SiCl_4$, $GeCl_4$, $SnCl_4$ and $PbCl_4$ are all **predominantly covalent** chlorides which with the exception of CCl_4 are hydrolysed by water. The dichlorides $GeCl_4$, $SnCl_2$ and $PbCl_2$ are solids, unlike the tetrahalides which are liquids. $GeCl_2$ and $PbCl_2$ are **predominantly ionic** in structure but $SnCl_2$ is **predominantly covalent**. The existence of covalent chlorides shows the elements to have some non-metallic character. However an ionic chloride for lead shows the **metallic character** of that element.

Table 55.4 The group IV elements

Element	*Structure*	*Chloride*	*Bonding*	*Oxide*	*Nature*	*Hydride*	*Bonding*	*Oxidation states*
Carbon	Diamond (giant) Graphite (layer)	CCl_4	Covalent	CO_2 CO	Acidic Neutral	CH_4	Covalent	+4,
Silicon	Giant lattice	$SiCl_4$	Covalent	SiO_2	Acidic	SiH_4	Covalent	+4,
Germanium	Metallic	$GeCl_4$ $GeCl_2$	Covalent Ionic	GeO_2	Acidic	GeH_4	Covalent	+2, +4
Tin	Metallic	$SnCl_4$ $SnCl_2$	Covalent Covalent	SnO_2 SnO	Amphoteric Amphoteric	SnH_4	Covalent	+2, +4
Lead	Metallic	$PbCl_4$ $PbCl_2$	Covalent Ionic	PbO PbO_2 Pb_3O_4	Basic Amphoteric Mixed	PbH_4	Covalent	+2, +4

The hydrides of the elements, CH_4, SiH_4, GeH_4, SnH_4 and PbH_4 are all covalent but the stability of these hydrides decrease down the group, PbH_4 being **particularly unstable**. The existence of covalent hydrides is also a sign of non-metallic character.

The oxides of the elements are (a) CO and CO_2, of which CO is **neutral** and CO_2 is **acidic**, (b) SiO_2 which is acidic, (c) GeO_2 which is weakly acidic, (d) SnO and SnO_2, of which SnO is **amphoteric** and SnO_2 is also amphoteric but more acidic than SnO, and (e) the oxides of lead are Pb_3O_4, PbO_2 and PbO. PbO_2 is amphoteric and PbO is **predominantly basic**, the other oxide of lead, Pb_3O_4 is a **mixed oxide** containing PbO and PbO_2 in a ratio of 1:2.

(vi) This change in properties from an acidic to a basic character for the oxides, exhibits the change in properties from non-metallic to metallic character.

The change in properties on decending the group shows that the elements become more metallic with increasing atomic number, and although they are not considered here, the lower elements in groups III and V also show increasing metallic character. Germanium, which is the middle element of the group is difficult to classify the formation of a predominantly covalent chloride, an amphoteric oxide, and the semi-conducting nature of the element indicates a non-metallic character. However, the bonding of the element is metallic which indicates a metallic character. This conflict of properties results in the classification of germanium as a **metalloid**. A consideration of these *p*-block element properties shows the diversity of the *p*-block elements.

The *d*-block elements

8 There are **three periods of *d*-block elements**, however the only elements considered in this book are those from **scandium to zinc**. Some of the properties are shown in *Table 55.5.* These elements occupy their position in the periodic table as a result of their electronic configurations. The **4*s* orbital is filled before the 3*d* orbital due to the lower energy of the 4*s* orbital**. Since the 4*s* orbital is completely filled, the size of the atomic radius of the ten elements which constitute the *d*-block elements are much more closely in agreement than the six *p*-block elements which follow them.

9 The chemistry of the *d*-block elements can be considered from the viewpoint of the characteristic properties of the elements; these are, (a) **variable oxidation states**, (b) **magnetic properties**, (c) **coloured compounds**, (d) **formation of interstitial**

Table 55.5 Some properties of the *d*-block elements

Element	*Atomic radius*, nm	*Valency states*	*Ions*	*Paramagnetism*	*Colour*
Scandium	0.161	3	Sc^{3+}	0	Colourless
Titanium	0.145	4 3 2	Ti^{3+}	1	Purple
Vanadium	0.132	5 4 3 2	V^{3+}	2	Green
Chromium	0.137	6 3 2	Cr^{3+}	3	Violet
Manganese	0.137	7 6 4	Mn^{3+}	4	Violet
		3 2	Mn^{2+}	5	Pink
Iron	0.124	6 3 2	Fe^{3+}	5	Yellow
			Fe^{2+}	4	Green
Cobalt	0.125	4 3 2	Co^{2+}	3	Pink
Nickel	0.125	4 2	Ni^{2+}	2	Green
Copper	0.128	2 1	Cu^{2+}	1	Blue
Zinc	0.133	2	Zn^{2+}	0	Colourless

compounds and (e) the **formation of complex ions**.

(a) The variable oxidation states of these metallic elements is a result of the **small difference in energy levels of the 3*d* orbitals, and the 4*s* and 4*p* orbitals**. The formation of positive ions rarely exceeds the removal of three electrons. The high oxidation states of the *d*-block elements are due to the formation of covalent or dative bonds in complex ions with non-metals (see *Table 55.5*).

(b) The ability of *d*-block elements to exhibit **paramagentism** is considered to be due to the **number of unpaired electrons** shown by the elements in its compounds. The ions with the greatest number of **unpaired** electrons, for example, Mn^{2+} and Fe^{3+} in compounds, show the **greatest magnetic moment**, as shown in *Table 55.5*.

(c) The five 3*d*-orbitals are not all of equal energy. **Two** of the orbitals are considered to be of **slightly higher** energy than the other **three**. Any electron in any of the **five orbitals can move from the lower energy level to the higher level** if it is supplied with **energy of the correct wavelength**. For many **electronic transitions**, the wavelength of energy required is within **the visible spectrum**. Thus the **absorption** of parts of the visible spectrum, and the consequent **reflection** of other parts, enables the **compounds to become coloured**.

(d) With the exception of Sc, Cu and Zn, the remaining *d*-block elements all form **interstitial compounds**. The size of the atoms is such that small elements like hydrogen, carbon and nitrogen can be enclosed within the metal crystal lattice when the molten metals solidify. The properties of the metals change considerably when other elements are enclosed within their structures.

(e) The formation of complex ions is dependent upon the **central metal ion** (or atom in the case of carbonyls) having **vacant orbitals** which can form **dative bonds** with surrounding **ligands**. The number of ligands which surround a central ion is either six or four, the six ligands form an **octahedral shape** and the four ligands form either a **square planar** or **tetrahedral shape**. Some examples of complexes showing isomerism are shown in *Figure 55.1.*

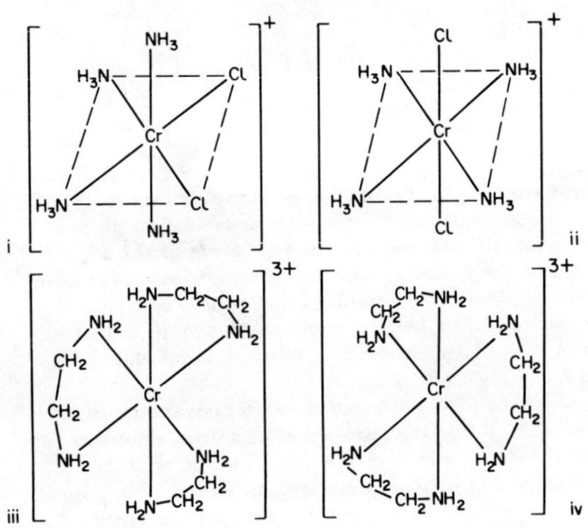

Figure 55.1

The structures of dichlorotetramine chromium (III) are shown in *Figure 55.1(i) and (ii)*. It is quite clear that there are two distinct **non-superimposable** structures *(i)* and *(ii)*. These structures are of the same molecular composition. In order to differentiate between the two structures the prefixes *cis* and *trans* are used to denote either similar groups on the same side of the structure, i.e. *cis* as shown in *Figure 55.1(i)*, or similar groups on opposite sides of the structure, i.e. *trans* as shown in *Figure 55.1(ii)*. Although the structures are different and called geometrical isomers, they both possess a plane of symmetry and are therefore not optically active.

The structures of *tris*(ethylenediamino) chromium are similarly shown in *Figure 55.1(iii) and (iv)*. In this case the structures are a pair of non-superimposable mirror images, neither of which has any element of symmetry. As a result these isomers are optically active, that is, they are capable of rotating the plane of plane-polarised light if it is passed through a solution of the isomer. Thus, depending on the constitution of the complex ion, geometrical or optical isomers can be formed.

10 It should be noted that the chemical reactions discussed in Chapter 50 act as a useful supplement to this brief comparative study of the elements.

Index

Newnes Technical Books